HANDBOOK OF
RELAY SWITCHING TECHNIQUE

HANDBOOK OF
RELAY SWITCHING TECHNIQUE

J. Th. APPELS

AND

B. H. GEELS

1966

Springer Science+Business Media, LLC

Translated from Dutch by R. H. Bathgate, Knegsel, The Netherlands

This book contains x + 321 pages, 390 illustrations

U.D.C. No. 621.3.06.001.1:621.318.5

Library of Congress Catalog Card Number: 66-23051

ISBN 978-3-662-39043-6   ISBN 978-3-662-40017-3 (eBook)
DOI 10.1007/978-3-662-40017-3

Original Dutch edition:

© N.V. Philips' Gloeilampenfabrieken, Eindhoven, The Netherlands, 1965

English edition:

© Springer Science+Business Media New York 1966

Originally published by N. V. Philips' Gloeilampenfabrieken, Eindhoven,
The Netherlands in 1966.

Softcover reprint of the hardcover 1st edition 1966

PREFACE

Science has taken enormous steps forward during the past 20 years. This has among other things resulted in the possibility of automating many processes previously carried out by manual labour. Automation, originally aimed at replacing manual labour by machines, is now also being used more and more to relieve men of a lot of mental routine work. The rapid development of electronic computers is one of the greatest achievements in this field.

A well-ordered society is unthinkable nowadays without extensive telegraph and telephone networks, and automatically controlled power stations. Automatic safety systems are widely used on railways, particularly at level crossings. The help of radar in safeguarding sea and air traffic has become indispensable.

The high standard of living is largely due to the ever-increasing automation of the production process.

All those who wish to make it their task to design, install or maintain electronic data processing equipment must have a thorough knowledge of the basic elements of switching techniques. Since the second world war, many firms engaged in the production of this kind of equipment have come into existence, which has given rise to a need for many skilled workers in this branch of industry. Unfortunately, there are still very few books which deal with relay switching techniques in simple language and without the use of too much mathematics.

The authors, who have had many years of teaching experience in this field, hope that this book will prove of assistance to those who wish to train themselves for work of this kind.

In fact, although this book deals with the basic elements of relay circuitry, much of the information given here is also necessary for the understanding of electronic switching techniques.

In many cases the theoretical information is illustrated by means of practical examples, while a number of questions are given at the end of each chapter to give the reader some practice in using the knowledge he has gained and to test his understanding of the material.

The elements of switching algebra are dealt with in chapter 3; this should be of considerable assistance in helping the reader to understand the reason behind various switching techniques. This branch of mathematics also serves as a basis for the study of electronic switching techniques.

A special chapter is devoted to a detailed treatment of the design of relays.

For those readers who are not yet acquainted with all the symbols normally used in switching techniques, a list of symbols with explanations is given in chapter 14.

The most important codes are dealt with, and their use in counter and calculating circuits is illustrated with the aid of many diagrams. Circuits for decoding, control, checking, recording, locking, translation, analysis and identification are discussed, each type in a separate chapter.

The authors welcome comments from readers, and will gladly make use of them in later editions.

May this book prove of use to many readers.

April 1966   The authors

CONTENTS

Chapter 1

CIRCUIT ELEMENTS

1.1 The neutral relay

A relay is a device for opening and/or closing electrical circuits by means of contacts attached to it. The relay is basically an electromagnet whose magnetic circuit is built up of a core, a yoke and a pivoted armature. A coil of one or more windings is placed round the core. The number of windings which can be placed on the core depends on the number of available terminals. This varies in the various designs from 2 to 6. Sometimes one terminal can be used as the common point of more than one winding, e.g. when the same voltage, or earth, can be used for several windings. Fig. 1 shows an example of a relay with 5 terminals and 3 windings.

Fig. 1

The number of turns of a winding varies from a hundred or less to tens of thousands, depending on the purpose for which the relay is to be used. When the current through a winding is large enough, a magnetic field will be produced which is sufficient to make the armature move towards the core. We then say that the relay is operated. The motion of the armature is transmitted to a set of contacts which can be used to open or close circuits, depending on the type of contact unit.

The following are the most commonly used contact units:

a) the make contact unit, with the symbol (m)

b) the break contact unit, with the symbol (b)

c) the change-over contact unit, with the symbol (c/o)

d) the make-before-break contact unit, with the symbol

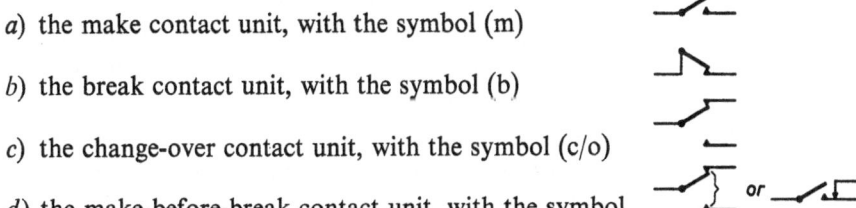

Many other sorts of contact units are possible, such as the make-before-make contact unit, etc.

There are various designs of relays, which all however serve to open or to close contact units and thus to change the constitution of the circuits in which these contact units are used. The number of contact units which can be operated by one relay depends on the construction of the relay and the purpose for which it is intended, and varies from one to twenty or more.

If the relay is to work reliably in various circuits, it must satisfy certain conditions. These can be divided into those concerning the contacts and those concerning the magnetic circuit.

In the production of a series of relays of the same type, small differences between individual relays are inevitable. This is a result of the spread in the magnetic resistance of the air gaps between the core and the yoke and between the yoke and the armature. The adjustment of the relay, i.e. the precise value given to the travel of the armature, the contact pressures, etc., also plays an important role. Because of this, not all relays of the same type will move the armature at the same minimum energizing current. These variations must be taken into account when designing relay switching circuits.

1.2   The properties of contacts in electrical circuits

A closed contact unit must have as low a contact resistance as possible. An open contact unit, on the other hand, must have as high a resistance as possible (insulation resistance).

The metal most used for contacts is silver – but not pure silver, as that is too soft. Other noble metals, such as gold or platinum, are also used when the constant resistance must be kept to a minimum.

Tungsten contact units are often used for switching large currents (e.g. 1 amp), but because the contact resistance of tungsten is many times that of silver or gold, it is not always possible to use it. In order to obtain as reliable a contact unit as possible, which still stays closed despite mechanical vibrations, one must apply a certain pressure between the two contact surfaces. However a high contact pressure leads to a high value for the energizing current, so a compromise has to be found which satisfies both conditions as far as possible. Depending on the contact material used, one adjusts the contact pressure to between 5 and 30 grammes. This relatively low value means that dirt (e.g. dust) between the contacts can give rise to insulation. A number of measures can be taken to prevent insulation by dust; e.g. the contact springs are normally split, giving essentially two equal but mechanically independent contacts. This greatly reduces the risk of insulation. Moreover, the contacts are in general placed vertically, so that any dust which may be present can fall off when the contacts move. A contact in the vertical position is shown on the right in Fig. 2.

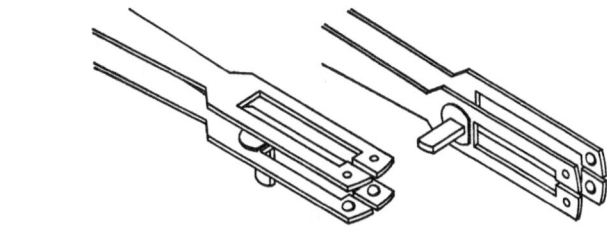

Fig. 2

In order to make the insulation resistance of an opened contact unit as large as possible, the distance between the contacts must be made sufficiently large. The insulation material used to pack the contact springs together in a spring set must also have a high resistance. The contact distance is also a quantity which has a strong influence on the sensitivity of the relay. A large contact distance means that the armature has a long way to travel; but a large distance from the armature to the core of the relay means that a large magnetic field is needed to attract the armature. Here too, a compromise must be made to arrive at a useful relay. A contact distance of 0.3 mm is much used.

1.3 Values of the magnetic flux for operating and releasing relays

A certain magnetic flux is needed to attract the armature of a relay. According to Hopkinson's law, this flux is given by

$$\Phi = \frac{magnetomotive\ force}{magnetic\ resistance}$$

The magnetic resistance of a relay will be mainly determined by the distance between the armature and the core. It is given by

$$\frac{l\,(\text{cm})}{\mu \times O\,(\text{cm}^2)}$$

The relative permeability (μ) of air is 1, while that of iron is several thousands of times greater.

In the manufacture of the relay one will try to make the air gaps as narrow and as well defined as possible, since any deviation from the chosen magnetic resistance will affect the necessary magnetic flux. If we assume a constant magnetic resistance, the magnetic flux needed will be proportional to the magnetomotive force, which is given by $0.4\pi ni$, where n is the number of turns (T) of the coil and i the current in ampères through the coil. The product ni is called the number of ampère-turns (which we shall here abbreviate to AT).

A certain number of AT (depending on the magnetic resistance) will thus cause a magnetic flux sufficient to attract the armature, given by:

$$\Phi = \frac{0.4\pi ni}{l/\mu O}$$

When the armature is attracted, the air gap is reduced. The magnetic resistance is thus also reduced, whence Φ is larger. The number of AT needed to reach the desired value of Φ is thus less. In other words, once the relay has been operated, it can be "held" with much less AT.

After the current through the winding is switched off, a certain remanent magnetism will remain, which will give rise to a certain magnetic flux. If the armature fits closely against the core, the air gap will be very narrow, and the Φ due to the remanent magnetism might be big enough to keep the armature attracted to the core. If however the armature does not fit so well against the core (manufacturing tolerances), the remanent magnetism will *not* be able to keep the armature attracted, and it will be "released". In order to get round this difficulty with the manufacturing tolerances, and to give the magnetic resistance of an energized relay a fairly accurate value, a platelet or rod of a non-magnetic material is applied to the armature, which prevents the armature from being pulled right against the core. This is known as the residual pin. This measure, together with the pressure on the armature of the contacts when they are switched over, guarantees that the armature will be released when the current is switched off.

The relay will operate at a certain value of the number of AT. If we now reduce the current, we will arrive at a value of the number of AT at which the relay just stays safely "operated". If the current is reduced further, the relay will "release". A fourth important value of the number of AT of a relay is the maximum value at which we can be sure that the relay will not operate.

Summarizing, we may state that there are four characteristic values of the number of AT:
a) the operate value (pick-up value),
b) the hold value,
c) the release value (drop-off value),
d) the non-operate value.
The smaller the spread of the above-mentioned values as a result of the

manufacturing tolerances, the more useful will the relay be. Here a compromise is necessary between the demands made by the circuitry and the manufacturing possibilities. It will be clear that the manufacturer of the relays would like to make his manufacturing tolerances as wide as possible, as he can then make his relays more cheaply.

The design engineer, on the other hand, wants accurate data in order to ensure that his circuit will work as reliably as possible. For this purpose it is necessary to determine the number of AT at which the worst relay will still operate; the better relays may however be expected to operate at a lower value of the AT. The same holds for the release value of the relay. This can be illustrated in a graph. At e.g. 175 AT, the worst relay still operates. At $120-1$ AT the best relay does not operate yet. At $60+1$ AT the worst relay will still hold, and at 30 AT the best relay will release (see Fig. 3).

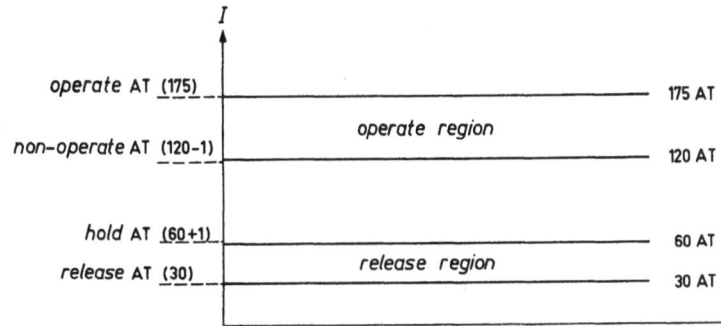

Fig. 3

1.4   The operate and release times of relays

A relay will operate and release when the current has reached such a value that the number of AT is sufficient to allow the armature to move. We may divide the time which elapses between the switching on of the current through the relay coil and the switching over of the relay contact units into two parts:

1. the time needed to reach the right number of AT, so that the armature can move.
2. the "mechanical" time needed for the relay contact unit to switch over.

The design engineer cannot do much about the mechanical time, which depends on the construction of the relay. The contact distance and the masses of the armature and the contact springs play a role in determining this quantity. It is however possible to influence the rise of the current and hence the time needed for the right number of AT to be attained.

Depending on the purpose for which a relay is intended, it may be necessary to make certain demands on both the operate time and the release time.

To a first approximation, we can regard a relay coil as a resistance and a self-inductance in series. If we let the number of turns be T and the resistance R, then the relationship between the current and the time from the switching on of the current is given by the formula:

$$i \cdot T = \frac{E}{R} \cdot T \left(1 - \varepsilon^{-\frac{R}{L}t}\right)$$

At the operate time t_a the current is i_a, so $i_a \cdot T$ is the operate value of the number of AT (AT_a). The term $(E/R)T$ represents the maximum possible value of the number of AT (AT_b). The equation

$$i \cdot T = \frac{E}{R} \cdot T \left(1 - \varepsilon^{-\frac{R}{L}t}\right)$$

then becomes

$$AT_a = AT_b \left(1 - \varepsilon^{-\frac{R}{L}t_a}\right)$$

$$\varepsilon^{-\frac{R}{L}t_a} = \frac{AT_b - AT_a}{AT_b}$$

$$\frac{R}{L}t_a = \ln\frac{AT_b}{AT_b - AT_a}, \qquad t_a = \frac{L}{R}\ln\frac{AT_b}{AT_b - AT_a}$$

The term L/R is called the *time constant* (τ in s), where $L = c \times T^2$ so $\tau = c(T^2/R)$.

The inductance L is proportional to T^2/R, which for our purposes we shall refer to as the *S-value*, and c is a constant depending on the materials and design of the relay as regards the magnetic circuit. The term:

$$\ln\frac{AT_b}{AT_b - AT_a}$$

indicates a multiplication factor p (e.g. 0.3), so we may write

$$t_a = \tau \times p.$$

Substituting this equation in

$$AT_a = AT_b \left(1 - \varepsilon^{-\frac{R}{L}t_a}\right)$$

gives:

$$AT_a = AT_b \left(1 - \varepsilon^{-p}\right)$$

or:

$$\frac{AT_a}{AT_b} = 1 - \varepsilon^{-p}$$

The operate time can now be given by the general formula:

$$t_a = c \frac{T^2}{R} \ln \cdot \frac{AT_b}{AT_b - AT_a}$$

where: t_a is the operate time in ms

c is a constant depending on the design of the relay

T is $\dfrac{\text{number of turns}}{1000}$

R is resistance in kΩ.

We can plot the term $1 - \varepsilon^{-p}$ in a graph, giving the "current-time curve" (Fig. 4). As $p \to \infty$, AT_a/AT_b tends to 1. The percentage AT_a of the total $AT_b (= 100\%)$ is plotted on the vertical axis. The horizontal axis then gives $x \cdot \tau$.

Example
1000 Ω relay coil 6000 turns. $AT_a = 100$,

$$E = 60V$$

$$AT_b = \frac{60}{1} \cdot 6 = 360$$

$$\frac{AT_a}{AT_b} = \frac{10^2}{360} = 28\% \equiv 0.33\tau \text{ (from current-time curve, Fig. 4)}$$

$$\tau = c \cdot \frac{6^2}{1} = 36c \text{ ms}$$

t_a is thus $0.33 \times 36c = 11.88c$.

If we assume $c = 0.3$ for this relay, the operate time becomes $0.3 \times 11.88 = 3.6$ ms.

In the above example we have only taken the S-value of the winding itself into account. We may regard the S-value as a measure of the magnitude of the time constant. There are however other factors which can influence the time constant, such as eddy currents in the magnetic circuit. The time constant can be very considerably increased by placing a short-circuited winding or a cylinder of a conducting material round the core. This gives rise to counter-voltages, which oppose the rise of the current. We can arrange the various possibilities in the following order:

a) *S-value of the magnetic circuit*

The eddy-current losses depend strongly on the total time found. The eddy currents in the core form as it were a conducting cylinder round the core, but the thickness of this cylinder is variable, since the eddy currents depend on the rate of change of the magnetic flux $(\mathrm{d}\Phi/\mathrm{d}t)$, which depends on the design of the relay.

b) *Short-circuited winding*

The *S*-value of the short-circuited winding must be added to that of the active winding (T^2/R) to get the total *S*-value of the windings:

$$S_{\text{total}} = \frac{T_a^2}{R_a} + \frac{T_k^2}{R_k}$$

c) *Conducting sleeve (slug) round the core*

This sleeve is usually made of copper or brass. It is made in various sizes, to give various *S*-values. Generally applicable *S*-values cannot be given in this case, as the area of the coil body determines the resistance. We give here some values for a particular make of relay.

         S-value 1 mm brass   =   25
         S-value 1 mm copper=   90
         S-value 2 mm copper=180
         S-value 3 mm copper=250

As with the short-circuited winding, this *S*-value has to be added to

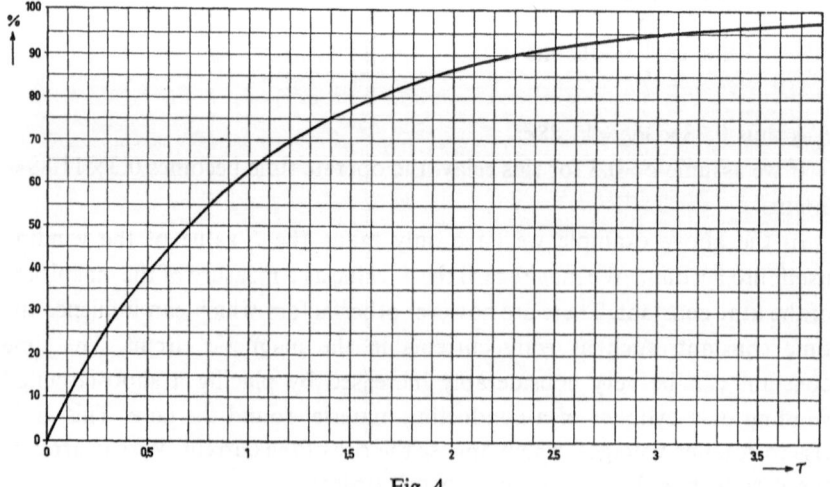

Fig. 4

that of the active windings to give the total value

$$S_{\text{total}} = \frac{T_a^2}{R_a} + S_{\text{copper}}$$

d) *Winding of another relay in series with the relay to be calculated*

The circuit now consists of 2 inductances in series, so that the time constant becomes:

$$\tau_t = \frac{L_1}{R_1 + R_2} + \frac{L_2}{R_1 + R_2} = \frac{L_1 + L_2}{R_1 + R_2} \quad \text{or:} \quad S = \frac{T_1^2 + T_2^2}{R_1 + R_2}$$

If the relay connected in series has already been energized via another winding, we may neglect L_2, when we find:

$$S = \frac{T_1^2}{R_1 + R_2}$$

If the relay connected in series has a short-circuited winding or a copper sleeve, its inductance will be reduced, and it may be partly or wholly (depending on the amount of copper) considered as a resistance.

Example (Fig. 5)
Relay C: winding 1–2 =   50 ohm   2400 turns
           winding 4–5 = 1000 ohm   7000 turns
           1 mm copper sleeve; $AT_a = 100$, $E = 60$ volt
Relay D: winding 1–5 = 2000 ohm   18200 turns
   The problem is to find the operate time of the relay C in ms, it being supposed that c for the relay is 0.265, and the S-value of the core is 30.

Solution:

$$\tau = c\frac{T^2}{R} = 0.265\left(90 + 30 + \frac{2.4^2}{0.050} + \frac{7^2}{1+2} + \frac{18.2^2}{1+2}\right) = 95 \text{ ms}$$

$$AT_b = \frac{60}{1+2}\cdot 7 = 140\,\text{AT}$$

$$\frac{AT_a}{AT_b} = \frac{100}{140} = 71.5\% \equiv 1.25\tau \quad \text{(from current-time curve, Fig. 4)}$$

whence $t_a = 1.25 \times 95 = 119$ ms.

Example (Fig. 6)
The energizing of a relay via a voltage divider.

Relay D: winding $1-5 = 2000$ ohm 18200 turns

$$AT_a = 100 \qquad E = 60 \text{ V}$$

According to Thévenin's theorem, the open voltage of point m is:

$$E_m = \frac{2}{2+1} \cdot 60 = 40 \text{ V}$$

with an internal resistance of

$$R_i = \frac{2 \times 1}{2+1} = 670 \ \Omega$$

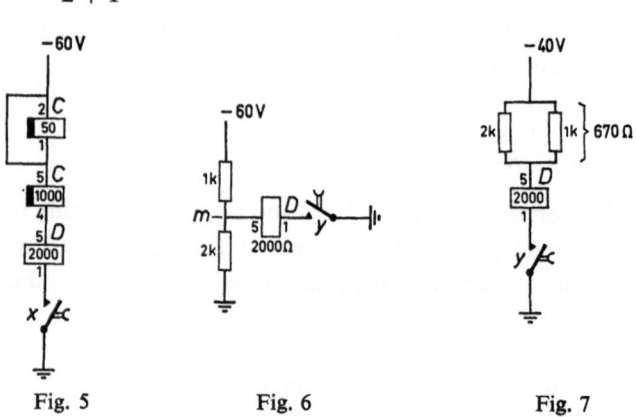

Fig. 5   Fig. 6   Fig. 7

The original circuit may thus be replaced by that of Fig. 7. Reckoning with an S-value of the core of 30, we find:

$$\tau = 0.265 \left(\frac{18.2^2}{2 + 0.670} + 30 \right) = 40 \text{ ms}$$

$$AT_b = \frac{40}{2 + 0.67} \cdot 18.2 = 272 \text{AT}$$

$$\frac{AT_a}{AT_b} = \frac{100}{272} = 37\% \equiv 0.47\tau \ \text{(from current-time curve, Fig. 4)}$$

whence

$$t_a = 0.47 \times 41 = 19 \text{ ms}$$

Release times

The formula

$$i \cdot T = \frac{E}{R} \cdot T \varepsilon^{-\frac{R}{L}t}$$

gives the relationship between the field strength (expressed in AT) and the time after the energizing current is cut off. The release time t_r will now be the time which gives a current i_r, such that $i_r \times T =$ release value AT_r. The term $(E/R) \cdot T$ represents the number of energizing AT.

The equation

$$i \cdot T = \frac{E}{R} \cdot T\varepsilon^{-\frac{R}{L}t}$$

then becomes:

$$AT_r = AT_b \cdot \varepsilon^{-\frac{R}{L}t_r}$$

$$\varepsilon^{-\frac{R}{L}t_r} = \frac{AT_r}{AT_b}$$

$$\frac{R}{L}t_r = \ln \frac{AT_b}{AT_r} \qquad \text{or:} \qquad t_r = \frac{L}{R} \ln \frac{AT_b}{AT_r}$$

In this equation L/R is again the time constant τ; it should be remembered however that the relay is now energized, so that the inductance $L = h \cdot T^2$, where the constant h is larger than the constant c used to find the operate time.

In order to arrive at a general formula for the release time, we must also take the $B-H$ curve of the material used into consideration. It will be clear that an increase of H above the saturation value for the iron will have very little or no influence on the value of B. This corresponds to increasing the number of energizing AT above the saturation point of the relay. We must then replace AT_b by the reduced AT, which is obtained from the $\Phi - AT (= B - H)$ curve of the magnetic circuit in question. This comes down to determining the active part of AT_b.

We may then write the following general formula for the release time:

$$t_r = h\frac{T^2}{R} \cdot \ln \frac{AT_{red}}{AT_r}$$

We can again determine the logarithmic term from an exponential current-time curve; the curve used for this purpose is the mirror image of the previous one with respect to the axis of abscissae.

In the equation

$$AT_r = AT_b \varepsilon^{-\frac{R}{L}t_r}$$

we must again replace AT_b by AT_{red}:

$$AT_r = AT_{red} \varepsilon^{-\frac{R}{L}t_r}$$

We may again write $-(R/L)t_r$ as $-p$, as we did for the operate time. Our equation now becomes $AT_r = AT_{red}\,\varepsilon^{-p}$, whence

$$\frac{AT_r}{AT_{red}} = \varepsilon^{-p}$$

If we plot the curve $y = \varepsilon^{-p}$ (Fig. 8), we see that it is indeed the mirror image of the former current-time curve. For $p=0$, $y=1$, which corresponds to AT_{red}.

If we mirror the curve $y = \varepsilon^{-p}$ back to the normal current-time curve, we may write

$$1 - y = 1 - \varepsilon^{-p} \qquad \text{or} \qquad 1 - \frac{AT_r}{AT_{red}} = 1 - \varepsilon^{-p}$$

or

$$\frac{AT_{red} - AT_r}{AT_{red}} = 1 - \varepsilon^{-p}$$

We may thus make use of the operate time calculations by considering $AT_{red} - AT_r$ as a percentage of AT_{red}, which gives us $p = x \cdot \tau$ (Fig. 9).

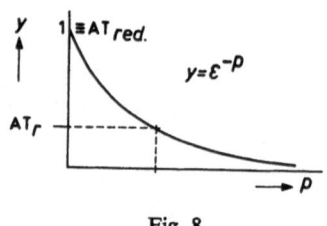

Fig. 8   Fig. 9

S-value for release times

Eddy-current losses etc. give the same trouble in calculations of the release time as in calculations of the operate time. No exact value can be given for general application, because the S-value again depends on the switching times and the construction. If we assume an S-value of 15 for the release time, then this value alone is taken into account when we consider the unretarded release of the relay. When the release is retarded (copper or brass sleeve or short-circuited winding), the S-value of the shorted winding or sleeve must be added.

Since the calculation of the release time is strongly dependent on the relay construction, the manufacturer usually gives a number of curves from which the release time can be read off as a function of the number of release AT, with the S-value as a parameter.

Figures 10*a*, *b* and *c* show a number of such curves for a given relay design, with residual pins of 0.1, 0.2 and 0.3 mm.

Making use of the data given on page 8, we see that a relay with a 2-mm copper sleeve round its core and an intrinsic *S*-value of 15 will have a total *S*-value for the release of 180 + 15 = 195. In table I of section 1.5 is given, among other things, the number of release AT for relays with different contact loads. For example, it may be seen from this table that a relay with 3 change-over contact units and 3 make contact units, and a 0.2 mm residual pin, will release between 39 and 76 AT. The release time can now be read off from the graph of fig. 10*b*. If the relay releases at 76 AT and the *S*-value is 195, the release time is 38 ms. If release does not start until 39 AT, the release time is 96 ms. All relays of this sort will thus have a release time lying between the limits 38 and 96 ms.

1.5 Calculation of relays for a given circuit

In order to calculate a relay we must know the winding data, the number of AT for operation, holding, release and possibly non-operation with a given contact load. This contact load depends on the number of contacts used, and their nature. For example, the contact load of a break or change-over contact unit will be greater than that of a make contact unit for the operation of the relay.

Another variable is the thickness of the residual pin to be used. For a constant travel of the armature (measured from the core to the residual pin), the thickness of the residual pin is one of the factors determining the magnetic resistance of the circuit. The number of AT will thus also be influenced by this variable. In the energized state, the thickness of the residual pin has even more influence on the magnetic resistance, and hence on the hold and release AT values.

Since the AT values also depend on the relay construction, it is not possible to give a general table for all sorts of relays.

Table I must thus be considered as valid only for one given construction. The three values given for the various numbers of AT refer to 3 sizes of the residual pin, viz. 0.1 mm, 0.2 mm and 0.3 mm.

It may for example be seen from Table I that a relay with 6 change-over contact units and a 0.3 mm residual pin will certainly operate at 228 AT, while only 131 AT are needed to hold it in the energized position.

A relay with 3 change-over contacts and a 0.1 mm residual pin will be sure to hold at 36 AT, and will release at 15 AT. This means that 16 AT

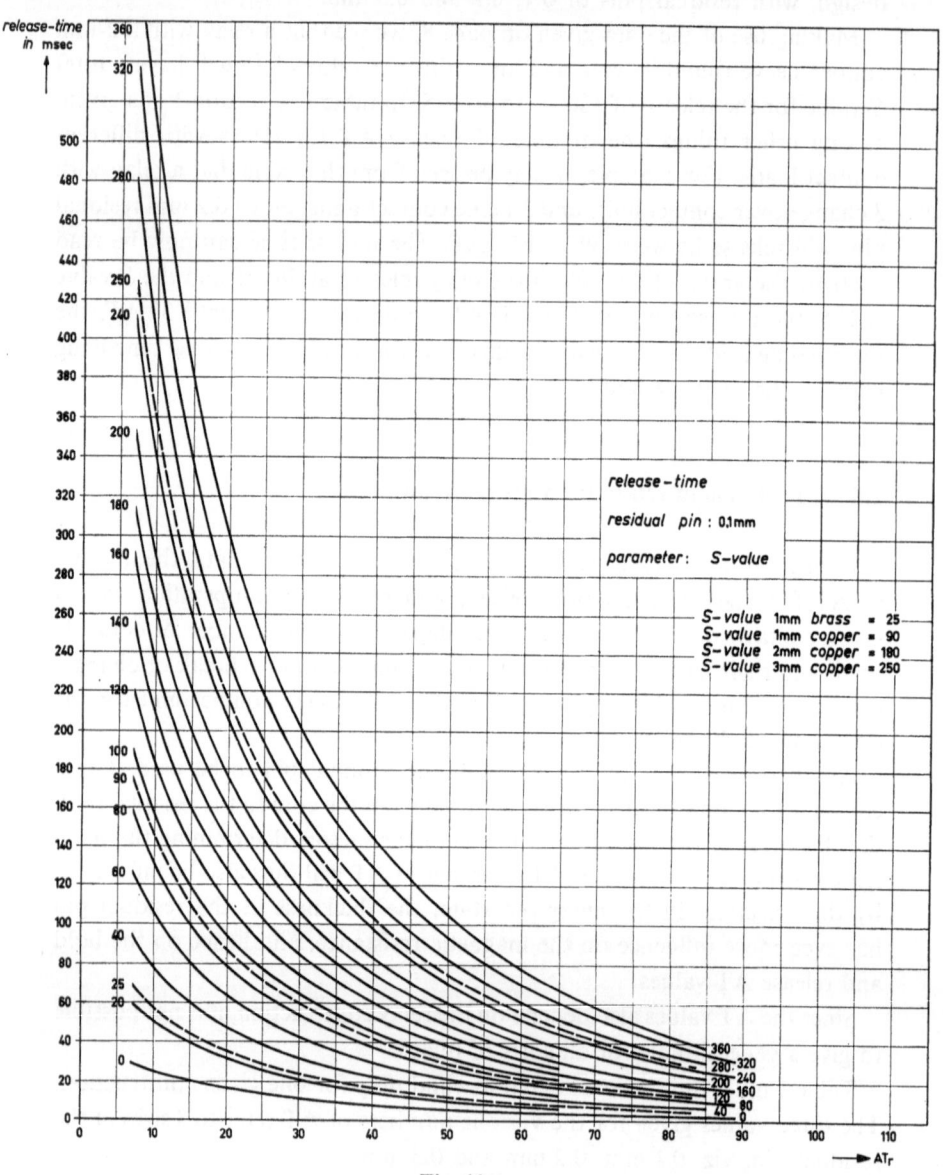

Fig. 10a

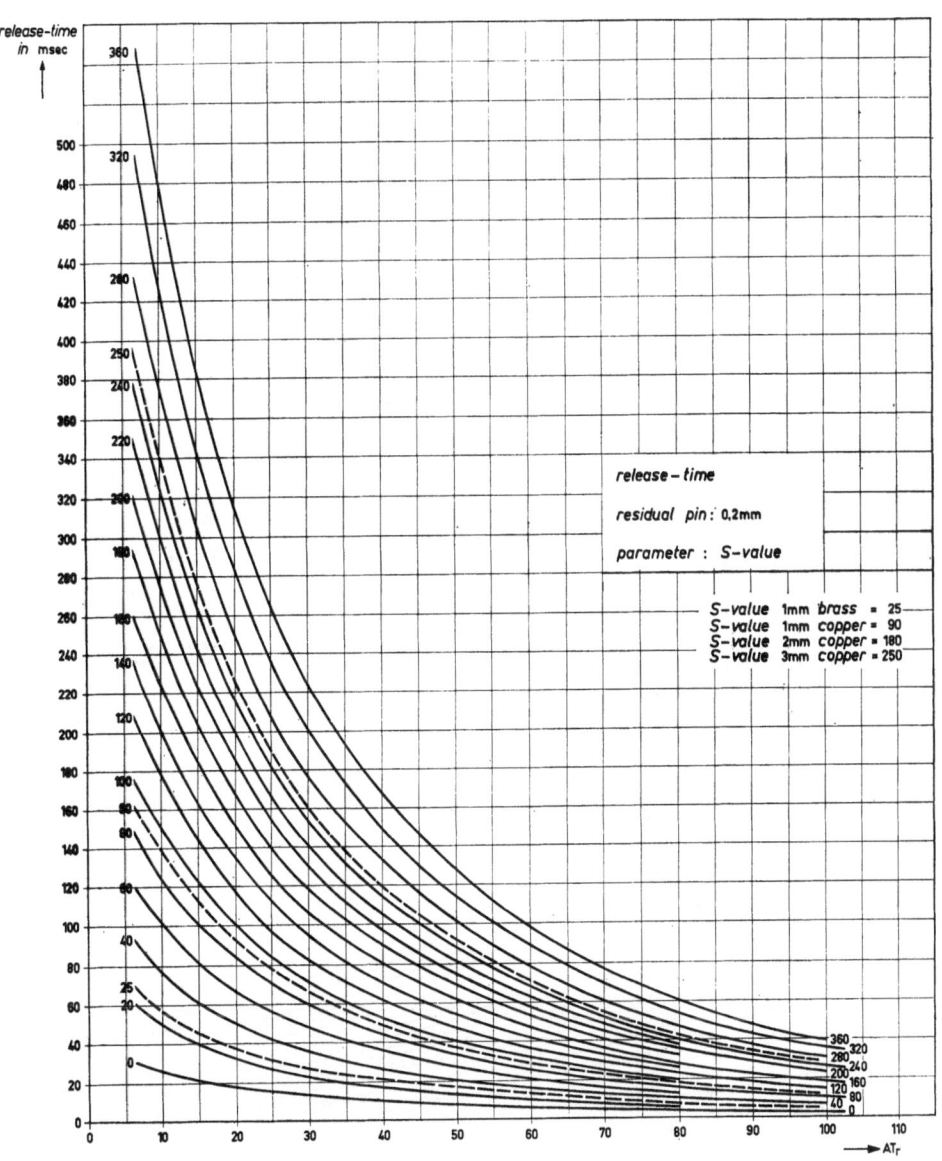

Fig. 10b

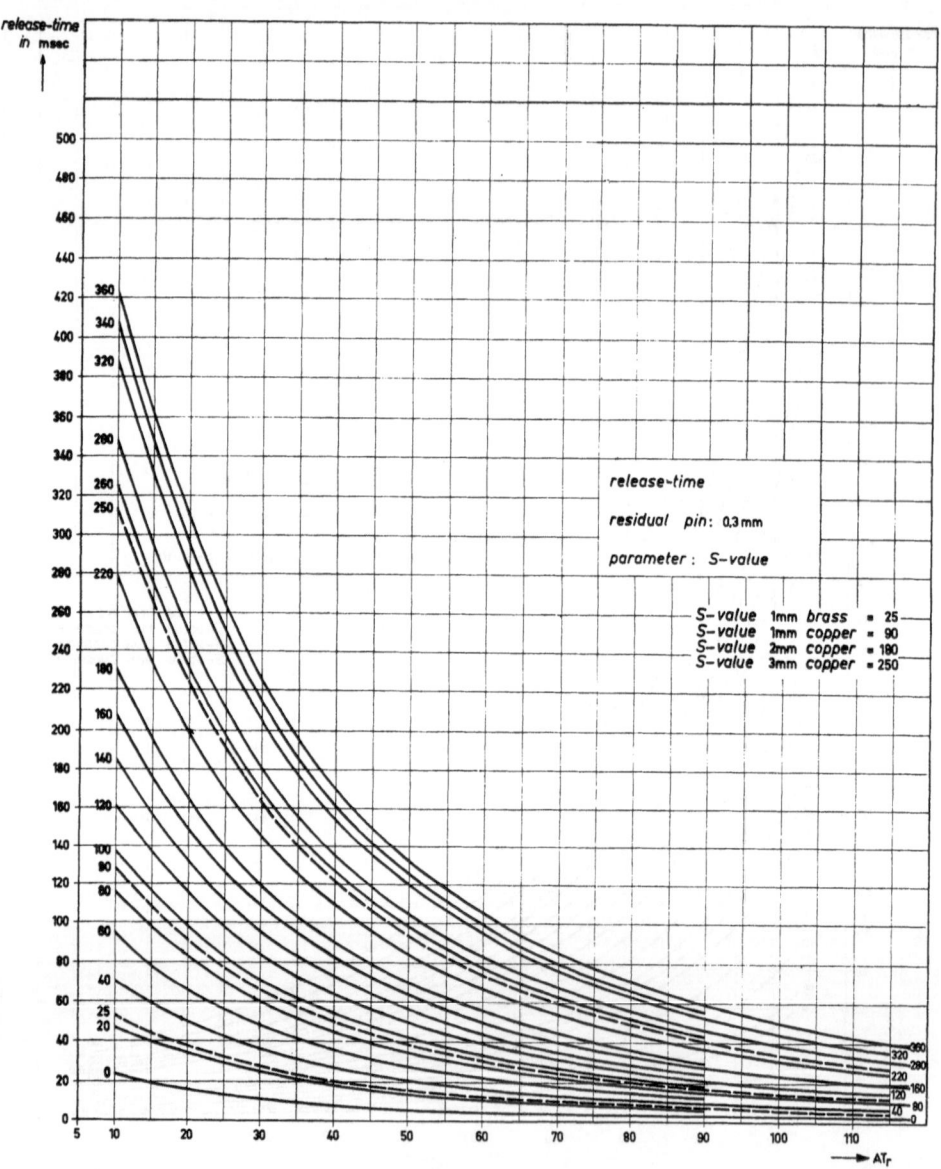

Fig. 10c

may be enough to hold this relay. This is however by no means certain, and the relay with the least favourable tolerances as regards holding will certainly release at this number of AT; one cannot be sure that the relay will hold until the above-mentioned value of 36 AT.

TABLE I

contact units	AT operate			AT non-operate			AT hold			AT release		
	0.1	0.2	0.3	0.1	0.2	0.3	0.1	0.2	0.3	0.1	0.2	0.3
3 make	97	105	116	53	59	64	28	42	61	12	22	33
3 change-over	127	140	152	71	78	85	36	53	76	15	28	42
6 make	130	142	154	73	79	85	49	70	101	22	37	53
3 change-over + 3 make	132	145	157	74	81	87	52	77	109	24	39	57
6 change-over	188	207	228	105	115	127	62	92	131	28	46	68
3 break + 6 make	147	162	179	75	83	92	50	75	107	23	38	57

Other important data are:

a) the maximum number of turns which can be accommodated on the coil bobbin, depending on the diameter of the wire used.

b) the winding resistance of the fully wound coil bobbin, which also depends on the wire diameter.

c) the data mentioned under a) and b) for coil bobbin which are not fully wound.

It will be clear that the length of a turn depends on its position in the coil: the length of a turn near the core will be much less than that of a turn near the outside of the coil. It follows that the resistance of a turn near the core will also be much less than that of a longer turn. These data, which like the above are given in tabular form, also depend on the form and dimensions of the coil bobbin.

The space available for the coil cannot be fully used, because of the space left between the various turns. Moreover, the wire used is insulated with cotton, silk, lacquer or some other insulating material. A "space factor" is therefore normally given for the various wire diameters. Table II gives the number of turns, the corresponding nominal resistance and the space factor for a given coil bobbin, when this is fully wound. The wire used in this case is enamelled copper wire.

The use of Table II will be illustrated later with reference to an example.

Table III gives the relationship between the extent to which a coil bobbin is wound (turns in % with respect to the fully wound coil bobbin) and the percentage resistance. We may see from this table for example that the first 40% of the turns (0–40%, as close to the core as possible) only provide 29.7% of the resistance of the entire coil.

TABLE II
Table for fully wound coil bobbins

wire diameter (mm)	number of turns	nominal resistance	space factor
0.05	68300	26400	1.11
0.06	47400	12750	1.11
0.07	35800	7050	1.10
0.08	28200	4260	1.09
0.09	22630	2700	1.08
0.10	18600	1790	1.07
0.11	15400	1230	1.07
0.12	13150	881	1.06
0.13	11390	650	1.06
0.14	10100	495	1.05
0.15	8900	383	1.05
0.16	8050	302	1.05
0.17	7180	241	1.05

If on the other hand we take the 40% furthest from the core (60–100%), we find that the percentage resistance is $100 - 49.7 = 50.3\%$ of the total. The relationships for all other percentages may be read off from the table in the same way.

TABLE III

% turns	% ohm	% turns	% ohm	% turns	% ohm
1	0.6	35	25.2	65	55.2
5	3.–	36	26.1	70	61.–
10	6.2	38	27.9	75	66.9
15	9.6	39	28.8	80	73.1
20	13.2	40	29.7	85	79.5
25	17.–	41	30.6	88	83.4
30	21.–	45	34.4	90	86.1
31	21.8	50	39.3	91	87.5
32	22.7	55	44.4	95	92.9
33	23.5	60	49.7	100	100.–

If the coil is wound on a coil bobbin with a copper sleeve, the space available for winding is reduced, depending on the thickness of the sleeve. Extra insulation in the form of a piece of paper or the like is normally placed between each layer of a coil. This insulation takes up $2\frac{1}{2}\%$ of the winding space.

The total reduction in the winding space due to use of a copper sleeve and paper insulation is given in Table IV.

TABLE IV

Cu sleeve	% turns	% ohm
1 mm	11.7	7.28
2 mm	28.3	19.6
3 mm	45.–	34.4

Further, when designing a relay we must take into account the possible deviations from the given nominal values.

The dc voltage used (say 48 V) depends on how well the accumulators used are charged, or if rectifiers are used on the mains voltage among other things. One often reckons with a deviation of 10% from the nominal value; this deviation may be positive or negative.

The resistance of the wire will also have a tolerance. For wire up to 0.06 mm in diameter, we may take this as $\pm 15\%$, and for thicker wire as $\pm 10\%$.

The resistance quoted for the wire is that at 20 °C; if the temperature is increased to 45 °C, we have:

$$R_{t2} = R_{t1}\{1 + 0.004(t_2 - t_1)\} = R_{20°}\{1 + 0.004(25)\} = 1.1\,R_{20°}.$$

An increase of 25 °C in the temperature thus increases the nominal resistance by 10%.

If we take other resistances which may be present in the circuit, with their tolerances, into account we are now in a position to design a relay with the aid of the data given in Tables I to IV. The object is to ensure that the relay designed will meet the requirements made even with the voltages etc. at the most unfavourable values.

If in connection with the data of Section 1.4 we state that we will always take about:

$1.2 \times$ operate value of AT

$0.8 \times$ non-operate value

$1.3 \times$ hold value

$0.7 \times$ release value,

we are in a position to determine the relay e.g. for the circuit of Fig. 11.

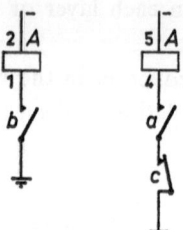

Fig. 11

We want a relay with 6 change-over contact units, 0.3 mm residual pin (to ensure that the release is as rapid as possible). Let us assume operate winding 1–2 = 2000 Ω and hold winding 4–5 = 4000 Ω. Voltage 48 ± 5 V. The operate AT must be 1.2 × 228 = 274 AT (see Table I). Resistance of winding 1–2 + 10% = 2200 Ω (neglecting any possible temperature rise). The number of turns must then be

$$\frac{274 \times 2200}{43} = 14000$$

For holding we need 1.3 × 131 = 170 AT. The number of turns for this winding is thus

$$\frac{170 \times 4400}{43} = 17400$$

The relay data are thus 2000 Ω (winding 1–2) 14000 turns
4000 Ω (winding 4–5) 17400 turns

Is it now in fact possible to manufacture such a relay with the material at our disposal?

The choice of wire is already more or less determined by the above calculation, since we have reckoned with a resistance tolerance of 10%. This means that we cannot use 0.06 mm wire, as this has a tolerance of 15%. Let us try a wire diameter of 0.07 mm, for which the resistance of the fully would coil bobbin is 7050 Ω (see Table II). The 2000 Ω of winding 1–2 corresponds to

$$\frac{2000}{7050} \times 100 = 28.4\%$$

of the total possible resistance, i.e. to 38.5% of the total possible number of turns (see Table III). The number of turns needed for this winding is thus 0.385 × 35800 = 13800 (see Table II). This differs by less than 1% from the value of 14000 turns stipulated above, which is certainly acceptable.

We can set out the results obtained so far in tabular form (Table V).

TABLE V

wire diam.	% turns	turns	% ohm	ohm
0.07 mm	38.5 2.5 insulation 41.–	13800	28.4 30.6	2000

The extra 2.5% turns is for the insulation between windings 1–2 and 4–5. The 4000 Ω for winding 4–5 (also with 0.07 mm wire) represents

$$\frac{4000}{7050} \times 100 = 57\% \text{ of the resistance}$$

Taking into account the results obtained so far and summarized in Table V, we find that this 57% of the resistance corresponds to 50% turns, i.e. 17900 turns (Table VI).

TABLE VI

wire diam.	% turns	turns	% ohm	ohm
0.07 mm	38.5 2.5 41	13800	28.4 30.6	2000
0.07 mm	50 91	17900	57 87.6	4000

We have thus used up 91% of the turns, of which $91 - 2.5 = 88.5\%$ are wound with 0.07 mm wire. Taking the space factor of 1.10 for 0.07 mm wire into account, we find that $1.1 \times 88.5 = 97.35\%$ of the coil bobbin is occupied. Adding the 2.5% for the insulation, we get a total of 99.85% of the available space: a tight fit, but just possible. The result is a relay which satisfies the stipulated conditions, though not exactly:

winding 1–2 = 2000 Ω 13800 turns
winding 4–5 = 4000 Ω 17900 turns

Another example is the design of a relay with two windings, each of which must have the same resistance and the same number of turns. Such relays are sometimes demanded when the inductance of the two windings must be about the same; another application is the differential relay described in Chapter 2, Section 2.3. Fig. 12 shows the half cross-section of the coil body, with the available space divided into three parts. One of the windings is wound in parts I and III, and the other in part II. It may be seen from the figure that the average radius of the turns in parts I and III $((r_1 + r_3)/2)$ is equal to the average radius of the turns in part II (r_2).

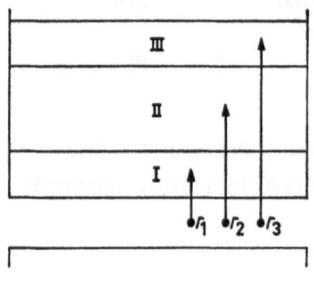

Fig. 12

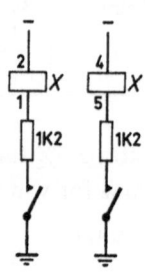

Fig. 13

In fact, the relay consists of 3 windings, of which windings I and III are connected with one another. The calculation for such a relay is given below. The wire diameter is the same for all windings. Relay X (Fig. 13) has a springset load of 3 change-over contact units, 3 make contact units and a 0.3 mm residual pin. The voltage is $60V \pm 10\%$. The resistance in series with each winding has the value $1200\,\Omega \pm 10\%$. The relay must not operate if both current circuits are closed; but if one current circuit is broken, the relay must operate.

The number of AT must be $1.2 \times 157 = 188$ AT (see Table I).

Let us assume the resistance of one winding to be about $1500\,\Omega$ and let us again reckon with a possible resistance tolerance of 10% due to variations in the thickness of the wire etc. (here again, temperature variations are neglected).

The total maximum resistance in one branch of the circuit is thus $1500\,\Omega$ from the relay winding $+1200\,\Omega$ from the resistance $= 2700\,\Omega$. In order to calculate the number of turns needed per winding, we must moreover add 10% to the total resistance, as stated above. The possible total resistance is thus $2700 + 270 = 2970\,\Omega$.

The number of turns per winding must then be

$$\frac{188 \times 2970}{54} = 10400$$

Here again the choice of wire is determined by the stipulation that the resistance tolerance of the winding should be 10%, as in the above calculation.

If we find in the course of the calculation that we have to use thinner wire after all, it will be necessary to revise the calculation given above.

Let us again start by trying 0.07 mm wire. Half of the total number of turns of one winding, i.e. 10400/2 must be placed in part I of Fig. 12, then the whole of the second winding (10400 turns) must be placed in part II, and finally the rest of the first winding (5200 turns) in part III.

The 5200 turns of 0.07 mm wire represent $5200/35800 \times 100 = 14.5\%$ of the number of turns of a fully wound coil (see Table II). This 14.5% corresponds to 9.26% of the resistance (Table III). This is 652 Ω. If we tabulate the results, taking into account the 2.5% winding space taken up by the insulation, we get the following picture (Table VII).

TABLE VII

wire diam.	% turns	turns	% ohm	ohm
0.07 mm	14.5 2.5 insulation	5200	9.26	652
0.07 mm	17.– 29.– 46.–	10400	11.04 24.34 35.38	1715

It will be seen from Table VII that the resistance of the winding in part II, with the required number of turns of 0.07 mm wire, is 1715 Ω according to the calculation. This is too high. It is thus not possible to achieve the desired windings of 10400 turns with 0.07 mm wire. Let us now try 0.08 mm wire. Making appropriate use of the various tables, we arrive at the picture shown in Table VIII. The resistance per winding is now 1435 Ω, which does satisfy the conditions made above. This relay can thus be regarded as the one we are after.

This gives a total of 79% turns, of which $79 - 5 = 74\%$ are of 0.08 mm

TABLE VIII

wire diam.	% turns	turns	% ohm	ohm
0.08 mm	18.5 2.5 insulation	5200	12.12	515
0.08 mm	21.– 37 58 2.5 insulation	10400	13.96 33.62 47.58	1435
0.08 mm	60.5 18.5 79.–	5200	50.25 21.61 71.86	920

wire. The space factor for 0.08 mm wire is 1.09 according to Table II, so that the total space occupied is $1.09 \times 74 + 5 = 85.66\%$ of the coil body.

If it is not important that the self-inductance of the two windings should be the same, but it is stipulated that they should both have the same resistance and the same number of turns, one can use wire of different diameters. Naturally, the diameter of the wire for the second winding should be chosen greater than that of the first. The second winding will take up more of the available space than the first one, because the diameter of the wire is greater. The reason why such a relay is preferred to the one calculated above is that the manufacturing costs (and thus the price) are lower, since one now only need apply 2 windings, instead of 3 as above. We shall now give an example of the design of a relay with only 2 windings.

Let us suppose that we want to calculate a relay with 6 change-over contact units and a 0.2 mm residual stop. The resistance per winding must be about 1100 Ω. A resistance of 500 $\Omega \pm 10\%$ is connected in series with each winding. The voltage is 48 ± 5 V.

According to Table I, for the relay to operate we need $1.2 \times 207 = 248$ AT. The number of turns per winding must be:

$$\frac{248 \cdot (1600 + 160)}{43} = 10100$$

The determination of the right wire diameter is a matter of experience, which only comes with much practice. The calculation is summarized in

TABLE IX

wire diam.	% turns	turns	% ohm	ohm
0.08 mm	36 2.5 insulation ——— 38.5	10100	26.1 28.35	1100
0.10 mm	54.5 ——— 93	10100	61.85 ——— 90.2	1100

Table IX. The results shown there can be obtained with the aid of the data of Tables II and III.

The space factor for 0.08 mm wire is 1.09, and that for 0.10 mm wire is 1.07.

The total space occupied by the windings is thus $1.09 \times 36 + 2.5 + 1.07 \times \times 54.5 = 100\%$. The relay calculated in this way satisfies all the requirements.

1.6 The polarized relay

The type of relay discussed so far is not suitable for the transmission of certain signals. If the signals received are very weak (low currents) and of very short duration, this type of relay is not sensitive enough. In particular for the reception of e.g. telegraph signals with a pulse length of 5–20 ms one needs a receiver relay which is very sensitive. Normal relays would have the following disadvantages for the above-mentioned application in telegraphy:

1. the switching times would be too long to record short signals.
2. the switching times depend strongly on the energizing current and the mechanical trimming of the relay.
3. the switching times are not symmetrical, i.e. the operate time is not equal to the release time, which means that signals are not relayed with the same length (distortion).
4. the life of this type of relay is usually not enough to allow it to work for long periods in telegraph lines without needing replacement.
5. moreover the character received will be distorted because the contacts cannot close without a certain amount of 'bounce'.

Therefore, for the reception of telegraph signals particularly, use is made of the polarized relay. The relay is given this name becuase it is sensitive to the direction of the current by which it is energized.

Relatively few AT are needed to move the armature; depending on the type, the number of AT varies from 2.5 to 40. Special measures must be taken to achieve this. For one thing, the polarized relay normally contains a magnetic circuit built up of electromagnets and permanent magnets, an armature that can move under the influence of the magnetic flux and one or two transfer contacts, which may or may not be adjustable, operated by the armature. The polarized relay cannot have many contacts, because the mass of the armature must be kept low to allow it to move *fast*.

Different manufacturers make the magnetic circuits of their polarized relays in different ways, but they all work on the same principle. In Figures 14 to 17 we show successively a differential magnetic circuit, a magnetic bridge circuit, a decoupled magnetic bridge circuit with 2 magnets and a decoupled magnetic bridge circuit with one magnet. The armature of the relay moves either on a pivot or on a spring.

The calculation which we shall now give will show the operation of the polarized relay, and will also give an impression of its sensitivity and the contact pressure.

According to Maxwell's law, the force of attraction between two magnetic

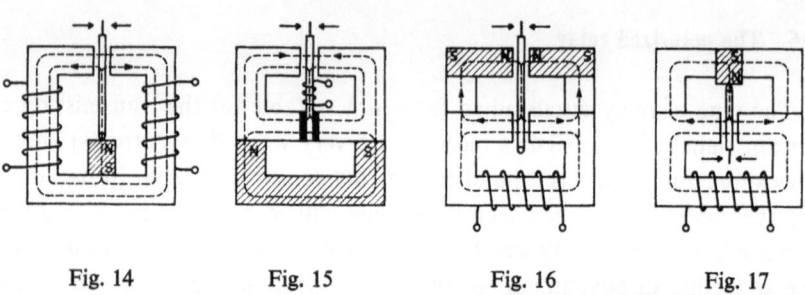

Fig. 14   Fig. 15   Fig. 16   Fig. 17

poles, between which a flux exists, is proportional to the square of the flux, i.e. $P = c\Phi^2$, where c is a constant. The force of attraction in a normal relay is independent of the direction of the magnetic flux, therefore independent of the direction of the energizing current through the coil. We shall show below that this is not the case with the polarized relay. Let us consider the magnetic circuit shown in Fig. 18, where Φ_1 and Φ_2 are magnetic fluxes caused by the permanent magnet and Φ_s is the magnetic flux caused by the energizing of one of the coils shown. If we imagine the armature to be precisely in the centre position, then $\Phi_1 = \Phi_2$, since the magnetic resistances are then equal. If the armature is displaced to the left (as in Fig. 18),

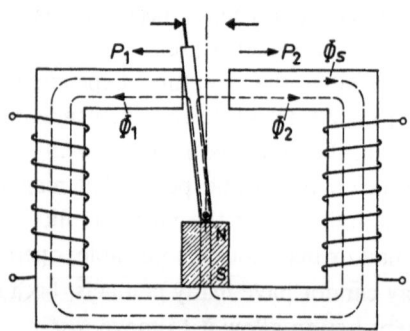

Fig. 18

Φ_1 will increase, since the corresponding magnetic resistance will decrease (smaller air gap). Φ_2 will decrease on the other hand, since the corresponding magnetic resistance will increase (larger air gap). $\Phi_1 + \Phi_2$, the total flux caused by the permanent magnet, may be assumed to be practically constant, since the total air gap does not change. Let us now also consider Φ_s, the flux caused by the energizing of one of the coils. If we suppose that the armature is initially over to the left, as in Fig. 18, what force is needed to make it move to the right?

The force exerted on the armature to the left is $P_1 = c(\Phi_1 - \Phi_s)^2$. The force exerted on the armature to the right is $P_2 = c(\Phi_2 + \Phi_s)^2$. The resultant force is thus

$$
\begin{aligned}
P_r = P_2 - P_1 &= c(\Phi_2 + \Phi_s)^2 - c(\Phi_1 - \Phi_s)^2 = \\
&\quad c(\Phi_2^2 + 2\Phi_2\Phi_s + \Phi_s^2) - c(\Phi_1^2 - 2\Phi_1\Phi_s + \Phi_s^2) = \\
&\quad c(\Phi_2^2 + 2\Phi_2\Phi_s + \Phi_s^2 - \Phi_1^2 + 2\Phi_1\Phi_s - \Phi_s^2) = \\
&\quad c\{(\Phi_2^2 - \Phi_1^2) + 2\Phi_s(\Phi_2 + \Phi_1)\} = \\
&\quad c\{(\Phi_2 + \Phi_1)(\Phi_2 - \Phi_1) + 2\Phi_s(\Phi_2 + \Phi_1)\} = \\
&\quad c(\Phi_2 + \Phi_1)(\Phi_2 - \Phi_1 + 2\Phi_s)
\end{aligned}
$$

Since $\Phi_1 > \Phi_2$, we must have $2\Phi_s > \Phi_1 - \Phi_2$ to make P_r positive, i.e. to make $P_2 > P_1$. If this is the case, the armature will move to the right. The magnitude of Φ_s is a measure of the sensitivity of the relay. The smaller the value of Φ_s at which the relay switches over, making and breaking a contact, the more sensitive is the relay.

In order to make the relay work with a small value of Φ_s, one must keep the difference between Φ_1 and Φ_2 as small as possible. Now this can be done in 2 ways. Both Φ_1 and Φ_2 can be made small by use of relatively weak permanent magnets or an extra air gap in the magnetic circuit of the permanent magnet; and if Φ_1 and Φ_2 are small, the difference between them will be even smaller. Another possibility is to ensure that the armature can only move a very small amount from its centre position (the "neutral line"). This will also serve to keep the difference between Φ_1 and Φ_2 small.

The contact pressure is determined by the force with which the armature is pulled to the left or the right. If we assume that Φ_s is no longer present, then the difference between Φ_1 and Φ_2 will be a measure of the contact pressure; the armature is held in the position into which it was brought by Φ_s by this differential force. In a balanced polarized relay, the armature thus remains where it was after the energizing current is switched off. With a weak permanent magnet, or a small deviation of the armature from the neutral line, however, the difference between Φ_1 and Φ_2 will be small, so that the armature will only be pressed against the contact with a small force. A strong permanent magnet on the other hand will give a high contact pressure.

The conditions for high sensitivity are thus opposed to those for a high contact pressure. One therefore tries to reach a compromise between the two.

One can use a more sensitive relay when the energizing current is present the whole time, so that the contact pressure does not depend only on the difference between Φ_1 and Φ_2. In such a case one speaks of double current.

Fig. 19 shows a circuit of this type, in which the coil is given a positive or a negative potential with respect to earth, and thus a corresponding current.

It is also possible to transmit signals with current in only one direction. During the transmission of a pulse, the circuit of the winding 1–2 of Fig. 20

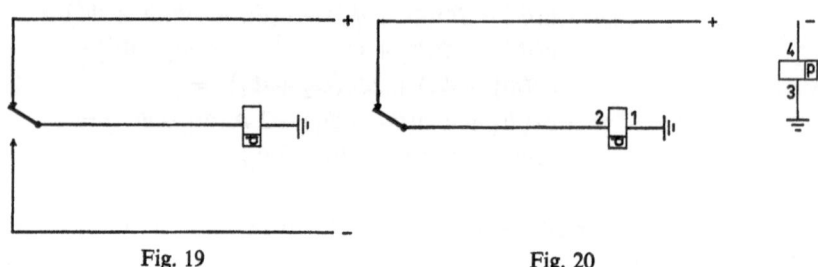

Fig. 19   Fig. 20

is open. The polarized relay reacts to this because a second winding (3–4) gives a Φ_s in the opposite direction, with half the number of AT of the "line winding" 1–2. In this case one speaks of single current, and here too a Φ_s is always present to ensure the necessary contact pressure.

It is however conceivable that in certain situations Φ_s must be absent for long periods, so that the contact is only maintained by the difference between Φ_1 and Φ_2. Mechanical vibrations can switch the relay over in such cases if a very sensitive relay is used. It is then better to choose a less sensitive relay (larger difference between Φ_1 and Φ_2).

1.7   Switches (selectors, finders)

In order to connect two given members of a group of instruments with one another, one can make use of a multi-position switch. By bringing the switch into the appropriate position, one connects the two instruments in question. If we consider by way of example a group of 10 telephone sub-scribers, any desired connection can be made with the aid of two 10-way switches. With more switches, it is possible to realize more than one con-nection at the same time. Fig. 21 shows such a system for ten subscribers, in which one connection can be made by means of switches I and II, and another by means of switches III and IV.

If subscriber 1 indicates that he wants to be connected with e.g. subscriber 8, switch I is put in position 1 and switch II in position 8. The wiper arm of each switch can rotate about an axis, coming into contact with each of the 11 contacts in turn. If no connection is required, the switch remains in

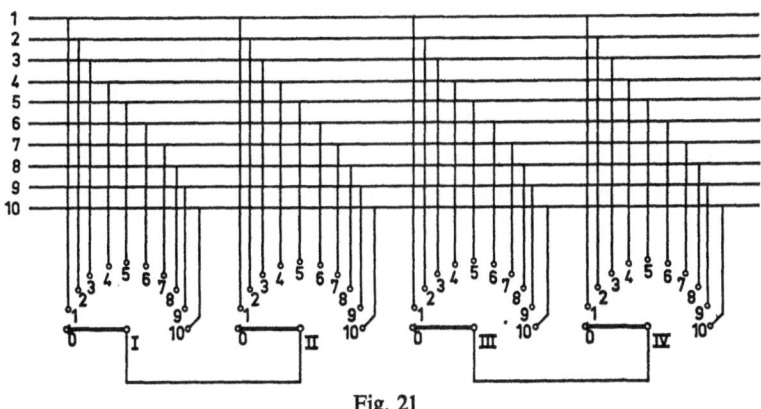

Fig. 21

position 0, which is not connected to any subscriber. If several wires have to be connected simultaneously to bring about one connection, the switch contains several wiper arms insulated from one another. Switches are made on this principle with an electromagnet to move the wiper to a desired position. Fig. 22 shows the principle of such a switch. If current flows

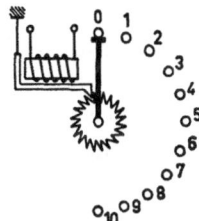

Fig. 22

through the electromagnet, the armature will move a ratchet (to which the wipers are fixed) over a certain distance. This distance is precisely that between contacts 0 and 1, so that a single energizing of the electromagnet will move the arms to position 1. By energizing the electromagnet once more, one moves the wipers to position 2, and so on.

Depending on the manner in which they are used, such switches are called selectors or finders. In Fig. 21, switch I finds the subscriber who is calling, while switch II selects the subscriber requested. Switch I is therefore called a finder and switch II a selector.

There are many different versions of electromechanical switches. All of them are however based on the same principle, viz the moving of one or more selector wipers in some appropriate way over a number of contacts. The moving part of the selector is called the rotor, and the stationary part (the contacts over which the wipers move) is called the contact bank or stator. The contacts are placed round the circumference of a circle. In different versions, the contacts occupy $\frac{1}{3}$, $\frac{1}{2}$ or a complete circumference. Fig. 23 shows a 180° contact bank. In order to do away with "dead time" while the contact wipers are moving over the half of the circle which is not provided with contacts, two sets of contact wipers are mounted diametrically opposite on the same spindle, in neighbouring planes.

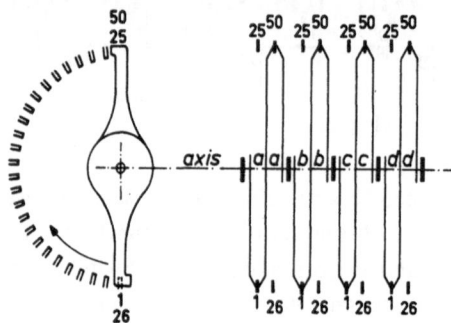

Fig. 23

Similarly, a selector bank which only occupies 120° of arc has 3 sets of contact wipers to bridge the "dead" space. The number of contacts of a selector varies, depending on the construction, from 10 to 500. The number of contact wipers can vary from 3 to 10.

Selectors are called uniselector when the contact wipers can only move in one direction. One also distinguishes between impulse driven selectors, such as the one sketched in Fig. 22, and continuously driven ones. In the latter case, the motion of the rotor is derived from a central drive system via a friction or cog-wheel coupling. Continuously driven selectors with a special little motor for each selector are also quite widely used; with such selectors, it is not so simple to move the contact wipers to a desired output by means of pulses.

Such selectors therefore normally make use of a "test method", in which the motion of the contact wiper is interrupted as soon as the indicated point on the contact bank has been reached. Centrally driven selectors generally have a coupling magnet by means of which the rotor is coupled to the drive mechanism, and a stop magnet, which is energized as soon as the desired output is reached and stops the motion of the contact wipers. In selectors with individual motors, the motor is switched on when the rotor has to start turning, and is switched off when the right contact has been reached. The method of ascertaining when the right contact has been reached is described in Chapter 9, Section 5.

The speed of these selectors, i.e. the number of contacts which can be scanned per second, depends on the purpose for which they are intended, and on their construction. It varies from 10 to 300 contacs/second.

Selectors are called two-motion selectors when the contact wipers can move in 2 different planes. Here again, we can distinguish between impulse and continuously driven models. Fig. 24 shows the principle of an impulse driven one.

When the vertical magnet L is energized, the corresponding armature

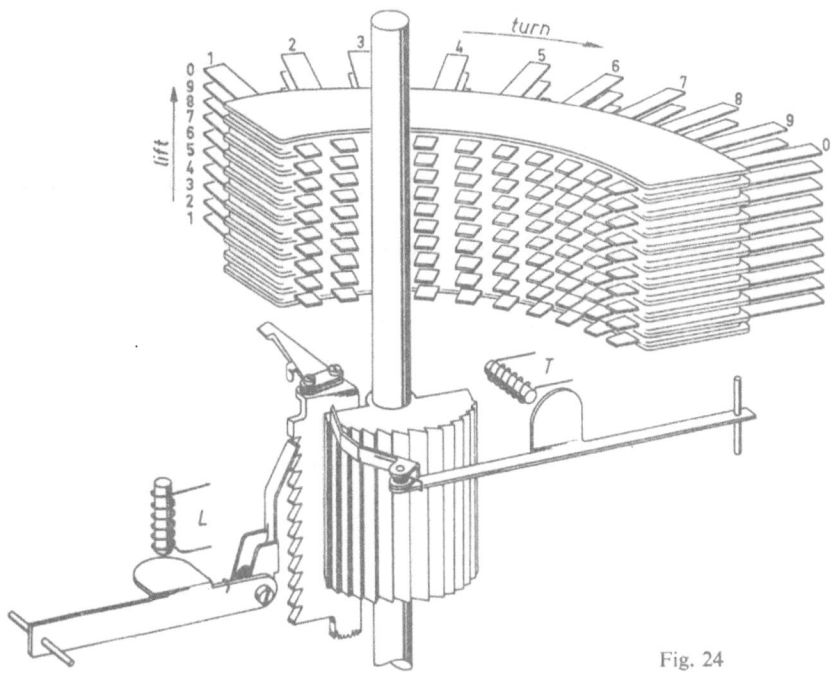

Fig. 24

comes up and the part on which the contact arm is mounted is raised. A locking spring (not shown) keeps the arm in this position, even though the energizing current of L is cut off for some time so that the armature falls back. A second energizing will lift the segment one step higher. If the rotary magnet T is now energized, the wiper assembly is caused to rotate in a horizontal plane by means of a ratchet attached to the armature of T. The wiper now reaches the first contact of the row to whose level it was brought by the lift pulses. The 100 contacts drawn in Fig. 24 can be reached in a minimum of 2 and a maximum of 20 steps, i.e. on the average

$$\frac{20 + 2}{2} = 11 \text{ steps}$$

A plain uniselector would need on the average

$$\frac{0 + 99}{2} = 49.5 \text{ steps}$$

The use of a two-motion selector thus reduces the number of steps needed to reach a given contact, on the average.

Selectors which are driven by pulses usually need to have a fixed starting position (zero position). Selectors also need a zero position if the construction does not allow the selector to reach any given position from an arbitrary starting position. In two-motion selectors, preparation is also necessary for changing from one dimension to the other. This is generally done by means of contacts mounted on the selector, so this type of selector also needs to start from a zero position.

If a selector has an essential zero position for one of the above-mentioned reasons, then the selector must be brought back to the zero position after the connection between 2 instruments (brought about by the selector) has been broken. This will give rise to more wear than in the case where the selector can remain where it is after the connection is broken (selector without essential zero position).

1.8 Semiconductors

The use of semiconductors can sometimes give important advantages in relay switching techniques. In this section we shall therefore discuss the most important properties of the diode and the transistor. In electrical practice one distinguishes between conductors and insulators. In general it may be stated that the resistivity of conductors, i.e. the resistance of a block of the material 1 cm long and 1 cm^2 in cross-section, is of the order of 10^{-6} ohm.

The resistivity of insulators is of the order of 10^6 to 10^8 ohm.

Semiconductors occupy an intermediate position, with a resistivity of about 10^{-3} ohm in the conductive state ("forward direction") and of about 10^7 ohm in the cut-off state ("reverse direction").

The structure of matter

All matter is built up of molecules, which are the smallest particles into which a substance can be divided without losing its specific properties.

Depending on the nature of the substance, a molecule consists of one or more atoms; if the molecule contains several atoms, these may be all the same or different.

A substance in which the molecules consist of one atom or a number of identical atoms is called an element. All other substances are called compounds.

There are roughly a hundred different elements. The number of known compounds amounts to several million.

An atom is defined as the smallest particle of an element which still possesses the characteristic properties of that element.

Further investigation of the structure of matter has shown that atoms, which were originally thought to be indivisible, are in fact composed of a nucleus of protons and possibly neutrons, circled (at a relatively great distance) by one or more electrons.

The electron has a certain negative charge and the proton an equal positive charge, while the neutron has no (external) electric charge.

Under normal circumstances, the number of electrons circling the nucleus of the atom is equal to the number of protons in the nucleus, so that the atom as a whole behaves as electrically neutral.

The motion of the electrons

The electrons do not move arbitrarily, but in certain well-defined "orbits" or "shells" round the nucleus.

These shells, which are normally denoted by the letters K, L, M, etc. (starting from the nucleus), can only contain a number of electrons equal to twice the square of the serial number of the shell in question $(2n^2)$. The maximum possible number of electrons in shells K, L, M, N, etc. is thus respectively 2, 8, 18, 32, etc.

According to quantum mechanics, a complete or fully occupied shell represents a completely symmetrical and thus very stable arrangement of the electrons round the nucleus. Moreover, 8 electrons in the outer shell also form a very stable distribution. no matter whether they are enough to fill the outer shell or not.

Very few elements have all their electron shells completely occupied. These elements are known as the inert gases, and include helium (complete

K shell), neon (complete K and L shells), argon $(2 + 8 + 8)$ and krypton $(2 + 8 + 18 + 8)$. In all the other elements, the outer shell is more or less incomplete.

All atoms try to achieve the stable inert-gas structure, because this has the lowest energy. This tendency is so strong that an atom with an incompletely filled outer shell will try to "borrow" electrons from other atoms. This can happen in several different ways. In the context of the present discussion, only one way is of importance, viz. that found in the covalent bond.

The covalent bond

The covalent bond is a bond between identical atoms, in which a number of electrons move round more than one nucleus, so that the outer shells of all atoms appear to be completely occupied.

For example, two hydrogen atoms, each of which has only one electron, can combine to give a situation in which both electrons move first round one nucleus and then round the other (see Fig. 25), giving both atoms an apparently complete K shell.

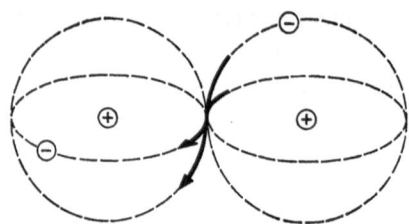

Fig. 25

A similar situation is found with fluorine, where the atoms have 9 electrons, i.e. they are one short of a complete L shell. Here again two electrons (one from each atom) will be shared, giving each atom an apparently complete L shell. If an atom has more than one electron too few in the outer shell, it will try to form covalent bonds with as many (identical) atoms as it lacks electrons. For example, an atom of the element carbon, which lacks 4 electrons in the L shell, will borrow electrons from 4 other carbon atoms to get an apparently complete L shell. Such an arrangement of electrons can be represented graphically as a regular tetrahedron (i.e. a body with four plane surfaces, see Fig. 26), in which the atom situated at the centre of gravity has borrowed one electron from each of the atoms situated at the corners. Another representation is the cubic form (Fig. 27), in which

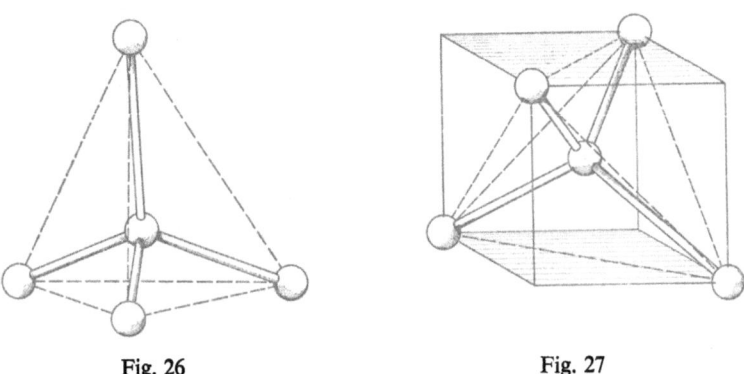

Fig. 26   Fig. 27

the atom at the centre of gravity has borrowed electrons from the diagonally arranged atoms at the corners.

When hydrogen or fluorine atoms form covalent bonds as indicated above, the two atoms which form a bond both get an apparent inert-gas structure and thus have no further tendency to form covalent bonds with other atoms. When carbon atoms form covalent bonds, there is such a tendency. As may be seen from Figures 26 and 27, the central atom has got an apparent inert-gas structure, but the atoms at the corners have only borrowed one electron each from the central atom, and thus still have a deficit of three electrons in the outer shell. They will thus in their turn try to borrow these 3 electrons from other carbon atoms, i.e. to form 3 more covalent bonds. The result is that the arrangement of atoms shown in Fig. 27 tends to extend itself systematically, as shown in Fig. 28. This process is known as the crystallization of the substance.

The pattern in which the atoms are arranged is known as the "crystal

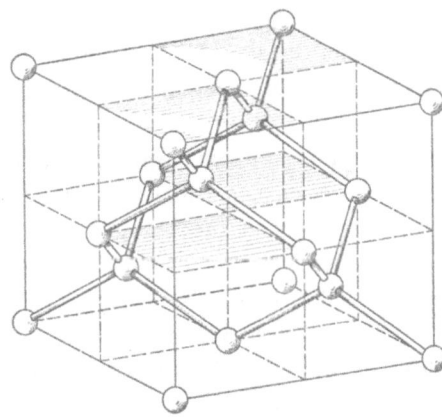

Fig. 28

lattice". The form of the crystal lattice depends on the number of covalent bonds which each atom forms with other atoms. Since the number of covalent bonds is equal to the number of electrons lacking in the outer shell, the form of the crystal lattice may also be said to be determined by the extent to which the outer shell is occupied.

The form of the crystal lattice of carbon, which is known as the diamond lattice (diamond is crystalline carbon), is thus characteristic for atoms which lack 4 electrons in the outer shell.

Germanium, whose nucleus is surrounded by 32 electrons $(2 + 8 + 18 + 4)$, needs 4 more electrons in its outer shell to reach the inert-gas structure of krypton $(2 + 8 + 18 + 8)$, and therefore crystallizes in the same way as carbon, i.e. in the diamond lattice. Another element with similar properties is *silicon*, with 14 electrons $(2 + 8 + 4)$ in the atom, i.e. 4 too few for the argon inert-gas structure $(2 + 8 + 8)$. Both elements belong to the class of semiconductors, that is a number of elements with the special electrical properties described below.

At the absolute zero $(- 273\ °C)$ the electrons which form the covalent bonds (the "valency electrons") are fixed firmly in place. There are thus no electrons which can carry an electric current, so the substance behaves as a complete insulator. As the temperature rises, the motion of the valency electrons increases, so that a (very small) number of these electrons escape from their orbits and move through the crystal as free electrons.

A positive charge (equal in magnitude to the negative charge on the electron) may be ascribed to the "hole" left by such an electron on its escape from the covalent bond, so that such a hole exerts a force of attraction on the valency electrons of other atoms. There will thus be a continual tendency for electrons to jump from their orbits to fill these holes, which increases the motion of electrons through the crystal even more. These free electrons give the element a certain electrical conductivity.

There exists a certain (thermal) equilibrium between the temperature of the crystal and the number of free electrons, this number increasing with the temperature.

Since the number of free electrons determines the electrical conductivity of the substance, elements like germanium and silicon, although they act as perfect insulators at the absolute zero of temperature, will acquire a certain (very slight) electrical conductivity at a certain temperature, and the conductivity will increase with any further increase in the temperature.

Since this increase in the conductivity corresponds to a decrease in the resistance, we may say that semiconductors have a *negative temperature coefficient of resistance*.

The very low electrical conductivity of a pure semiconductor can be very considerably increased by the introduction of certain irregularities into the crystal lattice.

If for example a germanium crystal is contaminated with a relatively small number of *antimony* atoms, which have five electrons in their outer shell, the antimony atoms will give up one electron each in order to be admitted into the crystal lattice. In this way the germanium acquires a large number of free electrons and thus a high electrical conductivity. Germanium contaminated in this way is called *n*-type germanium (*n* for negative, referring to the charge on the electrons).

Alternatively, one can introduce into the germanium crystal a relatively small number of atoms with 3 electrons in their outer shell, e.g. the element *gallium*. Such atoms can only form 3 covalent bonds, so that in the crystal all these atoms will lack one covalent bond. As has been mentioned above, a positive charge can be ascribed to a missing covalent bond ("hole"), which thus exerts a force of attraction on the valency electrons of neighbouring atoms. As a result of this, the introduction of the gallium atoms gives rise to an increase in the number of electrons jumping through the crystal to fill the holes. Since this means that a large number of electrons will continually be moving through the crystal, the material has an electrical conductivity which is considerably higher than that of pure germanium. A germanium crystal contaminated with 3-valent atoms is known as *p*-type (*p* for the positive charge ascribed to the holes).

The semiconductor diode

When a crystal of a semiconductor (e.g. germanium) consists partly of *p*-type material and partly of *n*-type (with a sharp boundary between the two), the free electrons of the *n*-type material will diffuse through the boundary layer to the *p*-type material, in order to eliminate the shortage of valency electrons there. This movement of electrons disturbs the equilibrium between the positive charge (protons) of the nuclei of the atoms and the negative charge of the electrons moving round the nuclei, so that the *n*-type material acquires a positive potential (too few electrons) and the *p*-type material a negative potential (too many electrons).

The *p*-type material will exert a force of repulsion on the electrons of the *n*-type material, as a result of its negative potential. This force increases as the potential difference across the boundary layer increases, i.e. as more electrons penetrate from the *n*-type region to the *p*-type region. As a result of this, the flow of free electrons from the *n*-type to the *p*-type material ceases as soon as a certain potential difference is reached. This equilibrium position is shown in Fig. 29 for a $p - n$ combination which is not connected to any voltage source.

If a voltage source is connected across the $p - n$ combination so that the positive pole is connected to the p-type material and the negative pole to the n-type (see Fig. 30), this voltage source will attract electrons from the p-type material and inject them into the n-type. This will reduce the potential difference existing across the boundary layer, so that the (repulsive) force exerted by the p-type material on the electrons from the n-type also derceases; it is now again possible for a number of free electrons to pass from the n-type material to the p-type. This transport of electrons by the

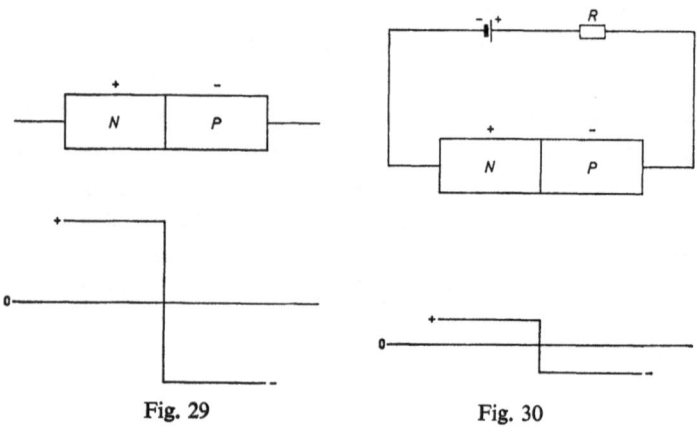

Fig. 29   Fig. 30

voltage source from the p-type material to the n-type, combined with the spontaneous passage of free electrons in the opposite direction via the $p - n$ junction, gives a continuous current of electrons from the p-type material via the voltage source to the n-type material and back via the $p-n$ junction. This gives rise to a voltage difference across the $p-n$ junction which is less than in the absence of an applied voltage. This situation is sketched in Fig. 30. The $p-n$ combination thus behaves as a conductor for this applied voltage. If the voltage source is connected to the $p-n$ combination the other way round (see Fig. 31), electrons will be removed from the n-type material, which already has a positive potential, an injected into the p-type. The potential difference present across the boundary layer will thus increase, so that no current can flow through the circuit formed by the $p-n$ combination and the voltage source; in other words, the $p-n$ combination behaves as an insulator for this external voltage. This is sketched in Fig. 31.

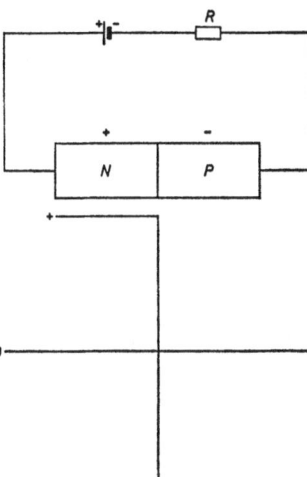

Fig. 31

It may be seen from the above that a *p–n* combination passes electric current from the *p*-type region to the *n*-type (the "forward direction"), but not from the *n*-type to the *p*-type (the "reverse direction"), thus showing a considerable resemblance to the 2-electrode electron tube, the diode. The *p–n* combination is therefore called a semiconductor diode (e.g. germanium diode or silicon diode).

Temperature dependence

In practice, the semiconductor diode is found not to act as a perfect insulator when a voltage is applied in the reverse direction. This is because the kinetic energy of the valency electrons in the covalent bond is increased by heat, light, etc., so that some valency electrons can escape from their bonds and move through the crystal as free electrons. This effect is found both in the *p*-type and in the *n*-type material, and also in the boundary layer. The electrons freed in the boundary layer will pass into the *n*-type material under the influence of the existing electric field. This gives on the one hand a continuous extra shortage of covalent bonds in the boundary layer, which attracts free electrons from the *p*-type material, and on the other hand an increase in the number of free electrons present in the *n*-type material; as a result, the voltage across the boundary layer decreases. This finally gives a state in which just as many free electrons diffuse into the boundary layer from the *p*-type material as valency electrons are freed from the boundary layer and transported to the *n*-type material, i.e. there is a slight continuous current from the *p*-type material to the *n*-type (leak current).

Since the number of electrons freed from the covalent bonds increases with the temperature, the leak current of a semiconductor diode will also increase with the temperature.

Transistors

The usefulness of semiconductor elements was very considerably increased when a combination of p- and n-type germanium was found which possessed amplifying as well as rectifying properties. This combination, which has been given the name of "transistor", consists of two regions (layers) of one type of germanium separated by a very thin layer of the other type. This combination can be either p–n–p or n–p–n; in both cases, the component parts of the transistor are called "emitter", "base" and "collector" respectively.

We shall now describe the operation of the transistor, with reference to the p–n–p combination.

In the absence of an external voltage, the free electrons will diffuse from the n-type to the p-type germanium, just as in the semiconductor diode, in order to annul the lack of covalent bonds in the p-type material. As a result of this, the emitter and the collector acquire a negative potential, and the base a positive one. Since the negative potential of the emitter and the collector (the p-type germanium) repels the free electrons of the base (n-type), the passage of free electrons from the base to the emitter and the collector will cease as soon as a certain potential difference is reached. This equilibrium position is sketched in Fig. 32.

The p–n junctions in the transistor have an incidental rectifying effect,

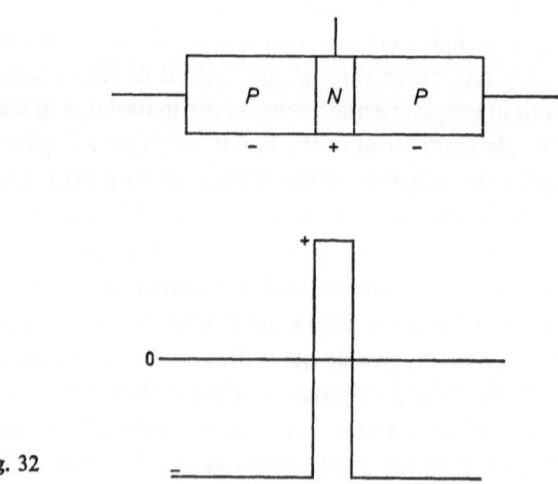

Fig. 32

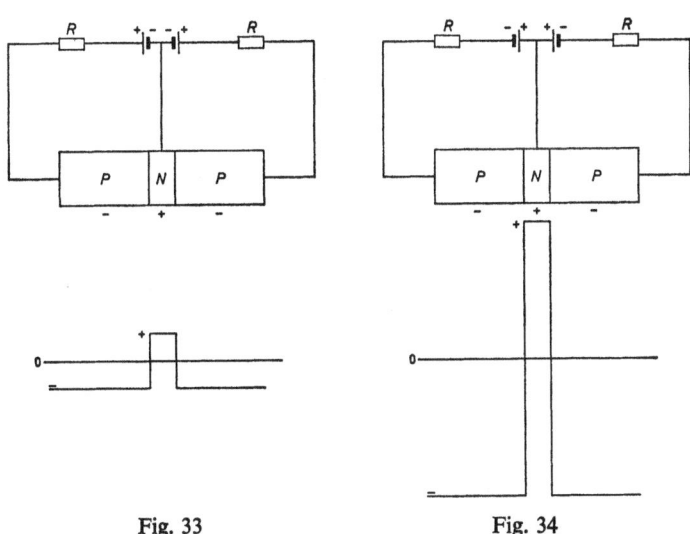

Fig. 33   Fig. 34

i.e. when an external voltage is applied across a junction, current will in general only flow when the negative pole of the voltage source is connected to the n-type germanium (the base); we neglect the leak current. This behaviour of the p–n–p combination is sketched in Figs. 33 and 34.

As long as the voltages across the two junctions are either both in the forward or both in the reverse directions, only rectification occurs. The real function of the transistor, the current amplification, is only found when one of these voltages, in practice that across the base-emitter junction, is in the forward direction, while the other (across the base-collector junction) is in the reverse direction, the latter voltage being large compared to the former (see Fig. 35).

The external voltage applied across the base-emitter junction extracts electrons from the emitter and injects them into the base. This causes the voltage across the junction to fall, so that a number of free electrons can diffuse from the base into the emitter; a continuous electron current thus

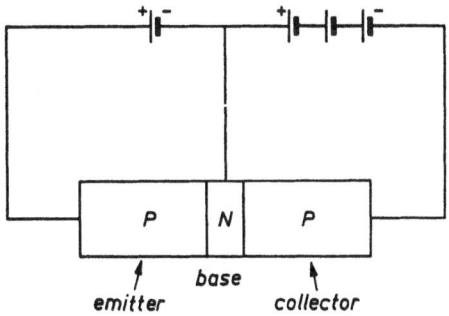

Fig. 35

flows through the circuit formed by the external voltage source and the base-emitter diode. This is the normal rectifying effect of a semiconductor diode, as described above.

While the electron current flowing through the base-emitter junction consists mainly of electrons freed by the contamination of the germanium with 5-valent atoms, some electrons will also occur in this current which have been freed from a covalent bond by thermal or other forms of energy supplied to the base. The latter electrons will leave a "hole" behind in the base, which as we have already mentioned can be regarded as possessing a positive charge equal in magnitude to the charge on the electron.

The external voltage applied across the base-collector junction extracts electrons from the base and injects them into the collector. The ratio of the number of free electrons to the number of holes thus decreases in the base and increases in the collector; in other words, the base has relatively more holes than the collector, so that a number of free electrons will diffuse from the collector into the base (giving rise to a current in what would normally be the reverse direction). Since these electrons cross the collector-base junction with a certain velocity and the base only consists of a very thin layer, a large number of these electrons will arrive in the vicinity of the emitter before they manage to "recombine" with a hole.

The emitter has even more holes than the base, because of its contamination with 3-valent atoms; an electron arriving in the vicinity of the emitter is therefore immediately attracted by the latter. As a result of this, fewer electrons recombine with holes than are freed under the influence of the voltage applied between the base and the emitter. This causes the current across the collector-base junction to be maintained.

The effect described above is stronger as the base layer is made thinner, since then there is less chance that the electrons will recombine with holes in the base: the electrons from the collector come under the influence of the emitter sooner.

If the external voltage applied between the base and the emitter is increased, more electrons will be extracted from the emitter and injected into the base, so that the potential difference across the base-emitter junction decreases and more electrons can cross this junction from the base to the emitter. Since this electron current consists for a certain part of electrons freed from covalent bonds, an increase (or decrease) in this current will cause a corresponding change in the number of covalent bonds permanently lacking in the base (the number of holes).

As the number of holes in the base increases, a greater force of attraction will be exerted on the free electrons of the collector, so that more of these will pass to the base. Because of their velocity and because of the thinness of the base, most of these electrons will also come under the influence of the emitter, and only a few will recombine with holes.

Consideration of the above shows that a change in the voltage applied between the base and the emitter causes not only a change in the number of electrons passing from the base to the emitter, but also a parallel change in the electron current across the collector-base junction.

Since this latter current change is dependent on the former, current amplification occurs. Fig. 36 shows how the current from the emitter is divided over the base and the collector. (So far we have been speaking of the electron current *to* the emitter; Fig. 36 shows the "conventional" current from + to —, i.e. *from* the emitter.)

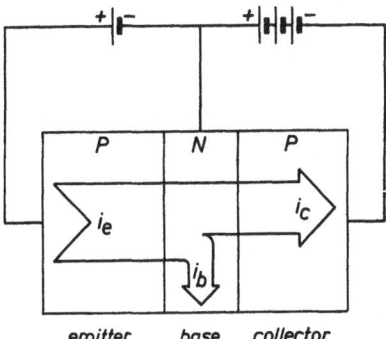

Fig. 36   emitter   base   collector

The relationship between the voltages applied to a transistor and the currents flowing through it is usually given graphically in the form of "characteristics". The voltages in question are:

V_{be} – the base-emitter voltage;
V_{ce} – the collector-emitter voltage;
V_{cb} – the collector-base voltage.

The following relationship always exists between these voltages:

$V_{ce} = V_{cb} + V_{be}.$

The currents flowing in the transistor are (see also Fig. 36):

I_c – the collector current, i.e. the current flowing across the collector-base junction;

I_e – the emitter current, i.e. the total current flowing across the base-emitter junction;

I_b – the base current, i.e. the current flowing across the base-emitter junction as a result of the voltage applied between base and emitter, less the part of the collector current which does not go to the emitter.

These currents are related as follows: $I_e = I_b + I_c.$

One of the most important characteristics is that giving the relationship between the collector voltage V_{ce} and the collector current I_c for a given base current. By way of example, the right-hand side of Fig. 37 shows for an arbitrary transistor a number of these characteristics for various values of the base current at the values A and B of the collector-emitter voltage V_{ce}. It follows from the nearly horizontal course of these lines that the collector current (after a very short initial period) is practically independent of the collector voltage, which means that the collector-emitter circuit of the transistor has a very high internal resistance.

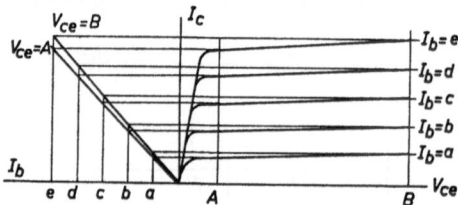

Fig. 37

We can derive directly from a family of I_c–V_{ce} output characteristics a second, also very important family of characteristics, viz the I_c–I_b forward transfer characteristics, which give the collector current I_c as a function of the base current I_b for various values of the collector voltage V_{ce}. The left-hand side of Fig. 37 gives an example of this.

Since the I_c–V_{ce} output characteristics are nearly horizontal, the I_c–I_b forward transfer characteristics for various values of the collector voltage are nearly the same. These, the static I_c–I_b forward transfer characteristics, are only valid when there is no resistance in series with the collector, so that the collector voltage is constant.

When there is a (load) resistance in the collector circuit, one should not use these characteristics, but the dynamic characteristics, obtained by drawing the load line corresponding to this resistance in the I_c–V_{ce} diagram. This is shown in Fig. 38.

A third, less important characteristic is that giving the relationship between

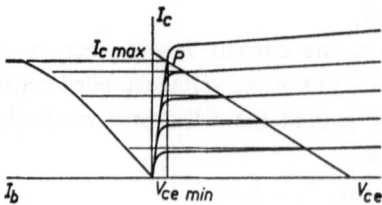

Fig. 38

I_b and V_{be} (input characteristic) of which an example is shown in Fig. 39. It may be seen from this characteristic that the base-emitter circuit of a transistor has a very low internal resistance.

There are many other characteristics apart from the three mentioned above, e.g. the I_c-I_e and I_b-V_{ce} characteristics, but these are in general of even less importance.

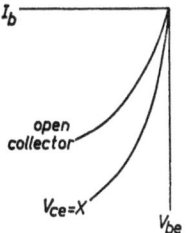

Fig. 39

The current-amplification factor

The current amplification given by the transistor is usually expressed by means of the current-amplification factor α, i.e. the ratio of the change in the collector current to the change in the emitter current causing it:

$$\alpha = \frac{\Delta I_c}{\Delta I_e}$$

Since the change in the collector current is practically equal to the change in the emitter current, α will be slightly less than 1, and will approach 1 even more closely as the current amplification increases.

In practice, therefore, it is usually easier to work with another amplification factor β (formerly α'), the ratio of the change in the collector current to the corresponding change in the base current:

$$\beta = \frac{\Delta I_c}{\Delta I_b}$$

Since $\Delta I_b = \Delta I_e - \Delta I_c$ and $\Delta I_c = \alpha \cdot \Delta I_e$, we can derive a relationship between α and β:

$$\beta = \frac{\Delta I_c}{\Delta I_b} = \frac{\Delta I_c}{\Delta I_e - \Delta I_c} = \frac{\alpha \Delta I_e}{\Delta I_e - \alpha \Delta I_e} = \frac{\alpha \Delta I_e}{(1 - \alpha) \Delta I_e}$$

i.e.

$$\beta = \frac{\alpha}{1 - \alpha}$$

1.9 Problems

1. What current is needed to operate a relay with 6 change-over contact units, 0.1 mm residual pin and 188 turns?
   What current is needed if the coil of the relay has 2000 turns and the residual pin is 0.3 mm thick? And what is the current needed to hold the latter relay? (Table I).

2. Consider a relay with 3 break and 6 make contact units, a 0.3 mm residual pin and a 1000-ohm winding in series with a 1000-ohm resistance, with a 48 V voltage supply.
   Find:
   a) how many turns the relay must have to guarantee that it will operate with these data.
   b) whether it is possible to apply this number of turns, and if so what wire thickness must be used.
   c) the percentage of the possible space taken up by the coil body. Make use of Tables I, II and III.

3. The relay in the circuit shown below has 6425 turns and the resistance tolerance of the winding is $\pm 5\%$.

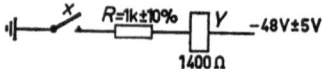

What contact load can be used with this relay so that the relay still operates all the time? And what residual pin is used in this case?

4. Calculate the nominal value of the resistance R at which the relay shown below will still be sure to hold. The relay has 3 change-over and 3 make contact units and a 0.2 mm residual pin.

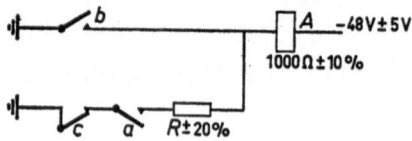

Determine:
a) the number of AT for operation and hence the required number of turns.
b) the maximum possible value of the resistance R at which the relay will be sure to hold.

5. What are the two functions of the residual pin?

6. The relay A of the circuits shown below has a winding of 1200 Ω and 6000 turns and a 0.3 mm residual pin. The relay must work at a voltage of 48 ± 5V and the resistance tolerance of the winding is 10%.

   The operate AT value as given in Table I must be multiplied by 1.2, in connection with the operate time, which must remain within reasonable limits (see Section 1.4).

   For which of the two circuits shown below (which give identical results) is the relay A suited and why?

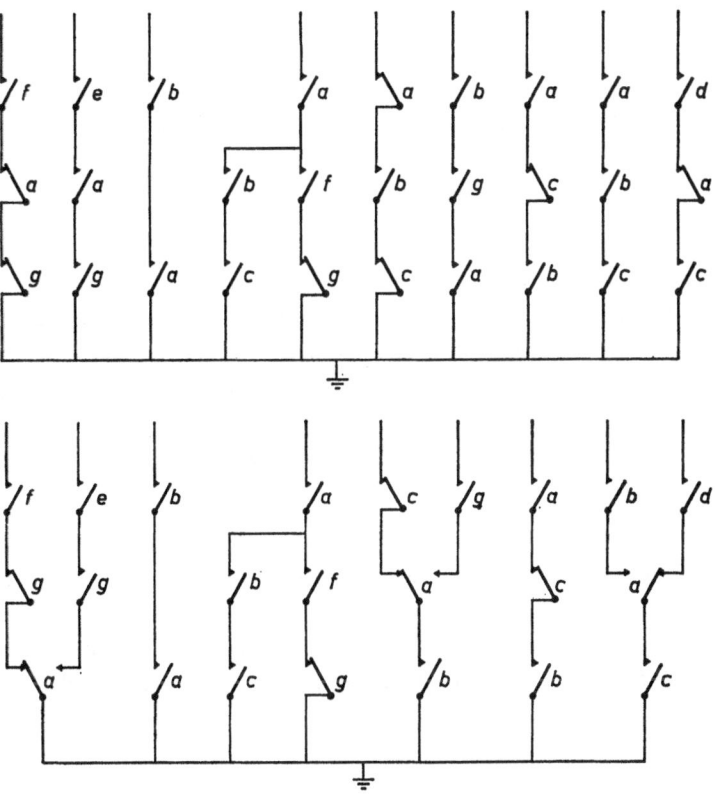

7. What methods do you know for increasing the release time of a relay? Do any of these methods have any influence on the operate time?

8. What is the effect on the operation and construction of a relay if a copper sleeve 2 mm thick is placed round the core?

Chapter 2

ELEMENTARY CIRCUITS

In order to operate a relay, one must produce the requisite number of AT. There are various ways of switching the relay on or off. The most usual elementary circuits will be discussed in this chapter.

2.1   Switching a relay on via a make contact unit

The simplest way of switching a relay on is shown in Fig. 40; the closing of the make contact unit y causes the relay X to operate.

Fig. 40

It is also possible to switch X on as shown in Fig. 41. This ensures that X will not be switched on at the wrong time by an accidental earthing in the connection between X and y. This makes use of the fact that an accidental earthing is more likely than the presence of a battery voltage $(-)$

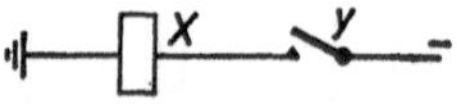

Fig. 41

at the wrong time. The fact that a contact unit y connected to a negative voltage source is available may also be a reason for choosing the circuit of Fig. 41.

In both cases, the opening of the make contact unit y causes the relay X to release: the relay follows the make contact unit y completely.

2.2   Switching a relay on via a break contact unit

If only a break contact unit is available for switching a relay on, the methods of Fig. 42 may be used.

A current will only flow through the winding of the relay X when the short-circuit across X is broken by the contact unit y. The resistance R serves to prevent the negative voltage source from being shorted to earth.

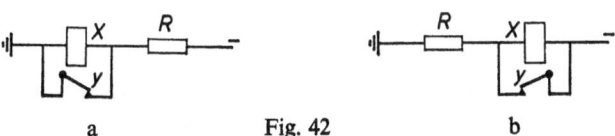

a   Fig. 42   b

In Fig. 42a the contact unit y is connected to earth and in Fig. 42b to the supply voltage $(-)$. The choice of circuit depends on the contact units available (connected to earth or to a negative voltage source) and on other functions which the contact unit y may have to perform (we shall be going into that in more detail later).

When the break contact unit y returns to its original position, X is again shorted and releases. In this case, however, unlike the cases shown in Figs. 40 and 41, the release of X will be retarded by the shorting caused by the contact unit y.

2.3 Circuits with opposed (differential) windings

If a relay carries two windings with the same number of turns, and if these two windings are traversed by a current of the same magnitude in opposite directions, the resultant magnetic field will be zero. It is easy to see from a circuit diagram whether two windings of a relay are connected in the same or opposite directions: the coils are always wound in the same direction, and the beginning of a winding ia always connected to the lowest numbered of the two points used.

For example, in the circuit of Fig. 43 the winding X 1–2 gives a positive AT, and the winding X 5–4 a negative. In the quiescent state, a current flows through both these windings. If this current is the same in both cases, there are no AT to make the relay operate.

The closing of the make contact unit z short-circuits the winding 5–4 of X, so that the current through X 1–2 can supply the number of AT needed to make the relay X operate. If the contact unit z is opened, X will release again. If the break contact unit y is opened, the winding 5–4 can supply

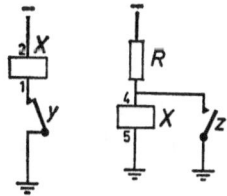

Fig. 43

the AT needed for X to operate, while if y is closed X will again release.

If X is energized because the contact unit z is closed, the opening of y will also cause X to release. In this case, neither winding is energized, so there are no AT at all. If z is now opened, X will operate again.

This circuit thus provides various possible ways of making the relay X operate or release.

One disadvantage of such a circuit is however that the resistance R in series with X 5–4 and the resistances of the X windings can never be made so exact (tolerances) that the current through X 1–2 and X 5–4 is exactly the same. There will thus always be a number of residual AT, due either to X 1–2 or to X 5–4. The circuit of Fig. 44 does not have this drawback. Here the current through X 1–2 and X 5–4 is precisely the same. Closing the contact unit z makes X operate, and opening it makes X release again.

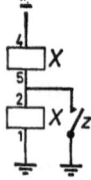

Fig. 44

2.4   Memory circuits (locking circuits)

The purpose of a memory circuit is to indicate that a certain switching operation has been carried out, even after the action in question has ceased. For example, when a patient in a hospital presses a button, this may light a lamp on a panel in the staff quarters. This signal must remain there, even if the patient no longer presses the button, until a nurse has reacted to it. Once the signal has been noticed, it can be erased by pressing another button. Many different memory circuits are known, of which we shall discuss a few in this section.

The simplest method is that in which a relay X, energized via a make contact unit y, itself closes another make contact unit x, which serves to hold X until the reset button T is depressed (Fig. 45).

A method in which a relay holds itself by opening a break contact unit is shown in Fig. 46. The operating contact unit y is also a break contact unit in this case. When the contact y is broken, X is no longer shorted and therefore operates. When y closes again, X does not release, because the break contact unit x now prevents X from being shorted. T makes X release again, thus restoring the short-circuit across X, which will therefore not operate even when T is released.

A circuit which is often used involves the use of a second winding on X. The second winding need only have enough AT (hold value) to hold X. Another possibility is to have one coil, and to reduce the current to the hold value. Fig. 47a shows a circuit with two windings, while Fig. 47b shows how

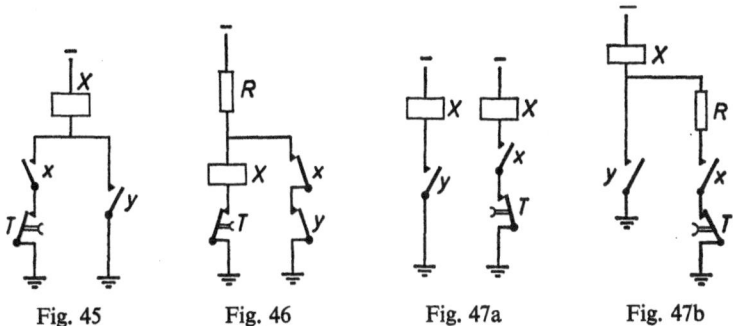

Fig. 45   Fig. 46   Fig. 47a   Fig. 47b

the same result can be obtained with one winding by switching in an extra resistance.

The differential circuits mentioned above can also be used to make memory circuits. In Fig. 48, closing contact unit y shorts the winding X 1–2, so that X operates via the winding 5–4. The contact unit x "remembers" that X has operated by keeping X 1–2 shorted, independent of the contact unit y. When T is opened, X releases, x is thus opened, and we return to the original situation.

2.5   AND and OR circuits

If several conditions have to be fulfilled before a relay X is switched on, one speaks of an AND circuit. Relay X of Fig. 49a can only operate if contact unit a and contact unit b are closed.

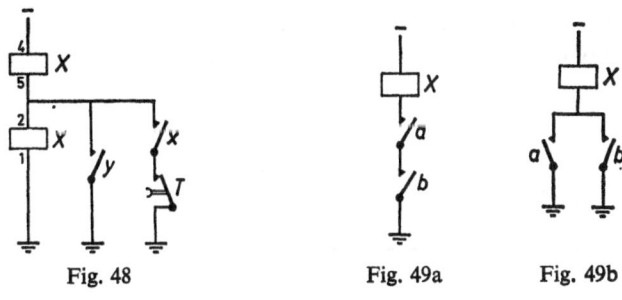

Fig. 48   Fig. 49a   Fig. 49b

An OR circuit is obtained when only one of several conditions has to be fulfilled to make the relay operate. In Fig. 49b, X will operate when the contact unit a or the contact unit b is closed.

Combinations of AND and OR circuits are also possible, as may be seen from Fig. 50a and b.

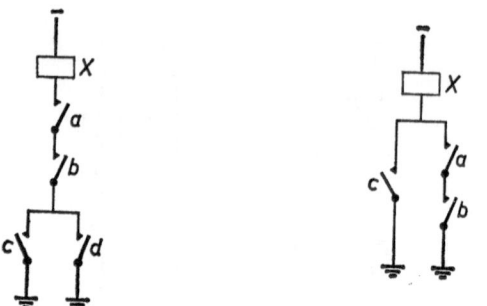

Fig. 50a.   (c *or* d) *and* a *and* b.   Fig. 50b.   c *or* (a *and* b).

2.6   A normal relay connected to operate as a polarized relay

A polarized relay has the advantage that it is sensitive for current flowing in one direction only, but the disadvantage that it has only one or sometimes two change-over contact units. A normal relay, with several contacts, can however also be connected so that it operates when current flows in one direction, but not in the other. This can be achieved by ensuring that the relay winding is shorted when the current flows one way, but not when it flows the other. A diode has the property of only passing current in one direction. In Fig. 51, switching over the change-over contact unit y causes current to flow from earth via the contact unit z, the diode, the contact unit y (which is now in the opposite position to that shown in Fig. 51) and the resistance R to a negative voltage source. The relay winding of X is shorted by the diode for this direction of the current. If, however, z is switched over starting from the position shown in Fig. 51, current will flow from earth via y, the relay winding and z to the negative voltage source. In this situation,

Fig. 51

the diode does not pass any current, so that the relay operates. The diode will give X a longer release time when connected in this way, but a break contact unit operated by X in series with the diode can solve this problem.

2.7 Simplification of contact networks

If we consider the demands which a circuit must meet, it is often possible to simplify the circuit in question. This goal should always be aimed at as far as possible, because the simplest circuit is the cheapest and is also the least likely to go wrong. A break contact unit of a relay can be combined with a make contact unit of the same relay to give a change-over contact

Fig. 52   Fig. 53

unit, as long as they have or can be given one point in common. This saves one contact spring.

When a relay A must operate when the relays X and Y are energized, *or* when the relays X *and* Z are energized, one obtains the circuit of Fig. 52.

In this circuit the two contact units x can be combined to a single contact unit, giving Fig. 53, corresponding to the condition that the relays X *and* (Y *or* Z) are energized.

Another example is that a relay A must operate when the relays (X *or* Y) *and* (X *or* Z) are energized. Here again it is possible to save one x contact unit, by simplifying the circuit to X *or* (Y *and* Z). Both possibilities are shown in Fig. 54.

An important aid in the simplification of contact networks is the taking of the inverse of the contact network in question. One then simplifies the network obtained, and takes the inverse of the simplified network, which gives the simplified version of the original network.

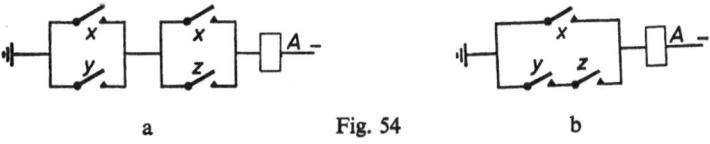

a   Fig. 54   b

2.8   Inversion of contact networks

Inverse circuits are circuits which do exactly the opposite of the circuit from which they are derived. The simplest example is a make and a break contact unit. When the make contact unit closes, the break contact unit opens. A make contact unit is thus the inverse of a break contact unit. The inverse of 2 make contact units in series is two break contact units in parallel. In the series combination, the circuit is only *closed* if both make contact units are closed (both relays energized), and in the parallel combination the circuit is only *open* if both break contact units are open (both relays energized).

The inverse of a given contact network is obtained by replacing all make contact units of the original circuit by break contact units, and vice versa, and all series connections by parallel connection, and vice versa.

A simple inversion is shown by way of example in Fig. 55.

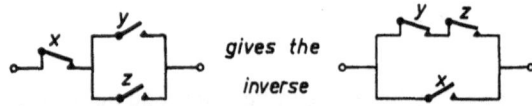

Fig. 55

Break contact unit x in series with parallel combination of make contact unit y and make contact unit z inverts to make contact unit x in parallel with series combination of break contact unit y and break contact unit z.

An example of the simplification of a contact network with the aid of inversion is shown in Figures 56–58. The original network (Fig. 56a) can be split up into the parallel combinations U, V, W and X, which are inverted to give U', V', W' and X'; the inverse of the series combination of U, V, W and X is thus the parallel combination of U', V', W' and X'. This inverse circuit is shown in Fig. 57, which gives a much clearer picture of the possible simplifications. Finally, we invert the simplified inverse circuit, giving the simplified version of the original circuit as shown in Fig. 58.

2.9   The use of capacitors in relay circuits

Capacitors are used as memory elements for storing information. They are only useful for this purpose if the information does not have to be stored for too long. The length of time for which the information can be usefully stored depends mainly on the internal leak resistance of the capaci-

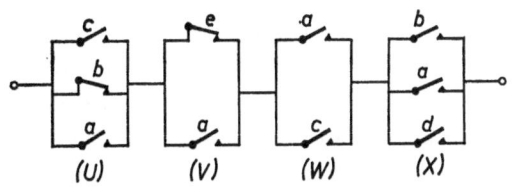

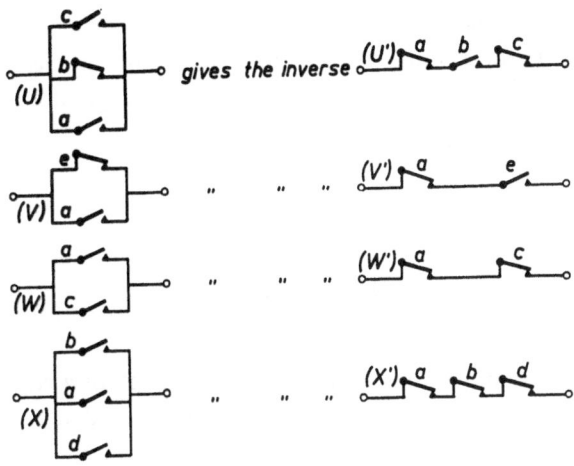

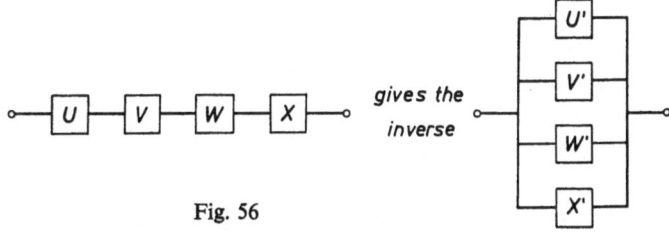

Fig. 56

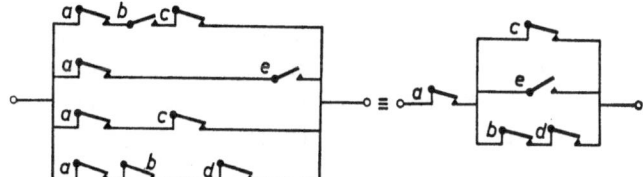

Fig. 57

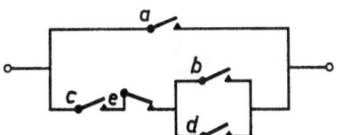

Fig. 58

tor in question. For example, a polystyrene capacitor can store information for 4 hours. If the original voltage is 48 V, at the end of this time it will have fallen to less than 24 V. Electrolytic capacitors have a lower leak resistance, so that they discharge faster and cannot retain information for as long.

Apart from the lower leak resistance, an electrolytic capacitor is considerably smaller than a paper or polystyrene capacitor of a comparable capacitance.

The capacitance of an electrolytic capacitor is inversely proportional to the applied voltage. If for example the nominal voltage for an electrolytic capacitor is 60 V and the capacitor is connected to a voltage of 30 V, the capacitance will be twice the nominal value.

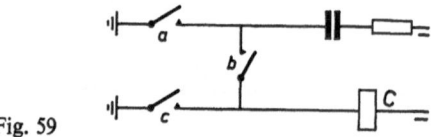

Fig. 59

Fig. 59 shows an example of the "writing" and "reading" of a capacitor. The closing of contact unit *a* causes the capacitor to be charged, after which *a* is opened again. As soon as the contact unit *b* is then closed, the capacitor discharges via the winding of relay *C*. Contact unit *c* closes a hold circuit and at the same time recharges the capacitor.

A capacitance of $10-50\ \mu F$ is needed to operate a relay; the exact value depends on the number of contact units which the relay has. These (electrolytic) capacitors are relatively large, so that a capacitor memory containing a large number of memory elements will take up a lot of room. The use of small polystyrene capacitors with a capacitance of $0.1-0.2\ \mu F$ is therefore very attractive: they are not only much smaller, but their leak resistance is so much higher that they can retain information for hours.

In order to enable such a small capacitance to *operate* a relay, use must be made of a transistor. The charge of the capacitor is only used to make winding 1-2 produce a sufficient flux change to induce a voltage in winding 4-5 which can make the transitor conduct. The circuit for this is shown in

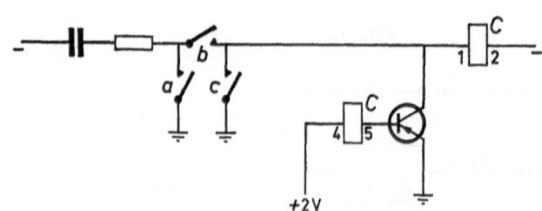

Fig. 60

+2V

Fig. 60. The quiescent base voltage of the transistor is positive ($+2\,V$), so
that the transistor is cut off. The capacitor is charged via contact unit a,
after which a is opened again.

When the capacitor has to be "read", contact unit b is closed, whereupon
the capacitor gives a discharge current through the winding 1–2. The current
pulse thus produced in winding 1–2 induces a voltage in winding 4–5 which
is enough to give the base a negative voltage, so that the transistor conducts.
The collector current flows through winding 1–2, inducing a higher voltage
in winding 4–5, so that the base current increases.

This phenomenon continues until the flux variation stops. This does not
however happen until relay C operates, causing contact unit c to close a
hold circuit. The circuit shown in Fig. 60 is not reliable, because even very
slight interference voltages or residual charges from the capacitor are enough
to cause the relay to operate. The circuit is therefore given a "threshold
voltage", which is much higher than the interference voltage or residual
charge. The complete circuit is given in Fig. 61. The quiescent voltage of
the capacitor C_1 is $-48\,V$ (not shown). The capacitor then has the binary
value 0.

The capacitor can be charged to a voltage of $+24\,V$ via contact unit a.
Its binary value is then 1. The voltage at point P is $-48\,V$. If contact unit b
were to close when the capacitor was at $-48\,V$, the voltage at P would
remain $-48\,V$. The closing of b thus has no effect, because of the diode in
the reverse direction. If the capacitor is charged to $+24\,V$, closing b gives
rise to a brief voltage pulse of $+24\,V$ at P. The diode then allows a current
pulse to pass through winding 1–2 of relay C. The voltage thus induced in
winding 4–5 makes the transistor conduct, as described above.

Even though the leak current of the capacitor allows its voltage to fall to
0, closing contact unit b will still give a pulse to C. On the other hand, a
capacitor with an initial voltage of $-48\,V$ can fall to $-24\,V$ and still no
pulse will be passed to C. The capacitor C_2 protects the circuit against

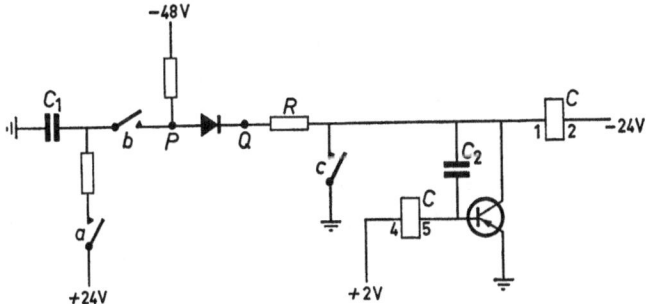

Fig. 61

brief positive interference pulses at point P or Q, by passing the positive pulses to the base of the transistor. Only when pulses with a duration of more than 10–15 μs are applied is the capacitor C_2 charged, and the positive pulse is no longer passed to the base. For further details of the use of capacitors as memory elements, see Chapter 11, Section 3.

Another application of capacitors in relay techniques is in delaying the operation or release of a relay. Fig. 62 shows a circuit in which the relay X, connected between earth and minus, will operate without delay when the contact unit a is closed. The closing of a will also cause the capacitor to be charged via the resistance R. The time needed to charge the capacitor depends on its capacitance and on the value of the resistance R. The current-time curve given in Fig. 4 of Chapter 1 represents the charging curve of the capacitor. The term RC is known as the time constant.

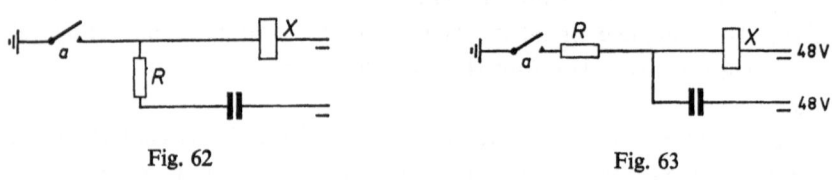

Fig. 62   Fig. 63

Here R is expressed in ohms, C in farads and τ in seconds. These units are impractical. If R is expressed in kohms and C in μF, τ will be in ms.

It may also be seen from Fig. 4 of Chapter 1 that at time 3τ the capacitor is 95% charged. If contact unit a is closed for long enough, R can be given a large value.

When a is opened, the capacitor discharges through the winding of relay X. The current again follows the exponential curve of Fig. 4, and both the value of the current and the discharge time depend partly on the resistance R. The object is to keep the current above the hold value of the relay for a certain time. It follows that R must be chosen as high as possible in order to make the RC time as long as possible; the value of R is limited by the number of turns on the relay winding. Delays of 0.5 to 20 s can be obtained with electrolytic capacitors, depending on the capacitance and the spring-set load of the relay.

The operation of a relay can also be delayed with the aid of a capacitor. The principle of this is shown in Fig. 63. When contact unit a is closed, the capacitor is charged via the resistance R.

The voltage of relay X is now determined by the difference between the voltage of 48 V and the voltage drop across the resistance R. Initially the current through the resistance R is high and the voltage of the relay X is hence low. As the charge current falls off (Fig. 4), the voltage of the relay increases. The operation of the relay is thus delayed by the presence of the capacitor. An application of this principle is shown in Fig. 64, where the relay is connected as a pulse generator. Both the pulse frequency and the operate time/release time ratio of the relay can be determined by the capacitance and the resistance.

Relay X in this figure is a differential relay (see Section 2.3). When contact unit a is closed, currents arise in both windings of relay X, in opposite directions. As soon as the capacitor has been charged to a certain level, the current through winding 4–5 is so low that the relay X can be energized by winding 1–2. This causes the contact unit x to be opened. The capacitor is now discharged through both windings of relay X. In this case however

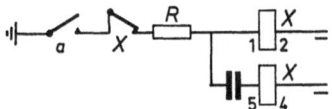

Fig. 64

the direction of the current in winding 4–5 is opposed to the direction in this winding during the operation of the relay, so that both windings have the same magnetization direction. The desired pulse frequencies and relative pulse lengths can be obtained by a suitable choice of R and C.

2.10   Problems

1. Give the inverse of the following circuits.

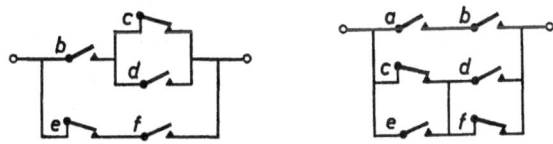

2. Combine and simplify the following network.

3. Given a relay X with one winding and one change-over contact unit. Also available are two non-locking buttons (A and B) and one resistance R. The relay X must operate when button A is depressed and stay energized when A is released. Pressure on button B should cause X to release again, and to remain so when B is released, until A is pressed again. The two push-buttons are not used simultaneously. Give a circuit which meets the above demands for each of the following cases:
   a) with a break contact unit on button A and a break contact unit on button B.
   b) with a make contact unit on A and a make contact unit on B.
   c) with a break contact unit on A and a make contact unit on B.
   d) with a make contact unit on A and a break contact unit on B.
   On each push-button, one contact must be connected to earth. Also give circuits for the above cases a–d if one contact of both push-buttons must be connected to $-$.
   A total of 8 circuits must thus be given.

4. Draw the inverse of the following circuit, after having removed two superfluous change-over contact units.

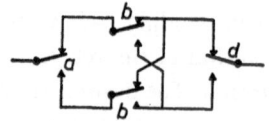

5. The relay X of the circuit given below is a differential relay, and must operate when one of the energizing circuits is broken.
   Both windings have a resistance of 200 ohms and 4200 turns.
   Relay X only needs one change-over contact unit, and has a 0.2 mm residual pin.
   The resistance tolerances of the resistor and the windings are taken into consideration if we assume independent voltages of 44 and 56V for the two windings.

The problem is: does this relay X work well in this circuit, and if not what measures can be taken to guarantee good operation? Make use of the data in the tables of Chapter 1, and take 1.2 × the number of operate AT and 0.8 × the number of non-operate AT.

The number of operate AT for a relay with 0.2 mm residual pin and 1 change-over contact unit is 90. The number of non-operate AT is 48.

6. Combine and simplify the following contact network (to 6 contact springs).

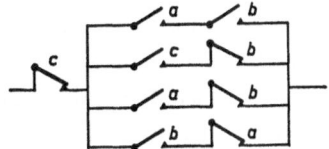

7. Draw the inverse of the following circuit.

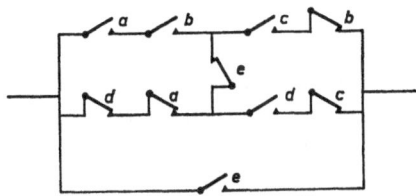

8. Combine and simplify the following circuit (to 14 contact springs), and give its inverse.

Chapter 3

SWITCHING ALGEBRA

For economical and practical reasons it is desirable when designing circuits to find the simplest possible form for a given circuit.

Switching algebra, which allows mathematical expression and manipulation of the form of a circuit, has become a useful aid in the design, simplification and combination of networks to the simplest forms.

Switching algebra relates to circuits built up of elements which can assume one of two states such as relays (energized or not), valves (conducting or cut off), transistors (ditto), switches (open or closed).

Switching algebra is also known under other names, e.g. relay algebra, binary algebra, etc.

Systematic use of switching algebra, whenever possible, gives a better insight into the structure of networks and stimulates logical thinking.

3.1 Symbols and theorems

In switching algebra, symbols can only have one of two values, viz 0 or 1. As in any other algebra, we make use of certain operations, namely addition $(+)$ and multiplication $(\times$ or $\cdot)$. With the aid of these operations we can form combinations such as $0+0, 0+1, 1+1, 1+0, 0\cdot0, 0\cdot1, 1\cdot1, 1\cdot0$, or more complicated combinations such as $0\cdot1+1+1\cdot0$.

If we introduce the $=$ sign, we may give the following definitions, which represent the rules of calculation in this algebra:

a	$0 \cdot 0 = 0$
b	$0 \cdot 1 = 1 \cdot 0 = 0$
c	$1 \cdot 1 = 1$
d	$0 + 0 = 0$
e	$0 + 1 = 1 + 0 = 1$
f	$1 + 1 = 1$

It will be seen that the only one of these six definitions which differs from those used in normal arithmetic is f. We will be returning to this definition below.

With the aid of these 6 rules of calculation, we can simplify complicated expressions. For example:

$$0 \cdot 1 + 1 + 1 \cdot 0 + 0 + 1 \cdot 1 = 0 + 1 + 0 + 0 + 1 = 1 + 0 + 1 = 1$$

If we now introduce symbols in this algebra as in normal algebra, we must remember that these symbols are "only" binary, i.e. they can only assume the values 0 and 1. In the following, a variable x must thus always be regarded as binary.

The way in which these two values for symbols are expressed will be given later.

We can now derive a number of theorems for the symbols introduced:

1	$1 \cdot x = x$
2	$0 \cdot x = 0$
3	$1 + x = 1$
4	$0 + x = x$
5	$x \cdot x = x$
6	$x + x = x$
7	$x \cdot y = y \cdot x$
8	$x + y = y + x$
9	$x + y + z \quad = (x+y) + z = x + (y+z) = (x+z) + y$
10	$x \cdot y \cdot z \quad = (x \cdot y)z = x(yz) = (xz)y$
11	$x + x \cdot y \quad = x(1+y) = x$
12	$x(x+y) \quad = x \cdot x + x \cdot y = x + x \cdot y = x(1+y) = x$
13	$x \cdot y + x \cdot z = x(y+z)$
14	$(x+y)(x+z) = x + xz + xy + yz = x(1+z+y) + yz = x + yz$

These theorems can be simply proved; many of them are in agreement with normal algebra. The proof of the theorems which do not agree with normal algebra is given below.

1. *Theorem 3* $1 + x = 1$

As mentioned above, x can only assume two values, from which it follows that:
if $x = 0$, $1 + x = 1 + 0 = 1$, according to definition e.
if $x = 1$, $1 + x = 1 + 1 = 1$, according to definition f.

2. *Theorem 5* $x \cdot x = x$

This can also be proved by substitution:
if $x = 0$, then $x \cdot x = 0 \cdot 0 = 0$ according to definition a.
if $x = 1$, then $x \cdot x = 1 \cdot 1 = 1$ according to definition c.
Thus $x \cdot x = x$ for both possible values of x, Q.E.D.

3. *Theorem* 6 $x+x=x$

If $x=0$, then $x+x=0+0=0$ according to definition d.
If $x=1$, then $x+x=1+1=1$ according to definition e.
It is thus always true that $x+x=x$.

4. *Theorem* 11 $x+xy=x$

This can be proved with the aid of theorem 3:

$$x+xy=x(1+y)=x(1)=x$$

because $1+y=1$ according to theorem 3.
It is also possible to prove theorem 11 by giving x and y all possible combinations of values.

To prove: $x+xy=x$

If $x=0$
$y=0$ $\Big\}$ then $x+xy=0+0\cdot0=0+0=0=x$ definitions a and d

If $x=0$
$y=1$ $\Big\}$ then $x+xy=0+0\cdot1=0+0=0=x$ definitions b and d

If $x=1$
$y=0$ $\Big\}$ then $x+xy=1+1\cdot0=1+0=1=x$ definitions b and e

If $x=1$
$y=1$ $\Big\}$ then $x+xy=1+1\cdot1=1+1=1=x$ definitions c and f

This thus also proves that $x+xy=x$ for both values of x and y.

5. *Theorem* 12 $x(x+y)=x$

This can be proved with the aid of theorems 3 and 5:

$$x(x+y)=x.x+x.y=x+xy=x(1+y)=x$$

since $x \cdot x=x$ according to theorem 5 and
$1+y=1$ according to theorem 3.

6. *Theorem* 14 $(x+y)(x+z)=x+yz$

This can also be proved with the aid of theorems 3 and 5. The proof will be left as an exercise to the reader.
Here again it is possible to prove the theorem by the substitution method. Since there are 3 variables, x, y and z, all of them binary, there are 2^3 possible combinations, viz:

	x	y	z
1st combination	0	0	0
2nd combination	0	0	1
3rd combination	0	1	0

4th combination 0 1 1
5th combination 1 0 0
6th combination 1 0 1
7th combination 1 1 0
8th combination 1 1 1

If we now calculate the values of $(x+y)(x+z)$ and $x+yz$ for all 8 combinations, we obtain:

	x	y	z		$(x+y)(x+z)$	$x+yz$
1st combination	0	0	0	gives	0	0
2nd combination	0	0	1	gives	0	0
3rd combination	0	1	0	gives	0	0
4th combination	0	1	1	gives	1	1
5th combination	1	0	0	gives	1	1
6th combination	1	0	1	gives	1	1
7th combination	1	1	0	gives	1	1
8th combination	1	1	1	gives	1	1

The above furnishes the proof that:

$(x+y)(x+z)=x+yz$ is true for both values of x, y and z.

The introduction of the operation of inversion into switching algebra allows us to derive a number of other very useful theorems. We will denote the inverse of x by x'. This symbol has the following significance:

If $x=0$, then $x'=1$ and *vice versa*. With the aid of this new operation we can derive the following theorems:

15	$x+x'$	$=1$
16	$x \cdot x'$	$=0$
17	x	$=(x')'$
18	$(x+y+z)'$	$=x'y'z'$
19	$(xyz)'$	$=x'+y'+z'$
20	$(x+y')y$	$=xy$
21	$xy'+y$	$=x+y$
22	$xz+x'y+yz$	$=xz+x'y$
23	$(x+y)(x'+z)(y+z)=(x+y)(x'+z)$	
24	$(x+y)(x'+z)$	$=xz+x'y$

Theorems 15 to 24 can be proved by substituting all possible values of the variables in each equation. It may be seen from theorems 18 and 19 that a number of terms between brackets can be inverted as a whole. This is indicated by placing the accent outside the brackets: $(...)'$.

We can prove theorem 18 by showing that $x+y+z$ is the opposite of

$x'y'z'$ for all possible combinations of values:

	$x+y+z$	$x' \cdot y' \cdot z'$
1st combination	$0+0+0=0$	$1 \cdot 1 \cdot 1 = 1$
2nd combination	$0+0+1=1$	$1 \cdot 1 \cdot 0 = 0$
3rd combination	$0+1+0=1$	$1 \cdot 0 \cdot 1 = 0$
4th combination	$0+1+1=1$	$1 \cdot 0 \cdot 0 = 0$
5th combination	$1+0+0=1$	$0 \cdot 1 \cdot 1 = 0$
6th combination	$1+0+1=1$	$0 \cdot 1 \cdot 0 = 0$
7th combination	$1+1+0=1$	$0 \cdot 0 \cdot 1 = 0$
8th combination	$1+1+1=1$	$0 \cdot 0 \cdot 0 = 0$

Theorem 20 can be proved as follows:

$$(x+y')y = xy + y'y$$

According to theorem 16, $yy' = 0$ and can thus be omitted.

Theorem 21 can be proved as follows. If we invert $xy'+y$ we find $(xy'+y)' = (x'+y)y'$ (from theorems 18 and 19). Multiplying out, we have $(x'+y)y' = x'y' + yy'$. Here again we can omit yy' (theorem 16), which leaves us with $x'y'$. We now invert again to get back to the original state: $(x'y')' = x+y$, i.e. $xy'+y = x+y$, Q.E.D.

This theorem may be stated in words as follows.

If a sum includes a certain quantity (combination of quantities), while the inverse of this quantity (combination) also occurs in the sum, multiplied by another quantity, we may omit the inverse of the first-mentioned quantity (combination). For example:

$$a(b+c) + b'c' = a + b'c'$$

This sum contains $b'c'$, together with the inverse of $b'c'$, i.e. $(b+c)$, multiplied by a.

This can also be proved by the method used for theorem 21. If we take the inverse of $a(b+c) + b'c'$ we obtain:

$$(a' + b'c')(b+c) = a'b + a'c = a'(b+c)$$

Inverting again, we find:

$$a + b'c', \quad \text{Q.E.D.}$$

In order to prove theorem 22, we multiply the term yz from $xz + x'y + yz$ by 1, which clearly leaves the expression as it is. However, $x + x' = 1$ (theorem 15). We can thus multiply yz by $x + x'$ without altering the above expression This gives:

$$xz + x'y + yz(x + x') = xz + x'y + xyz + x'yz = xz + xyz + x'y + x'yz =$$
$$xz(1 + y) + x'y(1 + z) = xz + x'y, \quad \text{Q.E.D.}$$

The proof of theorem 23 is similar to that of theorem 22. If we invert

$$(x + y)(x' + z)(y + z)$$

we obtain

$$\{(x + y)(x' + z)(y + z)\}' = x'y' + xz' + y'z'$$

According to theorem 22, this expression is equal to $x'y' + xz'$. Inverting back to the original state gives $(x + y)(x' + z)$. Consideration of theorems 21, 22 and 23 allows us to state the following generalization:

> *If a sum of products (product of sums) contains a quantity (combination of quantities) and its inverse, then the product (sum) of the quantities associated with the first-mentioned quantity may be omitted.*

Example 1: $abc + a'bd + bcd = abc + a'bd.$

The expression contains both a and a'. The product of the quantities associated with a, $bc \cdot bd = bcd$, may thus be omitted.

This can be proved by multiplying bcd from the above expression by $a + a' = 1$. We then get:

$$abc + a'bd + bcd(a + a') = abc + a'bd + abcd + a'bcd =$$
$$abc(1 + d) + a'bd(1 + c) = abc + a'bd$$

Example 2: $(a + b + c + d)(a + b' + d + e)(a + c + d + e) =$
$(a + b + c + d)(a + b' + d + e)$

The expression contains both b and b'. The sum of the quantities associated with b, $(a + c + d) + (a + d + e) = a + c + d + e$, may be omitted. If we invert the given expression we obtain:

$$a'b'c'd' + a'bd'e' + a'c'd'e'$$

According to the proof of example 1, we may omit $a'c'd'e'$ from the expression thus obtained. Inverting again gives the required expression,

$$(a + b + c + d)(a + b' + d + e)$$

Example 3: $(a + b)cd + a'b'ce + cde = (a + b)cd + a'b'ce$

The quantities a and b occur together as $a + b$, but also as the inverse of this, $a'b'$. The product of the quantities associated with a and b, $cd \cdot ce = cde$, may be omitted.

This can be proved by multiplying cde by $(a + b + a'b') = 1$ (if we regard $a + b$ as x, then $a'b'$ is x', and $x + x' = 1$).

This gives

$$(a+b)cd+a'b'ce+cde(a+b+a'b') \quad =$$
$$acd+bcd+a'b'ce+acde+bcde+a'b'cde =$$
$$acd(1+e)+bcd(1+e)+a'b'ce(1+d) \quad =$$
$$acd+bcd+a'b'ce=(a+b)cd+a'b'ce$$

This incidentally proves theorems 22 and 23.

If we multiply out the expression on the left-hand side of theorem 24, we obtain:

$$(x+y)(x'+z)=xx'+xz+x'y+yz=xz+x'y+yz$$

which according to theorem 22 is equal to

$$xz+x'y, \quad \text{Q.E.D.}$$

The simplest expression will be found directly by noting only the products wherein the term occurs, given in both states (x and x').

We can now try to simplify complicated expressions with the aid of the above.

Examples

1	Simplify $xy+xyz+yz$
2	Simplify $xy'+z+z(x+y')$
3	Simplify $a+a'b+a'b'c+a'b'c'd+a'b'c'd'e+\dots$
4	Simplify $wx+xy+x'z'+wy'z$

Example 1

$$xy+xyz+yz=xy(1+z)+yz=xy+yz=y(x+z)$$

Example 2

$$xy'+z+z(x+y')=xy'+z\{1+(x+y')\}=$$
$$xy'+z, \text{ since } 1+x=1 \text{ (theorem 3)}.$$

Example 3

$$a+a'b+a'b'c+a'b'c'd+a'b'c'd'e+\dots=$$
$$a+a'(b+b'c+b'c'd+b'c'd'e+\dots)= \quad \text{(theorem 21)}$$
$$a+b+b'c+b'c'd+b'c'd'e+\dots=$$
$$a+b+b'(c+c'd+c'd'e+\dots)=a+b+c+c'd+c'd'e+\dots=$$
$$a+b+c+c'(d+d'e+\dots)=a+b+c+d+d'e+\dots=$$
$$a+b+c+d+e+\dots$$

Example 4

Here we make use of theorem 22, the other way round: it is equally possible to replace $x'z'+wy'z$ by $x'z'+wy'z+x'wy'$. That this can finally also lead to simplification will be seen from this example.

$$wx+xy+x'z'+wy'z$$

may be written as

$$wx+xy+x'z'+wy'z+x'wy'$$

We may also add the term wy', since

$$wx+x'wy'=wx+x'wy'+wy'$$

We now have

$$wx+xy+x'z'+wy'z+x'wy'+wy'$$

We can put wy' outside brackets, which gives

$$wx+xy+x'z'+wy'(1+z+x')=wx+xy+x'z'+wy'=$$
$$xy+x'z'+wy'$$

This will be seen to be a simplification of the original form.

Problems

Simplify

1. $xy+xy'z+yz$
2. $(xy'+z)(x+y')z$
3. $(x+y')(y+z')(z+x')(xyz+x'y'z')$
4. $w'x'+x'y'+yz+w'z'$
5. Write the inverse of:
   $\{a+b'c'd\}\{a'd'+f(bc'+e)\}$

3.2 Application of switching algebra to relay circuits

A contact network may be regarded as a binary assembly. If the network forms a closed circuit, it is designated by 1, while an open circuit is represented by the symbol 0.

Whether the network is open or closed depends only on whether the contact units which make up the network are open or closed.

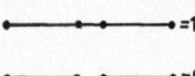

If we introduce contact units into the above circuits, we are in a position to assign a certain value (0 or 1) to the contact units. A closed contact unit is designated by the symbol 1 and an open one by 0.

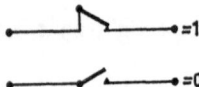

Two closed contact units in series are represented by $1 \cdot 1$, using the sign $\cdot$ to indicate the series connection.

Two closed contact units in series form a closed circuit, which we may indicate by $1 \cdot 1 = 1$ (definition c).

The other definitions given above in which the sign $\cdot$ occurs can be "translated" into relay circuits in a similar way.

Two open contact units in series are denoted by $0 \cdot 0$. This combination is an open circuit, i.e. $0 \cdot 0 = 0$ (definition a).

An open contact unit in series with a closed contact unit gives an open circuit: $0 \cdot 1 = 0$ (definition b).

The circuit is also open if a closed contact unit is connected in series with an open contact unit: $1 \cdot 0 = 0$ (definition b again).

Definition b shows that contact units connected in series can be interchanged without altering the properties of the circuit.

By using the operation $+$ to indicate parallel connections, we can "translate" the other definitions given above into contact circuits.

Two closed contact units in parallel give a closed circuit: $1+1=1$ (definition f).

$=1+1=1$

An open contact unit in parallel with an open contact unit gives an open circuit: $0+0=0$ (definition d).

$=0+0=0$

An open contact unit in parallel with a closed one, or a closed contact unit in parallel with an open one, gives a closed circuit (definition e): $0+1=1+0=1$.

$=0+1=1$　　　　$=1+0=1$

This shows that contact units may also be interchanged when they are connected in parallel.

In a network built up of relay contacts, the various relays are indicated by capital letters, and the contact units belonging to these relays by the corresponding small letters. Relay A thus has the contact units a. One distinguishes between make and break contact units by giving the latter an accent: a is a make contact unit, and a' a break contact unit.

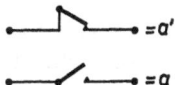

$=a'$

$=a$

The condition for the series combination of the contacts x and y to be a closed circuit is that both x and y should have the value 1.

$=x'.y'=1$

If the contacts x and y are in parallel, the circuit will be closed if x or y, or both, are closed: $x'+y=1$, $x+y'=1$, $x'+y'=1$.

Complicated networks can be represented by a formula, which we call the algebraic formula of the network in question. By making use of round and square brackets and braces, we can write a formula for even the most complicated network.

Examples

5

$=a+b.c$

6

$=a'(b'+c)$

7

$=(a+b')(c'+d)$

8

$=ab+c'(d+e'f)$

9

$=a\{bc+b'(cd'+ef)\}$

We can now simplify contact networks by "calculation", using the theorems and definitions given above on the algebraic formulae of the networks in question. Some examples will serve to illustrate this statement.

Example 10

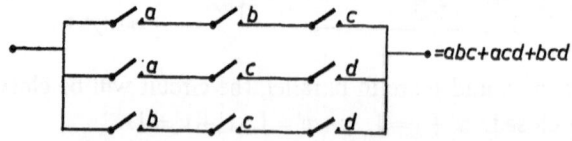

$=abc+acd+bcd$

The network shown above contains 9 contact units (18 contact springs). By placing e.g. c outside brackets, we obtain $c(ab+ad+bd)$. This expression can be further simplified by placing either a, b or d outside brackets. If we choose a, we find the expression $c\{a(b+d)+bd\}$. This corresponds to the network

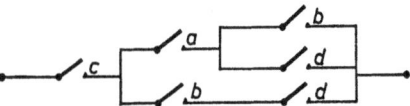

i.e. we have simplified the network to 6 contact units (12 contact springs).

If we put b outside brackets instead of a, we obtain $c\{b(a+d)+ad\}$, which gives the same network, but with different names for the various contact units.

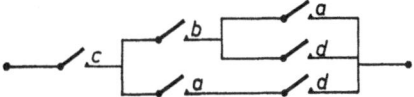

Another way of simplifying the original expression is to put ac outside brackets, which gives $ac(b+d)+bcd$. We can multiply the last term (bcd) by $b+d$ without altering the value of the expression, since $bcd(b+d)=bcd$.

This gives $ac(b+d)+bcd(b+d)=(ac+bcd)(b+d)=c(a+bd)(b+d)$. This gives a network different from that given above, but with the same number of contact units and the same result, viz

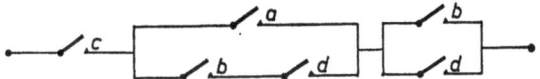

This shows that it is possible to get several *correct* solutions. It seldom happens that only one answer can be given.

Example 11

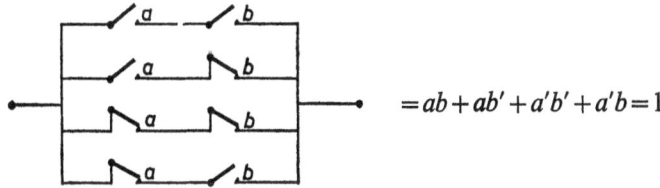

$=ab+ab'+a'b'+a'b=1$

$$ab+ab'+a'b'+a'b=a(b+b')+a'(b+b')=(a+a')(b+b').$$

Since both $a+a'$ and $b+b'$ are equal to 1, the whole expression boils down to $1\cdot1=1$.

In this case it is possible to show in a simpler manner that the circuit is always closed. All $2^2 = 4$ series combinations of the relays A and B occur in this circuit. It is not possible to imagine any combination in which one of the 4 parallel branches of the circuit is not closed. The result is thus always a closed circuit (1).

Example 12

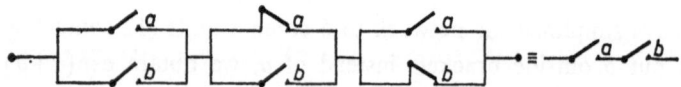

$(a+b)\,(a'+b)\,(a+b') = (aa'+ab+a'b+b)\,(a+b') = b(a+b')$, because aa' always gives an open circuit (0), and the presence of b allows us to omit the combinations ab and $a'b$. Further,

$$b(a+b') = ab+bb' = ab.$$

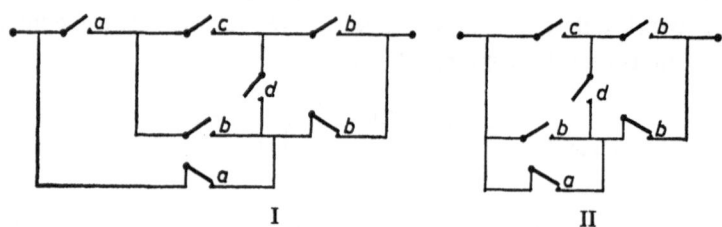

Example 13

Prove that the above two networks are equivalent.

The algebraic formula for network I is:

$$
\begin{aligned}
&acb + acdb' + abdb + a'b' + a'db + a'bcb = \\
&abc + ab'cd + abd + a'b' + a'bd + a'bc = \\
&bc(a+a') + bd(a+a') + ab'cd + a'b' = \\
&bc + bd + ab'cd + a'b' = bc + bd + b'(acd+a') = \\
&bc + bd + b'cd + a'b' \quad \text{(see theorem 21)}
\end{aligned}
$$

Network II gives the formula:

$$
\begin{aligned}
&cb + cdb' + bdb + a'b' + a'db = bc + b'cd + bd + a'b' + a'bd = \\
&bc + b'cd + bd(1+a') + a'b' = bc + bd + b'cd + a'b'
\end{aligned}
$$

The two networks are thus equivalent as regards their function.

Example 14

Simplify the following algebraic formula to 12 contact springs, and draw the simplified network.

$$(a+bc)\left[d'e'\{c'+a'(b+c)\}\right]+(b+d)(b'+d')+b'c'de$$

Multiplying out, we find:

$$(a+bc)\{d'e'(c'+a'b+a'c)\}+bd'+b'd+b'c'de=$$
$$(a+bc)(c'd'e'+a'bd'e'+a'cd'e')+bd'+b'd=$$
$$ac'd'e'+a'bcd'e+bd'+b'd=$$
$$ac'd'e'+bd'+b'd=d'(ac'e'+b)+b'd$$

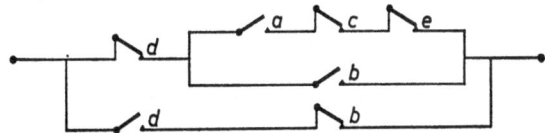

It is possible to make the contact units b and d change-over contact units, which gives a total of 12 contact springs in the final network.

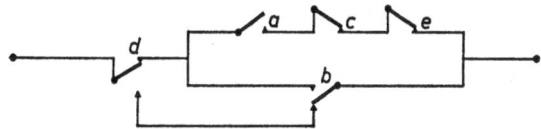

Problems

6. Simplify:
   $(a'+b)(b'+c)(c'+d)(a'+d')(ab'+bc'+cd'+a'd)$ to 8 contact springs.

7. Prove by switching algebra that:

$$a'b+b'c+ac'=ab'+bc'+a'c$$

8. Simplify to 9 contact springs and draw:

$$(ab'+a'c'+bc'+a'c)(ac+a'b+ac'+b'c)$$

9.

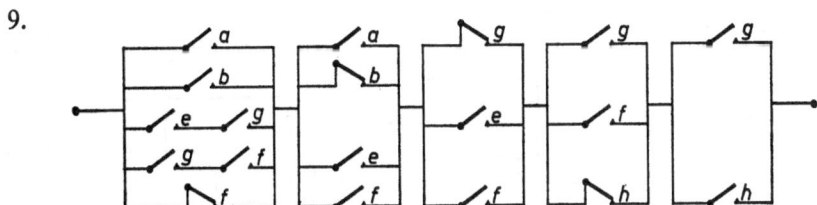

a) Give the algebraic formula for the above circuit.
b) Simplify this formula as much as possible.
c) Draw the simplified network (14 contact springs are possible).

10. Simplify $a(b+c+d)+b(a+c+d)+c'(a+b+d)+d'(a+b+c)$

11. Prove that $(a'+b+c+d')(a+b'+c'+d)$
is equal to $ab+b'c+c'd'+a'd,$
but is also equal to
$a'b'+bc'+cd+ad'.$

12. Show that the contact units a and d in the network shown below may be interchanged.

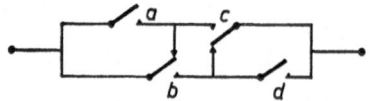

Invert the circuit and draw the inverse network.

13. Simplify and draw (possible with 10 contact springs):

$$\{b'+(d'+f)(e'+a+g)\}\{b+f(d+a+e')\}$$
$$\{a+g+c(d+e+f')\}(bd'f+df')$$

14.

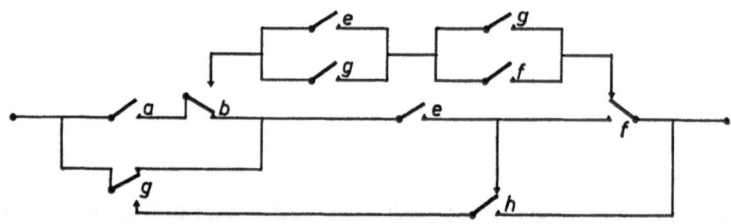

a) Give the algebraic formula for the above figure.
b) Simplify the expression found.
c) Draw the simplified network (14 contact springs).
d) Give the inverse of the expression a).

15.

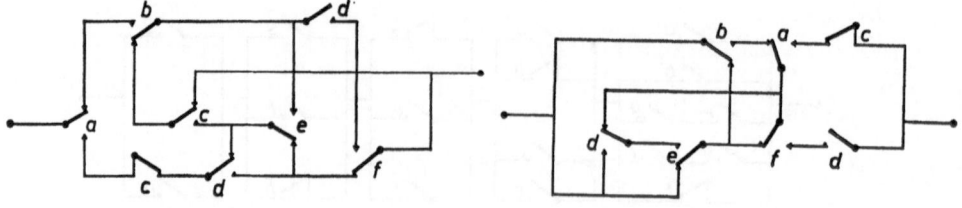

An attempt to simplify the left-hand network gave the right-hand one. Is this correct?

3.3 Functions and inverse functions

A function can be considered as a series of instructions. For example, $f(X, Y, Z)$ may mean that the behaviour of a circuit depends in the specified way on the contact units of the relays X, Y and Z.

An inverse function represents the exact contrary of the circuit from which it is derived. The simplest inverse functions are those of the make and break contact units of a relay: when the make contact unit is closed, the break contact unit will be open and *vice versa*. We have already mentioned that this process of inversion is indicated by an accent in switching algebra: the inverse of x is x'.

We have also mentioned that the series connection of a switching function with its inverse will always give an open circuit: $xx' = 0$.

The inverse of a series circuit is a parallel circuit, where all make contacts are replaced by break contacts and *vice versa*. Similarly, the inverse of a parallel circuit is a series circuit, with the contacts replaced in the same way.

A series connection of the *make contact units* of relays X and Y has the property that it is only closed when both relay X and relay Y are energized. The inverse of this circuit should therefore only be *open* under these conditions. A parallel combination of *break contact units* x and y does indeed fulfil this condition.

3.4 The sum of products and the product of sums

When designing and simplifying circuits, we may sometimes find it useful to regard a group of letters as a single letter.

This allows us to transform one circuit into another of quite a different form, but with the same overall function, for example to transform a parallel combination of series circuits into a series combination of parallel circuits.

In terms of switching algebra, this comes down to writing a sum of products (e.g. $yz + uv$) as a product of sums $(\cdots + \cdots)(\cdots + \cdots)(\cdots + \cdots)(\cdots + \cdots)$. This can be done as follows.

According to theorem 14, $yz + uv = (y + uv)(z + uv)$. To arrive at this result, we replace x in theorem 14 by uv, i.e. we regard the product uv as one letter.

Now

$$y+uv=(y+u)\,(y+v); \qquad z+uv=(z+u)\,(z+v)$$

whence:

$$yz+uv=(y+uv)\,(z+uv)=(y+u)\,(y+v)\,(z+u)\,(z+v)$$

If we draw this in terms of contact networks we find:

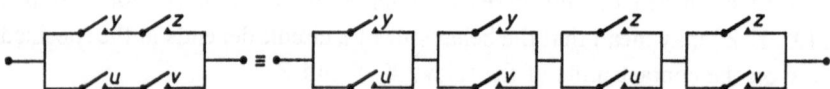

We have thus transformed our parallel combination of series circuits into a series combination of parallel circuits, as desired.

We will now give a practical example of the application of this method.

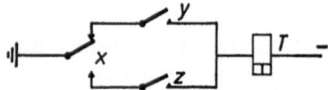

The relay T is a fast relay. In this circuit the transit time of contact unit x will be so large that the fast relay T will release if x switches over, even though contact units y and z may be closed.

Now the problem is to make a circuit with the same function, but in which the fast relay does not release when x switches over. The algebraic formula for the above circuit is $x'y+xz$. We may turn this into a product of sums as follows:

$$x'y+xz=(x'+xz)\,(y+xz)=(x'+x)\,(x'+z)\,(y+x)\,(y+z)$$

$x'+x=1$. We may also omit the factor $(y+z)$ (theorem 23), which gives:

$$x'y+xz=(x'+z)\,(x+y)$$

The same result can be obtained by use of theorem 22 the other way round: $x'y+xz=x'y+xz+yz$ (the addition of the term yz does not change the function of the circuit).

We may now add an open circuit to the parallel circuit thus obtained, since this will not alter its function either.

We therefore add xx', giving $x'y+xz+yz+xx'$. We may write $xz+yz$ as $z(x+y)$, and $x'y+xx'$ as $x'(x+y)$. The entire expression thus becomes:

$$x'y+xz+yz+xx'=z(x+y)+x'(x+y)=(z+x')\,(x+y)$$

Let us now draw this product of sums as a contact network:

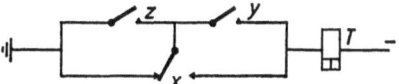

We can combine the make and break contact units x to one change-over contact unit, giving:

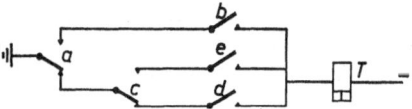

This does indeed have the same function as the original circuit, but also fulfils the condition that T should not release during the transit time of change-over contact unit x.

If we want to write an expression like $(abc+de)$ as a product of sums, we can use the same method we used for $yz+uv$, by:
1. first considering ab as one letter;
2. then considering a and b separately.

$$abc+de=(ab+de)(c+de)=(a+de)(b+de)(c+de)=$$
$$(a+d)(a+e)(b+d)(b+e)(c+d)(c+e)$$

Another way to change a sum of products into a product of sums is to invert twice. If we use this method on the previous example, we find on inverting once and multiplying out:

$$(a'+b'+c')(d'+e')=a'd'+a'e'+b'd'+b'e'+c'd'+c'e'$$

Inverting again gives the desired product of sums:

$$(a+d)(a+e)(b+d)(b+e)(c+d)(c+e)$$

Let us now try to change the circuit shown below, in which the relay T is again a fast one, so that it retains the same function but T is no longer dependent on the transit times of the contact units a and c.

The algebraic formula of this circuit is $ab+a'ce+a'c'd$. Inversion gives:

$$(a'+b')(a+c'+e')(a+c+d')$$

Multiplying out, we find:

$$(a+c'd'+ce'+d'e')(a'+b')=(a+c'd'+ce')(a'+b')=$$
$$ab'+a'c'd'+b'c'd'+a'ce'+b'ce'=ab'+a'c'd'+a'ce'$$

Inverting again, we obtain:

$$(a'+b)(a+c+d)(a+c'+e)$$

We may draw this as follows:

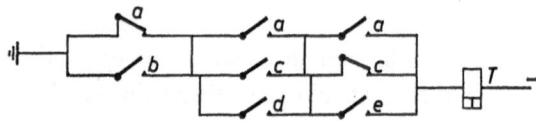

This can be further simplified by replacing the two make contact units of a by a single make contact unit.

This can also be shown algebraically, as follows.

$$(a+c+d)(a+c'+e)=a+ce+c'd$$

We may write $ce+c'd$ as:

$$(c+c')(c+d)(e+c')(e+d)=(c+d)(e+c')$$

The total expression $(a'+b)(a+c+d)(a+c'+e)$ thus becomes:

$$(a'+b)\{a+(c+d)(e+c')\}$$

Drawing this, we find:

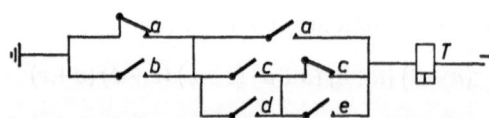

Combining make and break contact units to change-over contact units gives the desired circuit.

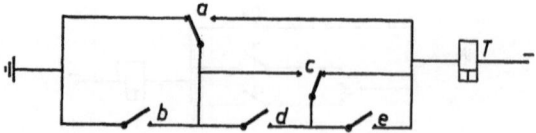

This has indeed the same function as the original circuit, but the contact units a and c no longer give trouble.

So far we have considered how to turn a sum of products into a product of sums. The reverse is also possible: we can turn a product of sums into a sum of products.

Let us take the following example.

Given the product of sums:

$$(a+b)\,(a'+b')\,(a+d')\,(a'+d)\,(b+c')\,(b'+c)\,(c+d)\,(c'+d')$$

Transform this into a sum of products, and find the equivalent network with the smallest possible number of contact springs.

The required sum of products can be obtained by multiplying out. The desired result is obtained most quickly if we start by multiplying pairs of sums in which the same letter occurs with the same value.

> a) the product of the 1st and 3rd sums is $(a+bd')$
> b) the product of the 2nd and 4th sums is $(a'+b'd)$
> c) the product of the 6th and 7th sums is $(c+b'd)$
> d) the product of the 5th and 8th sums is $(c+bd')$

The original expression has thus now become:

$$(a+bd')\,(a'+b'd)\,(c+b'd)\,(c'+bd')$$

Multiplication of a) and d) gives $(ac'+bd')$, while multiplication of b) and c) gives $(a'c+b'd)$, so that we have

$$(ac'+bd')\,(a'c+b'd)=ab'c'd+a'bcd'$$

This expression corresponds to the network:

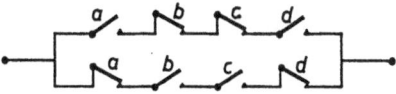

We have mentioned that we can interchange contact units which are connected in series. If we do so in the present case, we can obtain the network:

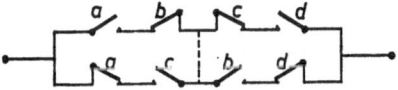

This corresponds to the formula $ab'c'd+a'cbd'$. If we can prove that:

$$ab'c'd+a'cbd'=(ab'+a'c)\,(c'd+bd')$$

then we may connect the points shown joined by a broken line in the above figure, and hence combine the contact units to give four change-over contact units.

The proof of the above statement is as follows. For:

$$ab'c'd+a'cbd'$$

we may write

$$ab'c'd+a'cbd'+ab'bd'+a'cc'd$$

(the products which we have added always represent an open circuit). Simplification gives:

$$ab'(c'd+bd')+a'c(c'd+bd')=(ab'+a'c)(c'd+bd')$$

Q.E.D.

If we draw this circuit, we find:

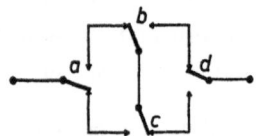

This completes the solution of this problem: we have transformed the product of sums into a sum of products, and have found and drawn the equivalent circuit with the fewest possible contact springs.

Theorem 25

When in the circuit shown below $f(X, Y, Z)$ and $f(A, B, X)$ are two switching functions which may or may not be related, then the connection indicated by the broken line may be added to the figure. This is only possible if the make and break contact units shown are from the same relay, or form (contain) inverse contact groups.

Proof

The algebraic formula for the figure without the broken line is $d \cdot f(X, Y, Z)+d' \cdot f(A, B, X)$. If we write this as a product of sums, we find: $(d+d')\{d+f(A, B, X)\}\{f(X, Y, Z)+d'\}\{f(X, Y, Z)+f(A, B, X)\}= \{d+f(A, B, X)\}\{f(X, Y, Z)+d'\}$.

Q.E.D.

Problems

16. Draw the contact networks represented by the following algebraic formulae:

    a) $a(b'c'+d)ef+(a'+b)(c+d+a)(b'c'+f')$
    b) $(a'+b'c'd)\{(a'b'+c)de+f'gh\}$
    c) $(vw+x'y')\{(v+x')(w+y+z'y')+vwx'(y+z)\}$

17. Write the algebraic formulae for the following networks.

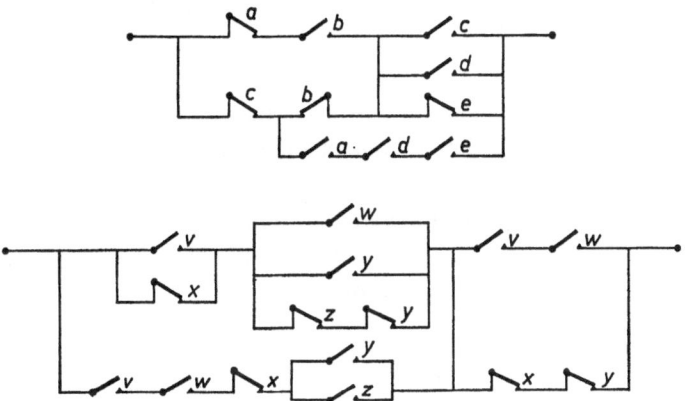

18. Simplify and draw each of the following formulae:

    a) $(a'd+b')(a+d)+(b+c)(bc'+d')+(a'+d')(a'd+c)+$
       $(a+bc')(b'+c')$ to 12 contact springs
    b) $(a+b')(b+c')(a'+c)(a'+b)(b'+c)(a+c')$ to 9 contact springs
    c) $(a+b+c')(a'+b'+c')(a'+b+c')(a+b+c)$ to 7 contact springs

19. Express the following as products of sums:

    a) $a'b+c'd'+e'f+de+b'c+af'=18$ contact springs
    b) $a'c'd+a'ce+ab$

20. Simplify:

    $$\{w'+x(y+z')\}\{w+y'(x'+z')\}+xy'z$$

Draw this function with as few contact springs as possible, so that no brief interruption occurs when the change-over contact unit included in the circuit switches over.

3.5   Expanding functions in series

We can represent the function $f(X, Y, Z)$ by the symbol:

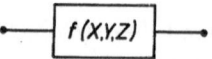

Let us consider this function in the cases that the relay X is energized and is released. When X is energized, the above "box" will behave as if all make contact units x are simply lengths of wire, and all break contact units x are cut through. The make contact units can thus be represented by a 1, and the break contact units by a 0.

When the relay X is released, the function behaves as if all make contact units were cut through and all break contact units were pieces of wire (make contact units=0, break contact units=1).

We can thus imagine the box representing $f(X, Y, Z)$ to be replaced by two boxes representing two different functions of Y and Z, the one being used only when the relay X is energized, and the other only when X is released. We may draw this as follows:

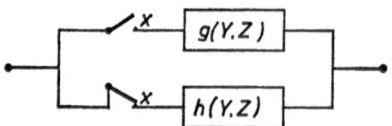

The function $g(Y, Z)$ is the result of replacing all make contact units x in $f(X, Y, Z)$ by 1, and all break contact units x by 0. This function may equally be denoted by $f(1, Y, Z)$.

Similarly, $h(Y, Z)$ is the result of replacing all make contact units x in $f(X, Y, Z)$ by 0 and all break contact units x by 1; this may be written $f(0, Y, Z)$. We may thus also draw:

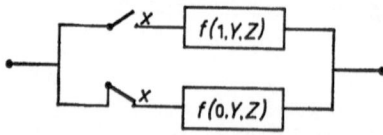

This is a parallel combination of the circuits $x \cdot f(1, Y, Z)$ and $x' \cdot f(0, Y, Z)$. In general, we may state:

Equation 1 $f(X, Y, Z) = x \cdot f(1, Y, Z) + x' \cdot f(0, Y, Z)$

According to what we said in the previous section (theorem 25), the expansion of the function can proceed as follows:

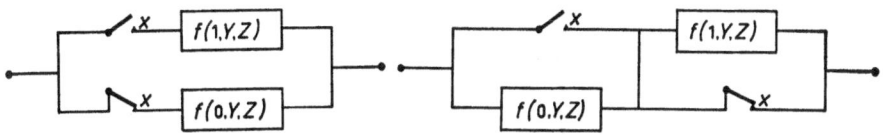

We can now derive the following general equation:

Equation 2 $f(X, Y, Z) = \{x + f(0, Y, Z)\}\{x' + f(1, Y, Z)\}$

All other equations can be derived from these two. The functions $f(0, Y, Z)$ and $f(1, Y, Z)$ can be expanded further in Y and Z. Equation 1 then becomes:

$$f(X, Y, Z) = xf(1, Y, Z) + x'f(0, Y, Z) =$$
$$xyf(1, 1, Z) + xy'f(1, 0, Z) + x'yf(0, 1, Z) + x'y'f(0, 0, Z) = \text{etc.}$$

Equation 2 becomes:

$$f(X, Y, Z) = \{x + f(0, Y, Z)\}\{x' + f(1, Y, Z)\} =$$
$$[x + \{y + f(0, 0, Z)\}\{y' + f(0, 1, Z)\}]$$
$$[x' + \{y + f(1, 0, Z)\}\{y' + f(1, 1, Z)\}] = \text{etc.}$$

The two following equations can be derived from equation 1.

Equation 3 $xf(X, Y, Z) = xf(1, Y, Z)$

Equation 4 $x'f(X, Y, Z) = x'f(0, Y, Z)$

These can be proved as follows:

$$f(X, Y, Z) = xf(1, Y, Z) + x'f(0, Y, Z) \quad \text{equation 1.}$$

For $xf(X, Y, Z)$ of equation 3 we may write:

$$\{x \cdot x \cdot f(1, Y, Z)\} + \{x \cdot x' \cdot f(0, Y, Z)\}$$

The second term is 0, because $xx' = 0$. We are thus left with:

$$x \cdot xf(1, Y, Z) = xf(1, Y, Z)$$

Q.E.D.

If we replace x by x', the same proof can be used for equation 4.

The following two equations can be derived from equation 2:

Equation 5 $x + f(X, Y, Z) = x + f(0, Y, Z)$

Equation 6 $x' + f(X, Y, Z) = x' + f(1, Y, Z)$

These may be proved as follows:

$$f(X, Y, Z)=\{x+f(0, Y, Z)\}\{x'+f(1, Y, Z)\}\quad\text{(equation 2)}.$$

For $x+f(X, Y, Z)$ in equation 5 we may write:

$$\{x+x+f(0, Y, Z)\}\{x+x'+f(1, Y, Z)\}$$

The last factor is equal to 1, since $x+x'=1$. We may therefore write:

$$x+x+f(0, Y, Z)=x+f(0, Y, Z)$$

Q.E.D. Replacing x by x' gives equation 6.

Equations 3 to 6 may be expressed in words as follows:

Equation 3

When a make contact unit x is in series with a contact network containing make and break contact units x, then all the make contact units x may be replaced by a conductor, and the break contact units x may be omitted.

Equation 4

If a break contact unit x is in series with such a network, the break contact units x can be replaced by a conductor and the make contact units omitted.

Equation 5

When a make contact unit x is in parallel with a contact network containing make and break contact units x, then the break contact units x can be replaced by a conductor and the make contact units x may be omitted.

Equation 6

If a break contact unit x is in parallel with such a network, the make contact units x may be replaced by conductors and the break contact units x omitted.

Example

Expand the function given below by means of equation 1, and then by means of equation 2.

$$f(A, B, C)=$$
$$(a'+b+c')(a+b+c)(a'+b'+c')(a+b'+c')(a'+b+c)(a+b+c')$$

Equation 1 states: $f(A, B, C) = af(1, B, C) + a'f(0, B, C)$, where $f(1, B, C)$ means that for all make contact units a a 1 may be read, and for all break contact units a a 0. Similarly, $f(0, B, C)$ means that all make contact units a must be read as 0, and all break contact units a as 1. Expanding the given function in this way, we find:

$$a(b+c')(b'+c')(b+c) + a'(b+c)(b'+c')(b+c')$$

Expanding further in terms of b, we find:

$$abc' + ab'c'c + a'bc' + a'b'cc' = abc' + a'bc' = \underline{bc'}$$

as the final result.

Equation 2 states: $f(A, B, C) = \{a + f(0, B, C)\}\{a' + f(1, B, C)\}$. Expanding in terms of a with this equation, we find:

$$\{a + (b+c)(b'+c')(b+c')\}\{a' + (b+c')(b'+c')(b+c)\}$$

Further expansion in terms of b yields:

$$(a+b+cc')(a+b'+c')(a'+b+c'c)(a'+b'+c') =$$
$$(a+b)(a+b'+c')(a'+b)(a'+b'+c') = b(b'+c') = \underline{bc'}$$

3.6 Eliminating contacts by tabular notation

One method of checking whether a number of contact combinations are worth using is to write them in tabular form. A make contact unit may be denoted by a 1 (relay energized gives a closed circuit) and a break contact unit by a 0 (relay energized gives an open circuit).

We can then represent the algebraic formula $abc'd'e + a'bc'de + ab'c'de + a'bcde'$ by:

```
A B C D E
1 1 0 0 1
0 1 0 1 1
1 0 0 1 1
0 1 1 1 0
```

Let us now suppose that we have to operate a relay by means of the contact units of two relays X and Y, and the following three conditions are known for this relay to operate: $x'y + x'y' + xy$. By applying the rules of switching algebra, we can simplify these conditions to $x' + xy = x' + y$.

We shall now show how these conditions can also be simplified with the aid of tabulation. We may write:

	X Y
1	0 1
2	0 0
3	1 1

Lines 1 and 2 show that as long as relay X is released, the circuit will be closed no matter what the state of the contact unit y (since y occurs as both 1 and 0 in combination with the 0 of x). The contact unit y is thus unnecessary, and the result of lines 1 and 2 will be $x=0$, giving the condition x' for the relay to operate. Similarly, we may see from lines 1 and 3 that here the contact unit x is unnecessary, which gives the condition y. The overall condition is thus $x'+y$, which agrees with the result obtained by switching algebra.

This allows us to state a rule for picking out the right combinations:
When all combinations of a number of different contact units (relays) occur together with only one combination of another number of contact units (relays), we may omit the first-mentioned combination.

Example

	A B C D
	1 1 1 0
	1 1 1 1
	1 1 0 0
	1 1 0 1

Here we have all four combinations of the relays C and D together with the same combination of A and B (1,1). These 4 conditions may thus be simplified to $a \cdot b$.

Example

Simplify the following formula:

$$(a+b'+c+d'+e')\,(a+b'+c+d'+e)\,(a'+b'+c'+d+e')$$
$$(a+b+c+d+e)\,(a+b+c+d+e')\,(a+b+c+d'+e')$$
$$(a'+b'+c'+d+e)\,(a+b+c+d'+e)\,(a+b'+c+d+e')$$
$$(a+b'+c+d+e).$$

Multiplying these ten factors out would doubtless give the right result, but it is clear that it is a lengthy business. Tabulation is definitely preferable in this case. We write:

	$A+B+C+D+E$				
	1	0	1	0	0
1	1	0	1	0	0
2	1	0	1	0	1
3	0	0	0	1	0
4	1	1	1	1	1
5	1	1	1	1	0
6	1	1	1	0	0
7	0	0	0	1	1
8	1	1	1	0	1
9	1	0	1	1	0
10	1	0	1	1	1

In rows 1, 2, 4, 5, 6, 8, 9 and 10, A and C always have the value 1 $(1, -, 1, -, -)$, while the relays B, D and E have in these rows all possible combinations of values for 3 relays $(2^n = 2^3 = 8)$. We may thus ignore the combinations of B, D and E here, leaving us with $(a+c)$. If we write this in tabular form again, we find:

$A+B+C+D+E$
1   –   1   –   –   from rows 1, 2, 4, 5, 6, 8, 9 and 10
0   0   0   1   0   row 3
0   0   0   1   1   row 7

Rows 3 and 7 can be simplified to 0, 0, 0, 1, –. The overall result:

$$(a+c)(a'+b'+c'+d)$$

can be still further simplified to:

$$a(b'+c')+c(a'+d).$$

3.7 Problems

21. Simplify $(a+b+c'+d')(a+b'+c+d')(a+b+c'+d)$
$(a+b+c+d')(a'+b+c'+d)(a+b+c+d)$. Draw the circuit.

22. Expand according to equation 2:
$a'b'+bc+c'd+d'e+e'f'+af$. Draw this circuit with as few contact springs as possible.

23. Prove that:

$$abx + bx'y' + a'b'x' + b'xy = a'b'y + axy + a'x'y' + aby'$$

24. Simplify the following formula, and draw with 12 contact springs:

$$(a+bc)\left[d'e'\{c'+a'(b+c)\}\right] + (b+d)(b'+d') + b'c'de.$$

25. Simplify and draw with 12 contact springs:

$$(a+b+cd+ed)(ab+ac'+ad+bc+cd+be+c'e+ed+bd'+c'd').$$

26. Design a contact network to light 2 lamps under the following conditions. Use only one change-over contact unit per relay.

Green light	Red light
$ab'c'$	$ab'c'$
$a'bc'$	$a'bc'$
abc	
ac	
bc	

27. Simplify the following expression and draw with 12 contact springs:

$$\{a(b+c')+b'c'\}\{d'(b+c')+bc\}\{c(a'+d)+a'd'\}\{b'(a'+d)+ad\}$$

28. Prove that:

$$bc' + b'd'e' + b'df'$$ is the inverse of $$bc + b'd'e + b'df$$

Chapter 4

CODES

The properties of the switching devices used in a circuit play a large part in the design of this circuit. One of these properties is the binary nature of by far the most elements of which a circuit can be built up.

By binary nature we mean that these elements can assume only two different states. For example, a relay is either energized or not, a contact unit is open or closed, a transistor or a diode is conducting or cut off. A result of this property is that alphabetic or decimal data can only be stored or reproduced by these circuits in coded form.

Since the switching devices are binary, all possible combinations of a given number of switching devices can be expressed in a binary code. Each one of these devices represents one "bit" (abbreviation of *binary* digi*t*).

In this chapter we will discuss a number of much-used codes by means of which decimal data can be represented in binary form. In order to determine the content of a piece of coded information, we make use of the binary system with the symbols (digits) 0 and 1 (relay energized $=1$, relay not energized$=0$, contact unit closed$=1$, contact unit open$=0$, etc.). We will start by determining how much scope the binary system allows us, depending on the code used.

4.1 Determination of the number of combinations of a given number of units m out of a total of n units

The number of different ways in which m equivalent units can be placed in n positions can be found as follows. One peg can be placed in only one manner in one hole; the number of choices in this case is thus factorial 1 (written 1!).

Two pegs can be placed in two holes in two different ways

$$
\begin{matrix} 1 \\ 2 \end{matrix} \qquad \begin{matrix} A\ B \\ B\ A \end{matrix} \ (2! = 1 \times 2)
$$

Three pegs can be placed in 3 holes in 6 different ways.

1	$A\ B\ C$
2	$A\ C\ B$
3	$B\ A\ C$
4	$B\ C\ A$
5	$C\ A\ B$
6	$C\ B\ A$

$(3! = 1 \times 2 \times 3 = 6)$

It will be clear that in general m pegs can be placed in m holes in $m!$ different ways.

If n holes are available, where $n > m$, then the first peg can be placed in any one of n different holes, and the second peg in any one of $(n-1)$ holes. If $m = 2$, the total number of ways in which 2 pegs can be placed in n holes is thus $n(n-1)$.

Let $m = 2$ and $n = 5$ (denoted by "2 out of 5"); we then get the following combinations, where the pegs are called A and B (Table X).

TABLE X

	Positions							Positions				
	1	2	3	4	5			1	2	3	4	5
1	A	B					11		A	B		
2	A		B				12		A		B	
3	A			B			13	B			A	
4	A				B		14		B		A	
5	B	A					15			B	A	
6		A	B				16			A		B
7		A		B			17	B				A
8		A			B		18		B			A
9	B		A				19			B		A
10		B	A				20				B	A

It may be seen from the table that the total number of combinations is indeed $n(n-1) = 20$. For $m = 3$, the total number of combinations is $n(n-1)(n-2) = 60$.

In general, we may state that the number of possible combinations of m out of n is given by $n(n-1)(n-2)(n-3)...(n-m+1)$. The last factor possibly requires some explanation: we have seen that for $m = 2$ we have $n(n-1)$, and for $m = 3$ we have $n(n-1)(n-2)$; in each of these cases, the last factor is $n-(m-1) = n-m+1$.

So far we have assumed that we can distinguish between the different units to be arranged. For example, possibilities 1 and 5 in Table X would be equivalent if we did not distinguish between A and B. If we are only interested in whether a given position is occupied or not, we must divide the number of combinations by $m!$.

This may be illustrated by considering 3 pegs in 3 holes. As we have seen, these can be arranged in 6 different ways, but if we are only interested in whether the holes are filled or not, we are only left with one possibility (all holes filled). In general, therefore, we have:

$$\binom{n}{m} = \frac{n(n-1)(n-2)(n-3)\dots(n-m+1)}{m!}$$

where $\binom{n}{m}$ is the number of possible combinations of m out of n.

In order to get this expression into a more convenient form, we multiply top and bottom by a number of factors so as to continue the series in the numerator right down to 1.

$$\binom{n}{m} = \frac{n(n-1)(n-2)(n-3)\dots(n-m+1)(n-m)(n-m-1)\dots 1}{m!(n-m)(n-m-1)\dots 1}$$

The value of the numerator is now $n!$, while that of the denominator is $m!(n-m)!$. We may thus write:

$$\binom{n}{m} = \frac{n!}{m!(n-m)!}$$

We may ask what value of m will in general give the most combinations for a given value of n. The fewest combinations are found with $m=0$ and $m=n$: use of the above formula shows that in both cases $\binom{n}{m}=1$ ($0!=1$).

$$\frac{n!}{0!(n-0)!} = \frac{n!}{n!(n-n)!} = 1$$

For $m=1$ and for $m=n-1$ we have:

$$\frac{n!}{1!(n-1)!} = \frac{n!}{(n-1)!\{n-(n-1)\}!} = n$$

and for $m=2$ and $m=n-2$:

$$\frac{n!}{2!(n-2)!} = \frac{n!}{(n-2)!\{n-(n-2)\}!} = \frac{n!}{2!(n-2)!} = \frac{n(n-1)}{2!}$$

If we plot the value of $\binom{n}{m}$ against the value of m in the form of a graph (Fig. 65), we see that a symmetrical figure is obtained, with the maximum value of $\binom{n}{m}$ at $m=\frac{1}{2}n$.

If n is even, then $\frac{1}{2}n$ will be an integer and will indeed give the maximum value of $\binom{n}{m}$.

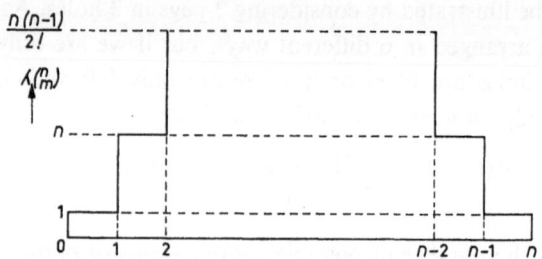

Fig. 65

If however n is odd and the maximum value of $\binom{n}{m}$ must also be an integer, then there will be two values of m which give a maximum value of $\binom{n}{m}$ namely $(\frac{1}{2}n+\frac{1}{2})$ and $(\frac{1}{2}n-\frac{1}{2})$. A practical example of the application of this statement may be found by consideration of the "two-out-of-five code". This code, in which one always takes 2 possibilities out of 5, gives 10 different combinations (Table XI), and this is the maximum possible. One would get an equal number of combinations by taking 3 out of 5 (Table XII). With the 1 out of 5 and 4 out of 5 codes (Tables XIII and XIV respectively) we get only 5 combinations, while 0 out of 5 and 5 out of 5 naturally give only 1 combination. All the combinations for m out of 5 are shown in Fig. 66.

<table>
<tr><td colspan="2" align="center">TABLE XI</td></tr>
<tr><td></td><td align="center">2 out of 5</td></tr>
</table>

	2 out of 5			
1	A	B		
2	A		C	
3	A			D
4	A				E
5		B	C	
6		B		D
7		B			E
8			C	D
9			C		E
10				D	E

TABLE XII

	3 out of 5			
1	A	B	C	
2	A	B		D
3	A	B			E
4	A		C	D
5	A		C		E
6	A			D	E
7		B	C	D
8		B	C		E
9		B		D	E
10			C	D	E

TABLE XIII

	1 out of 5			
1	A			
2		B		
3			C	
4				D
5					E

TABLE XIV

	4 out of 5			
1	A	B	C	D
2	A	B	C		E
3	A	B		D	E
4	A		C	D	E
5		B	C	D	E

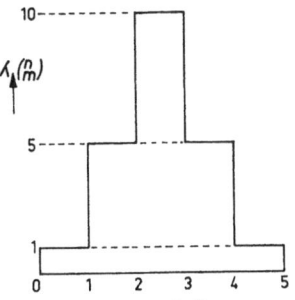

Fig. 66

We shall now determine the total number of combinations as m varies from 0 to n. Starting from:

$$\binom{n}{m} = \frac{n!}{m!(n-m)!} \text{ we find:}$$

$$\sum_{m=0}^{m=n} \binom{n}{m} = \frac{n!}{0!\,n!} + \frac{n!}{1!(n-1)!} + \frac{n!}{2!(n-2)!} + \ldots + \frac{n!}{n!\,0!} \qquad (1)$$

Now the binomial theorem states that:

$$(a+b)^n =$$

$$a^n + \frac{n}{1}a^{n-1}\cdot b + \frac{n(n-1)}{1.2}a^{n-2}\cdot b^2 + \ldots + \frac{n(n-1)\ldots 1}{1.2\ldots n}b^n =$$

$$a^n + \frac{n(n-1)\ldots 1}{1(n-1)\ldots 1}a^{n-1}\cdot b + \frac{n(n-1)(n-2)\ldots 1}{1.2(n-2)\ldots 1}a^{n-2}\cdot b^2 + \ldots$$

$$+ \frac{n(n-1)\ldots 1}{1.2\ldots n}\cdot b^n =$$

$$a^n + \frac{n!}{1!(n-1)!}a^{n-1}\cdot b + \frac{n!}{2!(n-2)!}a^{n-2}\cdot b^2 + \ldots + \frac{n!}{n!\,0!}b^n \qquad (2)$$

If we compare equations (1) and (2), we see that their right-hand sides become identical if we write $a=b=1$. Substituting these values into equation (2), we have:

$$(1+1)^n = 1 + \frac{n!}{1!(n-1)!} + \frac{n!}{2!(n-2)!} + \ldots + \frac{n!}{n!\,0!}\cdot 1$$

This proves that the total number of combinations $\binom{n}{m}$ for all possible values of m is $(1+1)^n = 2^n$.

$$\sum_{m=0}^{m=n} \binom{n}{m} = 2^n$$

The practical application of the above theory in switching techniques is found e.g. in the determination of the number of relays needed to obtain a certain number of combinations. For example, 1 relay gives $2^1 = 2$ "combinations":

>   relay off (0)
>   relay up (1)

Two relays give $2^2 = 4$ combinations:

>   A   B
>   0   0
>   1   0
>   0   1
>   1   1

As we have already hinted above, it is frequently practical to use slightly more than the minimum number of relays: the 2-out-of-5 code makes use of 5 relays, only 2 of which are up at any time, to represent 10 different combinations, while 4 relays fully used can already give $2^4 = 16$ different combinations.

4.2   The binary code

Various number systems have been thought out for the purpose of counting, of which the best known are Roman numerals, Arabic (decimal) numbers and the binary system.

During the past couple of decades, a large number of codes have been developed for counting in computers. We shall now describe the most important of these, starting with the binary code.

The decimal notation makes use of ten symbols (digits), 0, 1, 2, ..., 9, with 10 as the "base". The number 2087 is composed of the following:

$$7 \times 10^0 = \quad\;\; 7$$
$$8 \times 10^1 = \quad 80$$
$$0 \times 10^2 = \quad 000$$
$$2 \times 10^3 = 2000$$
$$\overline{2087}$$

The binary notation has only two digits, 0 and 1. The base of this notation is 2. We thus find the following relation between the binary and decimal notations:

binary	decimal
2^0	1
2×2^0	2
$2 \times 2 \times 2^0$	4
$2 \times 2 \times 2 \times 2^0$	8 etc.

Writing the exponents as Arabic numbers for the sake of simplicity, we find for example:

binary	decimal
$1 \times 2^0 =$	1
$1 \times 2^1 =$	2
$1 \times 2^2 =$	4
$1 \times 2^3 =$	8
1111 $=$	15

A property of this system is that the sum of successive powers of 2 is $1+2+4+...+n=2n-1$.

With the aid of this binary notation, in which successive digits represent the values 1, 2, 4, 8, 16, etc., one can represent any decimal number by means of only two different digits (0 and 1). For example:

			binary						decimal		
2^7	2^6	2^5	2^4	2^3	2^2	2^1	2^0		10^2	10^1	10^0
128	64	32	16	8	4	2	1		100	10	1
1	0	0	1	1	0	1	0		1	5	4

The digits 0 and 1 indicate from right to left whether the values $2^0, 2^1, 2^2, ..., 2^n$ are present (1) or absent (0).

The addition of two binary digits of the same value always gives 0 as result; if the digits in question are both 1, one also has to carry 1.

0	0	1	1
0	1	0	1
—	—	—	—
0	1	1	10

By way of example we give below the decimal numbers 365 and 287 in the binary notation (each with 9 bits having the value 0 or 1), and add them according to the above rules.

```
1 0 1 1 0 1 1 0 1 = 365
1 0 0 0 1 1 1 1 1 = 287
───────────────────
1 0 1 0 0 0 1 1 0 0   652
```

Relays can be used to represent binary digits if we define:

$$\text{relay energized} \quad = 1$$
$$\text{relay not energized} = 0$$

The binary code for representing or counting 16 different combinations can thus simply be realized with the aid of 4 counting elements (e.g. relays). Table XV shows the complete code.

TABLE XV

Value	Counting element of value				Value	Counting element of value			
	8	4	2	1		8	4	2	1
0	0	0	0	0	9	1	0	0	1
1	0	0	0	1	10	1	0	1	0
2	0	0	1	0	11	1	0	1	1
3	0	0	1	1	12	1	1	0	0
4	0	1	0	0	13	1	1	0	1
5	0	1	0	1	14	1	1	1	0
6	0	1	1	0	15	1	1	1	1
7	0	1	1	1					
8	1	0	0	0					

4.3 The reflected code (Gray code)

We have already proved above that n units can be combined in 2^n different ways. In the binary code sometimes more than one digit changes when one goes from a given number to the next higher number. For example, on going from combination 1 to combination 2 of Table XV, counter element 1 changes from 1 to 0, and counter element 2 from 0 to 1. When we go from 3 to 4, three counter elements already change over. This can be a nuisance in switching techniques, because undesirable intermediate states can arise. For example, the transition from 3 to 4 can give the value 7 for a moment, because the 4-element can assume the value 1 before the elements 1 and 2 have become zero. There is therefore a need for a code in which only one counter element changes per transition. Such a code is the "reflected code", which owes its name to the fact that all the counter elements apart from the right-hand one form a reflected image about an imaginary line (reflecting line), which divides the code into 2 equal parts.

Table XVI gives the code for 2, 3, or 4 counter elements.

TABLE XVI

	Counter element			
	A	B	C	D
0	0	0	0	0
1	1	0	0	0
2	1	1	0	0
3	0	1	0	0
4	0	1	1	0
5	1	1	1	0
6	1	0	1	0
7	0	0	1	0
8	0	0	1	1
9	1	0	1	1
10	1	1	1	1
11	0	1	1	1
12	0	1	0	1
13	1	1	0	1
14	1	0	0	1
15	0	0	0	1

reflecting lines 1, 2, 3

For two elements (A and B), with 4 combinations, reflecting line 1 applies. This divides the number of combinations into two equal parts, and we see that the values of element A above the reflecting line form the reflection image of those below the line.

Reflecting line 2 holds for three elements, and reflects both A and B.

Finally, reflecting line 3 divides all 16 combinations into two parts, forming reflection images of columns A, B and C.

The right-hand element in the code obeys a different law: above the reflecting line it is always 0, and below the reflecting line it is 1.

4.4 The biquinary code

The biquinary code is formed by 7 counter elements, with from left to right the values 5 0 | 4 3 2 1 0. The decimal digits from 0 to 9 are always represented by one of the 2 elements from the first group and one of the 5 elements from the second group. This gives the combinations shown in Table XVII.

The important advantage of this code is that each combination contains one element from the group of 2, and one from the group of 5, so that the occurrence of a single error can always be detected (self-checking code). When data has to be transmitted over large distances or via complicated

TABLE XVII

Value	Counter elements						
	5	0	4	3	2	1	0
0	0	1	0	0	0	0	1
1	0	1	0	0	0	1	0
2	0	1	0	0	1	0	0
3	0	1	0	1	0	0	0
4	0	1	1	0	0	0	0
5	1	0	0	0	0	0	1
6	1	0	0	0	0	1	0
7	1	0	0	0	1	0	0
8	1	0	0	1	0	0	0
9	1	0	1	0	0	0	0

equipment, it is sometimes absolutely necessary to have some method of checking the numbers received. The biquinary code offers a limited possibility of doing this.

4.5   The two-out-of-five code

We have already mentioned in this chapter that the total number of possible combinations of n units out of m is:

$$\binom{n}{m} = \frac{n!}{m!(n-m)!}$$

The 2-out-of-5 code gives 10 combinations, so that it can be used to represent all the decimal digits. To simplify the reading off, the counter elements are given the values 0, 1, 2, 4 and 7 (so that this could also be called the 0 1 2 4 7 code). This gives the code shown in Table XVIII, where the sum of the values of all the counter elements gives the value of the number represented, except for the digit 0, which is represented by $4+7$.

TABLE XVIII

Value	Counter element					Value	Counter element				
	0	1	2	4	7		0	1	2	4	7
1	1	1	0	0	0	6	0	0	1	1	0
2	1	0	1	0	0	7	1	0	0	0	1
3	0	1	1	0	0	8	0	1	0	0	1
4	1	0	0	1	0	9	0	0	1	0	1
5	0	1	0	1	0	0	0	0	0	1	1

Like the biquinary code, the 2-out-of-5 code can easily be checked, because if an error occurs in one bit, the resulting combination no longer forms part of the 2-out-of-5 code. In both cases one sacrifices a large number of possible combinations for the sake of a possibility of checking.

4.6 Hamming's self-correcting code

R. W. Hamming in 1950 published a code which allows automatic correction of an error. Each digit is here represented by 7 elements. When a digit coded in this way is received, 3 check additions are made, the result being expressed in the binary values 1, 2 and 4 (see Table XIX).

TABLE XIX

Value	Counter elements							Check		
	1	2	3	4	5	6	7	4	2	1
0	0	0	0	0	0	0	0	0	0	0
1	1	1	0	1	0	0	1	0	0	0
2	0	1	0	1	0	1	0	0	0	0
3	1	0	0	0	0	1	1	0	0	0
4	1	0	0	1	1	0	0	0	0	0
5	0	1	0	0	1	0	1	0	0	0
6	1	1	0	0	1	1	0	0	0	0
7	0	0	0	1	1	1	1	0	0	0
8	1	1	1	0	0	0	0	0	0	0
9	0	0	1	1	0	0	1	0	0	0

The following additions are made:

counter elements

        1, 3, 5 and 7 (1)
        2, 3, 6 and 7 (2)
        4, 5, 6 and 7 (4)

For each digit, these additions should give the result 0 0 0 in the check (the sum of an even number of "ones" = 0, the sum of an odd number of "ones" = 1). Now let us suppose that the digit 7 is received with an error, as follows:

        1 2 3 4 5 6 7
        ‾‾‾‾‾‾‾‾‾‾‾‾‾
        0 1 0 1 1 1 1

The three check additions give:

$$0+0+1+1=\quad 0$$
$$1+0+1+1=\quad 1$$
$$1+1+1+1=0$$

$$\overline{0\ 1\ 0}=2 \text{ in the binary notation.}$$

This indicates that the error is at position 2. The 1 at this position is therefore changed to a 0.

Another example is the reception of the digit 4 in the following form:

$$\overline{1\ 2\ 3\ 4\ 5\ 6\ 7}$$
$$1\ 0\ 0\ 1\ 0\ 0\ 0$$

Addition gives:

$$1+0+0+0=\quad 1$$
$$0+0+0+0=\quad 0$$
$$1+0+0+0=1$$

$$\overline{1\ 0\ 1}=5 \text{ in the binary notation}$$

The error has been detected: the 5th element should not be 0, and is therefore changed to 1.

Closer inspection of this code will show that the information concerning the digit to be transmitted is given by the bits 7, 6, 5 and 3, which have the binary values 1, 2, 4 and 8 respectively. The bits 1, 2 and 4 are added so that the number can be checked and corrected.

4.7 Protecting codes against error by the addition of information

We have already pointed out the advantage of codes in which all combinations are of the same kind. The biquinary and 2-out-of-5 codes are examples of this.

If however information must be transmitted in the binary code, the combinations are not all of the same form, since e.g. 0, 1, 2 or 3 out of 4 can occur. However, this difficulty can be got round by adding information. Let us suppose that the number 37296 has to be transmitted from A to B. At A, one bit (the "parity bit") is added after the transmission of each digit; this bit is given the value 1 if the sum of the "ones" occurring in the binary coded digit is even, and the value 0 if the sum of the "ones" is odd. This is illustrated in Table XX.

TABLE XX

digit	a binary value 8 4 2 1				b parity bit	check $a + b$
3	0	0	1	1	1	1
7	0	1	1	1	0	1
2	0	0	1	0	0	1
9	1	0	0	1	1	1
6	0	1	1	0	1	1

The digit 3 (0 0 1 1) has two "ones". A "1" is therefore added to this group. The digit 7, on the other hand, has three "ones", so a 0 is added. The sum of a and b is given in the last column. If this is "1", the information is accepted. If however it is "0", the information is not accepted, because there is an error somewhere. However, we have no means of knowing where in the row the error is.

By way of example we give in Table XXI the same number, in which however the digit 7 has been mutilated in the bit of value 2, so that 0 1 0 1 is received instead of 0 1 1 1.

TABLE XXI

digit	a binary value 8 4 2 1				b parity bit	check $a + b$
3	0	0	1	1	1	1
7	0	1	0	1	0	0
2	0	0	1	0	0	1
9	1	0	0	1	1	1
6	0	1	1	0	1	1

The sum of the "ones" in the second row now gives "0", which shows that this digit has been received wrong. We have however no means of knowing which bit is wrong. For example, the right digit might just as well be 4 or 1 as 7. One must therefore be content to reject the information transmitted, and to ask for it to be repeated. This however costs time and money.

A way has therefore been sought to make the code self-correcting by the addition of more information. Apart from the above-mentioned check bit, check bits are added at the end of each "block" of digits in the same way. These blocks must not be made too long, as the value of the check would then be greatly reduced. In the following example, the block of information consists of the number 438732, to be transmitted from A to B (Table XXII).

TABLE XXII

	digit	a binary value 8 4 2 1	b parity bit	check a + b
	4	0 1 0 0	0	1
	3	0 0 1 1	1	1
	8	1 0 0 0	0	1
c	7	0 1 1 1	0	1
	3	0 0 1 1	1	1
	2	0 0 1 0	0	1
d	parity bit	0 1 1 0		
	check c + d	1 1 1 1		

The check bits for the digits (column b) are determined as described above, and the check bits for the vertical columns are determined in exactly the same way. There is an odd number of "ones" in the 8 and 1 columns, so that a "0" is added under these columns. The number of "ones" in columns 2 and 4 is however even, so that a "1" is added here. The additional check consists in adding up all the columns $(c+d)$. The result should always be 1.

Let us suppose that the same block is transmitted, but that the digit 7 is mutilated, so that 0 1 0 1 ($=5$) is received instead of 0 1 1 1. The digit in question is transmitted correctly from A, as is the extra information (parity bit). At B, the digit is received mutilated as described above, but the check bit is received correctly. We thus get the result of Table XXIII.

TABLE XXIII

digit	binary value 8 4 2 1	parity bit	check	
4	0 1 0 0	0	1	
3	0 0 1. 1	1	1	
8	1 0 0 0	0	1	
7	0 1 0 1	0	0	←— error
3	0 0 1 1	1	1	
2	0 0 1 0	0	1	
parity bit	0 1 1 0			
check	1 1 0 1			

↑ error

The horizontal and vertical checks each detect one error. The position of this error is given by the intersection of the row and the column indicated. The correction is simple. If a "0" is found at this position, it should be replaced by a "1", and *vice versa*.

4.8 Binary-decimal code

Each decimal number can be represented in the binary notation. For example, we can write a number of 4 digits in one binary total:

1586 = 1 1 0 0 0 1 1 0 0 1 0

A number written in the binary notation in this way is however difficult to decode. If *each separate digit* is written in the binary notation, decoding becomes much easier. The number 1586 is then written:

1586 = 0 0 0 1   0 1 0 1   1 0 0 0   0 1 1 0

The highest value used in this way (the digit 9) is thus 1 0 0 1.

Only ten of the possible 16 combinations of 4 bits are used. The result is that an addition can give one of the values which is not used.

$$
\begin{array}{rl}
4= & 0\ 1\ 0\ 0 \\
9= & 1\ 0\ 0\ 1 \\
\hline
13 & 1\ 1\ 0\ 1
\end{array}
$$

The ten must be carried to the next group of 4 bits, leaving a result of 3.

In order to do this when a value greater than 9 is obtained, 6 (0 1 1 0) is added to the sum:

$$
\begin{array}{rl}
4 & 0\ 1\ 0\ 0 \\
9 & 1\ 0\ 0\ 1 \\
\hline
13 & 1\ 1\ 0\ 1 \\
6 & 0\ 1\ 1\ 0 \\
\hline
& 1\ 0\ 0\ 1\ 1
\end{array}
$$

The bit underlined is carried, and the remaining group of four bits has the value 3. It is easy to see that the addition of 6 leads to the desired result: this value is equal to the difference between the largest number which can be represented by 4 bits (1 1 1 1 = 15) and the highest decimal digit (9). The above addition gives the value 19, which is 3 more than the underlined bit of value 16. This bit is carried, leaving 3. This process is described in more detail in Chapter 8.

4.9 The 1242 code

In the binary notation the value of the successive bits in the binary-decimal code is from right to left 1 2 4 8. The sum of these values is 15. In the 1 2 4 2 code, the value of the successive bits is 1 2 4 2. The sum of these values is 9, so that all the decimal digits can be represented in this way. The code is given in Table XXIV. If we add the coded values of the numbers 0 and 9, 1 and 8, 2 and 7, 3 and 6 or 4 and 5, we get the result 1 1 1 1. These numbers are said to be complementary to each other. Use is made of this property in computers.

TABLE XXIV

digit	bits			
	2	4	2	1
0	0	0	0	0
1	0	0	0	1
2	0	0	1	0
3	0	0	1	1
4	0	1	0	0
5	1	0	1	1
6	1	1	0	0
7	1	1	0	1
8	1	1	1	0
9	1	1	1	1

4.10 Fractions in the binary notation

In all numerical notations, unity is represented by g^0, where g is the base of the notation.

Values greater than 1 are represented by a series of terms, each consisting of a digit which is considered to be multiplied by the base to the power 0 or to some positive integral power. In the decimal notation ($g=10$), for example, the number 109 represents $1 \times 10^2 + 9 \times 10^0$.

In the binary notation, the same value is represented by 1 1 0 1 1 0 1 ($2^6 + 2^5 + 2^3 + 2^2 + 2^0$).

Values <1 are represented by a sum of terms, each being considered to be multiplied by the base of the notation to a negative integral power. In the decimal notation, the number 0.875 represents $8 \times 10^{-1} + 7 \times 10^{-2} + 5 \times 10^{-3} = 8 \times \frac{1}{10} + 7 \times \frac{1}{100} + 5 \times \frac{1}{1000}$. In the binary notation, the same value is represented by $0.111 = 2^{-1} + 2^{-2} + 2^{-3} = \frac{1}{2} + \frac{1}{4} + \frac{1}{8}$.

Decimal $0.875 =$ binary 0.111.

The number 37.4375 represents:

$$10^1 \quad 10^0, \quad 10^{-1} \quad 10^{-2} \quad 10^{-3} \quad 10^{-4}$$
$$3 \quad 7 \,, \, 4 \quad 3 \quad 7 \quad 5 \; = 30+7+0.4+0.03+0.007+0.0005.$$

The corresponding number in the binary notation is given by:

$$2^5 \quad 2^4 \quad 2^3 \quad 2^2 \quad 2^1 \quad 2^0. \quad 2^{-1} \quad 2^{-2} \quad 2^{-3} \quad 2^{-4}$$
$$1 \quad 0 \quad 0 \quad 1 \quad 0 \quad 1 \, . \, 0 \quad 1 \quad 1 \quad 1 \; =$$
$$32 \quad + \; 4 \quad + \; 1 \; + \quad \tfrac{1}{4} + \tfrac{1}{8} + \tfrac{1}{16} \; = 37\tfrac{7}{16} = 37.4375.$$

Binary fractions can be obtained from decimal fractions by repeated multiplication by 2 (the base of the binary notation) until unity is reached.

Consider 0.5. Multiplication by 2 gives:

$$2 \times 0.5 \quad = \overline{1} \qquad \text{This number in the binary notation is 0.1.}$$

Consider 0.25:

$$2 \times 0.25 \; = 0.50$$
$$2 \times 0.5 \quad = \overline{1} \qquad \text{The binary notation is 0.01.}$$

Consider 0.375:

$$2 \times 0.375 \; = 0.75$$
$$2 \times 0.75 \quad = \overline{1}.50$$
$$2 \times 0.50 \quad = \overline{1} \qquad \text{The binary notation is 0.011.}$$

The decimal part of the above number 37.4375 can be converted as follows:

$$2 \times 0.4375 = 0.875$$
$$2 \times 0.875 \; = \overline{1}.75$$
$$2 \times 0.75 \quad = \overline{1}.50 \quad 0.4375 = \text{binary } 0.0111.$$
$$2 \times 0.50 \quad = \overline{1}$$

The value of a fraction in the binary notation always has a numerator which is equal to the sum of the binary values 1, 2, 4, 8, 16, 32, etc. For example, $0.4375 = 0.0111 = \tfrac{1}{4} + \tfrac{1}{8} + \tfrac{1}{16} - \tfrac{7}{16}$.

If the decimal fraction to be converted does not have a numerator of this form, the corresponding binary fraction will be repeating.

Consider the fraction 0.4.

$$2 \times 0.4 = \overline{0}.8$$
$$2 \times 0.8 = \overline{1}.6$$
$$2 \times 0.6 = \overline{1}.2$$
$$2 \times 0.2 = \overline{0}.4$$
$$2 \times 0.4 = \overline{0}.8$$

The binary equivalent is thus the repeating fraction $0.01100\dot{1}1\dot{0}$.

The conversion of a binary fraction to the decimal equivalent is based on the fact that the values after the "decimal point" are successively a factor 2 smaller ($\frac{1}{2}$, $\frac{1}{4}$, $\frac{1}{8}$, etc.).

If for example the binary fraction:

$$2^0 2^{-1} 2^{-2} 2^{-3} 2^{-4} 2^{-5} 2^{-6} 2^{-7}$$
$$0.\ 1\ \ 0\ \ 1\ \ 1\ \ 0\ \ 1\ \ 1$$

has to be converted to the decimal notation, we start by taking half the last term $1(2^{-7})$ plus the whole of the next to the last term $1(2^{-6})$:

$$(1/2)+1=1.5\ (=\text{sum of } 2^{-6} \text{ and } 2^{-7}, \text{ multiplied by } 2^{-6})$$

We now repeat this process with the 2^{-5} term:

$$(1.5/2)+0=0.75$$

with the 2^{-4} term:

$$(0.75/2)+1=1.375$$

the 2^{-3} term:

$$(1.375/2)+1=1.6875$$

the 2^{-2} term:

$$(1.6875/2)+0=0.84375$$

the 2^{-1} term:

$$(0.84375/2)+1=1.421875$$

and finally the 2^0 term:

$$(1.421875/2)+0=0.7109375.$$

The same result can also be obtained by direct addition of the values of the binary digits after the point:

$$\tfrac{1}{2}+\tfrac{1}{8}+\tfrac{1}{16}+\tfrac{1}{64}+\tfrac{1}{128}=\tfrac{91}{128}=0.7109375.$$

4.11 The 1245 code

Like the 1 2 4 2 code, the 1 2 4 5 code is intended for the coding of decimal information, so that no use is made of the values 10 to 15 which occur in the binary code.

In the 1 2 4 2 code, the sum of the values of the bits is $1+2+4+2=9$, so that the highest digit in the decimal notation is represented by the code group 1 1 1 1.

The starting point of the 1 2 4 5 code is that the ten digits in the decimal notation are divided into two groups:

> 0 1 2 3 and 4 without the bit of value 5
> 5 6 7 8 and 9 with the bit of value 5.

A motive for the code is that in certain applications of counter circuits it is not desired to indicate position 0 by having all relays in the non-energized state. One example of this is the counting of pulses during the dialing of a telephone number, when 10 pulses must be counted when the digit 0 is dialed.

The situation where all relays are not energized already exists before the production of the pulses starts. It is therefore necessary to choose another

TABLE XXV

	1	2	4	5
0	1	0	1	0
1	1	0	0	0
2	0	1	0	0
3	1	1	0	0
4	0	0	1	0
5	0	0	0	1
6	1	0	0	1
7	0	1	0	1
8	1	1	0	1
9	0	0	1	1

combination than 0 0 0 0 to represent the digit 0. The code is given in Table XXV.

It may be seen from the above table that the digits 1 to 4 are represented in the same way as in the binary code. The digit 5 is represented by the bit of value 5, and 6, 7, 8 and 9 are formed by the combination of 1–4 with the bit for 5.

The code groups are divided into two equal sets by the presence or absence of the bit of value 5. The combination 1 0 1 1 cannot be chosen for 0, as this belongs in the set of digits greater than 5. The only other remaining combination, 1 0 1 0, is therefore used to represent 0.

4.12 Problems

1. Add the following numbers in the binary notation:

   1 0 0 1 0 1 1
   1 1 1 0 1 1 0
   1 0 1 1 1 0 1
   0 1 0 1 1 1 1
   1 1 1 0 0 1 0
   ─────────────

2. How many different combinations does a 5-out-of-12 code offer?

3. Give the reflecting code with 32 combinations, using the relays A, B, C, D and E. You need not give the counter circuit.

4. Write the numbers 54 and 47 in the binary notation, and add them.

5. How many different combinations can be made with 7 relays?

6. How many combinations can be made with 8 relays, of which at least 3 must be energized?

7. Compose a 2-out-of-6 code in which only one relay operates and one relay releases on going from one combination to the next, even when the code is cyclic. What is the maximum number of combinations offered by this code?

8. Write the following fractions in the binary notation.
   Indicate which give a repeating binary fraction.

   a) 0.82666015625
   b) 0.975
   c) 0.703125
   d) 0.203125

9. Write the following binary fractions in the decimal notation.

   a) 0.1011101
   b) 1.1101101
   c) 0.1011001
   d) 0.001011

10. Determine the check information which must be transmitted with the number 49489 in the 0 1 2 4 7 code.

Chapter 5

COUNTER CIRCUITS

Digits are often transmitted in the form of groups of pulses. These pulses can be produced by the brief *opening* of a circuit containing a pulse relay (receiver relay), so that the armature and hence the contacts of this relay follow the pulses. A pulse can also be produced by the brief *closing* of a circuit containing a pulse relay. When digits are to be transmitted, the number of pulses in each group corresponds to the digit in question. These pulses must thus be counted, so that after the transmission the number in question is stored in a number of relays. The circuits which are capable of counting pulses and thus of storing digits are called counter circuits.

The numbers in question can be stored in a decimal (linear) form, or coded. In this chapter we shall describe a number of counter circuits using the codes given in Chapter 4.

Counter circuits are normally operated by a pulse relay, but it is also possible to design a self-operating counter circuit, which takes up all positions in turn at a predetermined rate, finally arriving at its starting point again. These "ring" counter circuits can be compared with the selectors described in Section 1.7. The selector wiper is here replaced by a contact tree, which will be described in more detail in Chapter 6.

If the number stored in the counter circuit has to be passed on uncoded, contact trees are also used. In many cases, however, the digit received can be passed on in coded form, when much fewer contacts are needed on the counter relays. This will be discussed in more detail in Chapter 11 (register circuits).

A number of successive signals may be counted with the aid of relays. These signals usually consist of pulses with a certain duration and a certain spacing. Fig. 67 shows a series of pulses, each of which is characterized by two changes of state. All the changes of state must be recorded by the counter circuit so that it can distinguish the identical pulses from one another and thus count them.

The circuit of Fig. 68 can be used to record both changes of state of a pulse. The start of a pulse fed in via contact unit i is recorded by relay 1. During this pulse, relay A remains shorted. *After* the pulse, when contact unit i is opened again, relay A is energized via the hold circuit of relay 1, thus recording the end of the pulse.

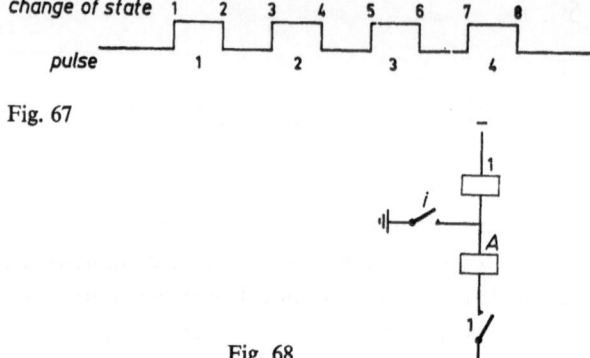

Fig. 67

Fig. 68

5.1 Linear counter circuits

With *n* pairs of relays, *n* pulses can be counted. These pairs of relays can be connected as shown in Fig. 69.

After the first pulse has been registered (relays 1 and *A* energized), the circuit is prepared to receive the second pulse by the switching over of change-over contact unit *a*. After the second pulse, relays 2 and *B* have been energized too, and the pulse contact unit *i* is connected with the third relay pair, 3 and *C*, and so on. The highest-numbered of the energized relay pairs indicates the number of pulses received.

A linear counter circuit where only the pair of relays corresponding to the number of pulses received is energized is given in Fig. 70. In this circuit the pulse contact unit *i* is connected with the first pair of relays via the change-over contacts *m...a*. After the first pulse, relays 1 and *A* are energized. The change-over contact unit *a* is now switched over to prepare the circuit for the reception of the second pulse. After the second pulse, when relays 2 and *B* are energized, relays 1 and *A* release again, thanks to the opening of a break contact unit of relay *B* in the hold circuit of these relays. The relay

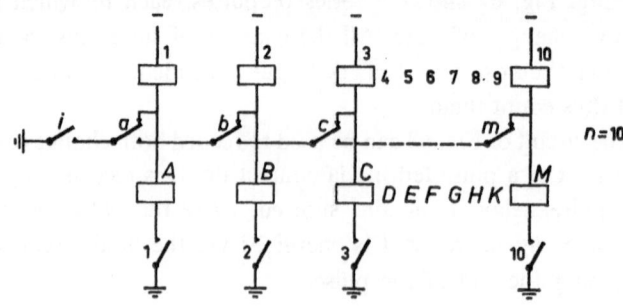

Fig. 69

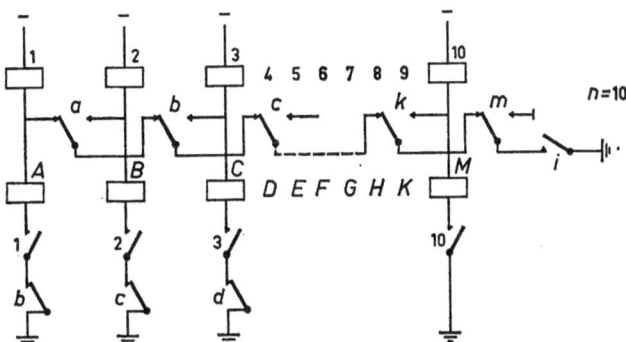

Fig. 70

pairs 2 to 9 are released in a similar way after the 3rd to 10th pulse respectively.

Fig. 71 shows how a ring counter can be constructed on this principle, using n (>2) pairs of relays. After the nth pulse has been recorded, the $(n+1)$th pulse is passed to the first pair of relays again. In Fig. 71, $n=3$.

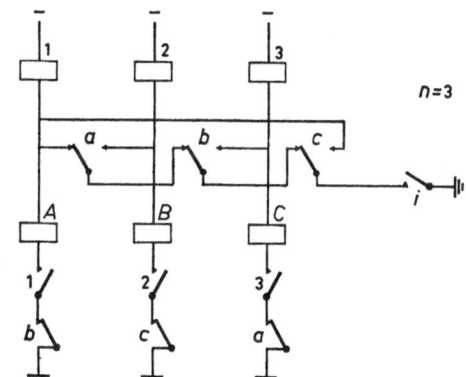

Fig. 71

5.2 Using relay pairs twice

The efficiency of a pair of relays of a counter circuit is:

$$\frac{number\ of\ pulses\ recorded\ per\ relay\ pair}{number\ of\ relay\ pairs}$$

For a linear counter circuit with n pairs of relays and P pulses ($P=n$), the efficiency per relay pair is $1/n$.

The efficiency can be improved by using the relay pairs 1 to $(n-1)$ twice during counting. Relay pair n "remembers" that n pulses have already been counted, so that pulses $n+1$ to $n+(n-1)=2n-1$ can be recorded by relay pairs 1 to $(n-1)$ again. Fig. 72 shows a circuit for doing this, for $n=6$. The first 5 pulses are recorded as in Fig. 70. However, as soon as relay 6 operates at the start of the 6th pulse, relay 5 releases.

Relay E, on the other hand, remains energized during the 6th pulse via a second winding, but then releases, so that the 7th pulse is received by the first relay pair, the 8th by the second relay pair, and so on, until the 11th pulse is recorded by the 5th relay pair. Fig. 72 also gives the counting code for this case, and the sequence diagram of the relays.

When each of n pairs of relays are used once, the efficiency per pair is $1/n$. If relay pairs 1 to $n-1$ are used twice, the efficiency is increased by the factor

$$\frac{n+(n-1)}{n}$$

and thus becomes:

$$\frac{1}{n} \cdot \frac{2n-1}{n} = \frac{2n-1}{n^2}$$

5.3 Multiple use of relay pairs

The efficiency can be improved even more, and thus the number of relays used considerably reduced, if some of the relay pairs are used *more* than twice.

With n relay pairs, one can count to n the first time round, to $n+(n-1)$ the second time round, to $n+(n-1)+(n-2)$ the third time round, and so on up to $n+(n-1)+(n-2)+...+1$.

The total number of pulses which can be recorded in this way can be found as follows:

$$
\begin{aligned}
P &= \quad n + (n-1) + (n-2)...+1 \\
P &= \quad\; 1 + \quad\;\; 2 + \qquad 3...+n \\
\hline
2P &= (n+1) + (n+1) + (n+1)...+(n+1) \\
P &= \frac{n(n+1)}{2}
\end{aligned}
$$

Number of pulses per relay pair is $n(n+1)/2n$, and the efficiency is

$$\frac{n(n+1)}{2n^2} = \frac{n+1}{2n}.$$

Ten pulses can be counted in this way with $n=4$.

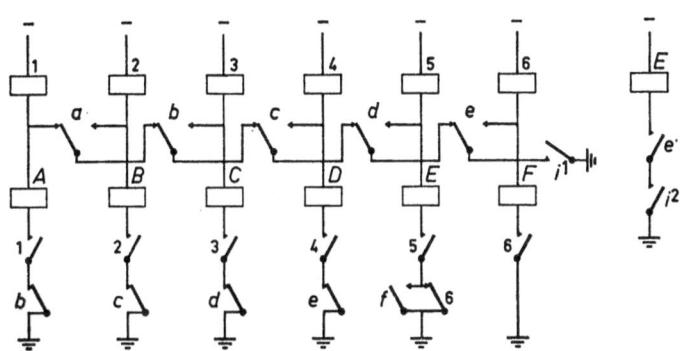

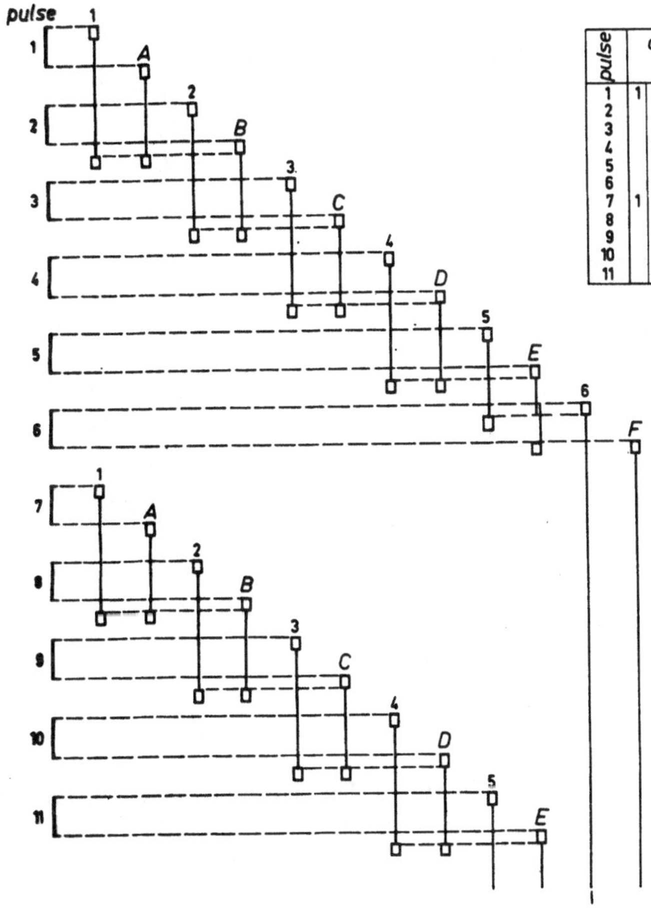

pulse	counting code					
1	1					
2		2				
3			3			
4				4		
5					5	
6						6
7	1					6
8		2				6
9			3			6
10				4		6
11					5	6

Fig. 72

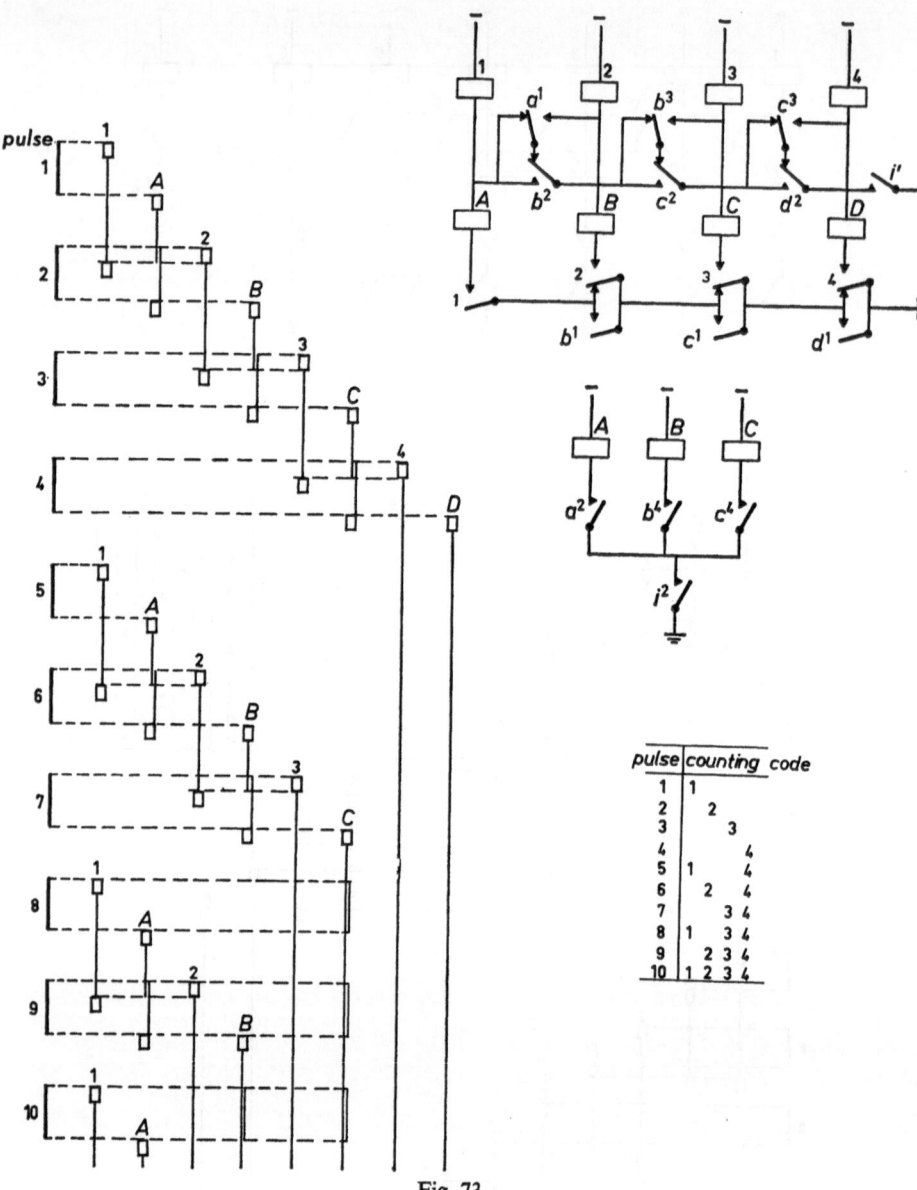

Fig. 73

The circuit for $n=4$ is given in Fig. 73. This figure also contains a sequence diagram, which shows the order in which the various relays operate and release. The circuit works as follows.

When the first pulse is received, the circuit for relay 1 is closed via contact unit i^1.

After the first pulse, relay A operates in the hold circuit of relay 1. As soon as the second pulse begins, relay 2 operates, but unlike the case with the circuits described above, the hold circuit of relay 1 is broken at precisely the same time, so that this relay releases. Relay A, however, remains energized during this pulse, via a second winding and the closed contact unit i^2. After the second pulse relay B operates, and relay 3 is prepared to record the third pulse. At the start of the third pulse relay 3 operates, which immediately causes relay 2 to release. Relay B remains energized during the third pulse. After the fourth pulse, relays 4 and D are energized, and all the others have released. There is no break contact unit in the hold circuit of relays 4 and D, so that these do not release.

The 5th, 6th and 7th pulses are recorded by the 1st, 2nd and 3rd relay pairs respectively. After the 7th pulse the relays 3 and C remain energized, because since contact unit d^1 is closed, the hold circuit of these relays is not broken any more. The 8th and 9th pulses are recorded by relay pairs 1 and 2 again. After the 9th pulse, the hold circuit of relays 2 and B remains closed, because contact unit c^1 remains closed. Finally, the 10th pulse is again recorded by the first relay pair.

5.4 Counter circuits with a common auxiliary relay

We have seen above how we can reduce the number of *relay pairs*. However, it is also possible to reduce the number of *relays* very considerably by recording the ends of all pulses with a single relay.

Fig. 74 shows a circuit for counting 10 pulses in this way. At the beginning of the first pulse (i contact unit closed), relay 1 operates and closes a hold circuit with contact unit 1^1. However, relay K, which is also included in this hold circuit, is still shorted. Relay K operates at the end of the first pulse.

The switching over of the change-over contact unit k causes the start of the second pulse to be recorded by relay 2, which is connected in parallel with a hold winding of relay 1. This is necessary because the first hold winding is broken by the switching over of contact unit 2^1. During the second pulse, relay K remains energized via contact unit i^1. The other winding of K is already broken by contact unit 2^1.

After the second pulse, relays K and 1 release. Contact unit 2^2 has already prepared relay 3 to record the third pulse. Subsequently, the counter relays are switched on in turn via the "even" or the "odd" side by contact unit k in a similar way. After each pulse, only the counter relay corresponding to the number of pulses received is energized.

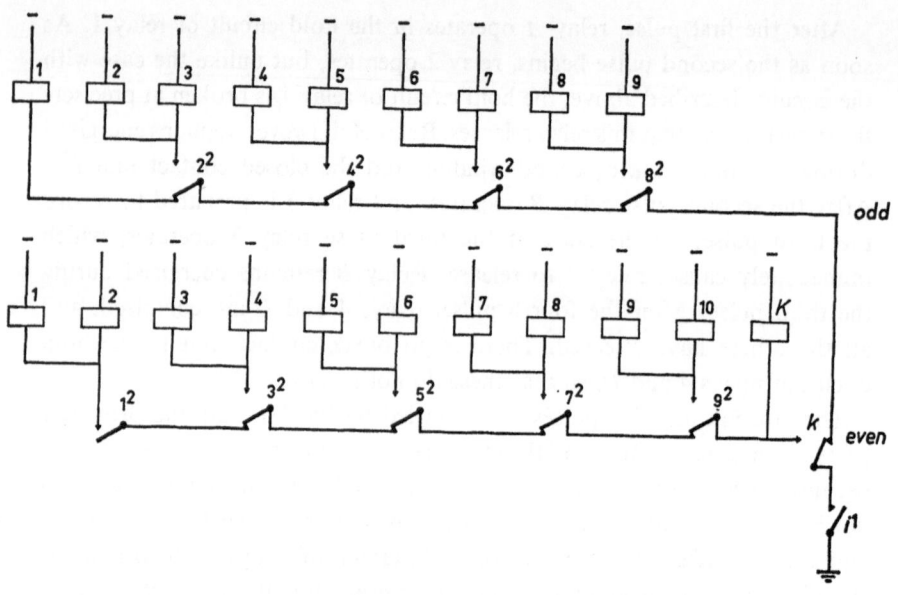

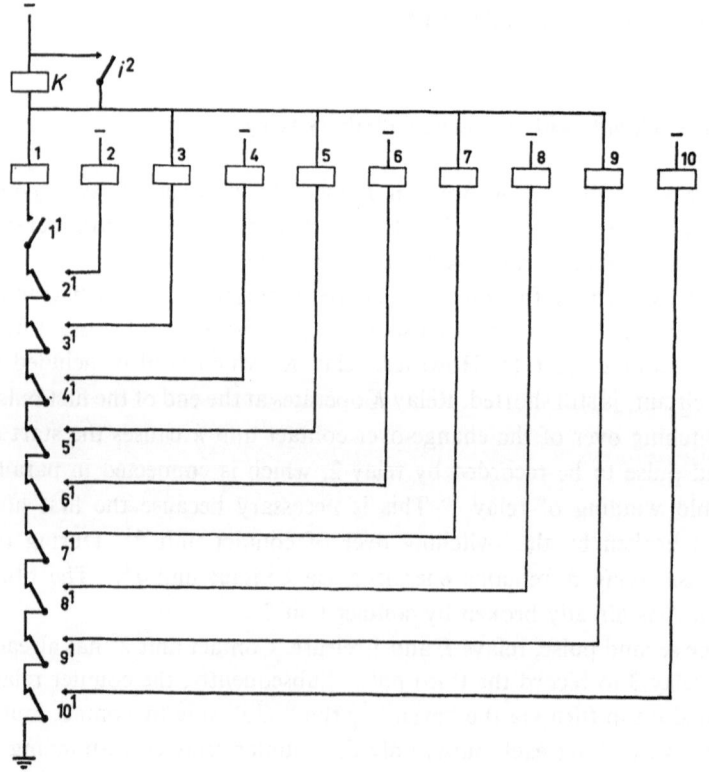

Fig. 74

5.5 Double use of counter relays with a common auxiliary relay

The number of relays used can be reduced even further by a combination of the methods of 5.2 and 5.4 above. Fig. 75 shows a circuit which can count up to 10 with 6 counter relays and one common auxiliary relay K. The counting up to 6 is carried out as described for Fig. 74. Then, however, relay 6 remains switched on. The 7th, 8th, 9th and 10th pulses are counted by relays 1, 2, 3 and 4 respectively. The sequence diagram and the code table indicate the operation of the circuit in full.

5.6 Counter circuits for the binary code

Elements of a binary counter circuit halve the number of changes of state received. Figure 76 gives the circuit of a binary counter element. When pulse contact unit i is closed, relay A operates. Contact unit a closes a hold circuit in preparation for the further counting procedure. Relay 1 does not operate yet, because both windings are shorted. As soon as contact unit i opens again after the first pulse, relay 1 operates in the hold circuit of relay A. At the beginning of the second pulse, the bottom windings of relays A and 1 are shorted again. The upper winding of relay A was already shorted by contact unit 1, so that relay A releases. Relay 1 still remains switched on via the pulse contact unit, contact unit 1 and the top winding of relay 1. After the second pulse, relay 1 releases too, so that we are back at the original state. The two pulses cause 4 changes of state, but relays 1 and A only undergo two changes of state each.

Fig. 77 shows a binary counter circuit with three elements, which can count $2^3 = 8$ pulses. After the 8th pulse, the circuit is back in its original state.

The first counter element (1 and A) is operated by pulse contact unit i. It undergoes only one change of state as a whole per pulse. The second element (2 and B) is operated by a contact unit of relay 1, and only changes state once in two pulses. Finally, the third element (4 and C) is operated by a contact of relay 2, and changes state only once in four pulses. The halving effect of the successive elements can be clearly seen from both the sequence diagram and the code table of Fig. 77.

The general structure of a binary counter circuit can be represented as shown in Fig. 78.

The counter elements can be realized in various ways. For example, relays A, B and C of Fig. 77 release as a result of *short-circuiting*, which

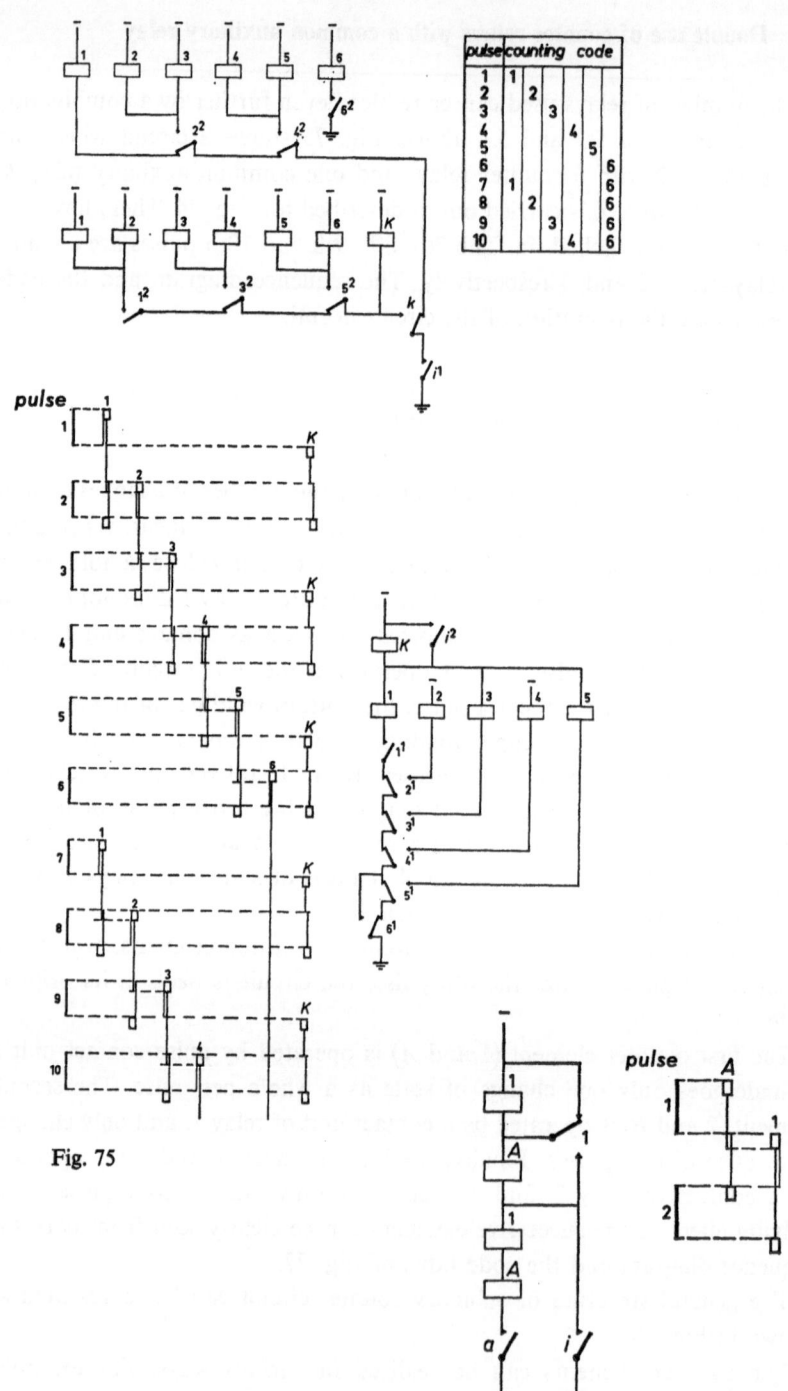

pulse	counting	code				
1	1					
2		2				
3			3			
4				4		
5					5	
6						6
7	1					6
8		2				6
9			3			6
10				4		6

Fig. 75

Fig. 76

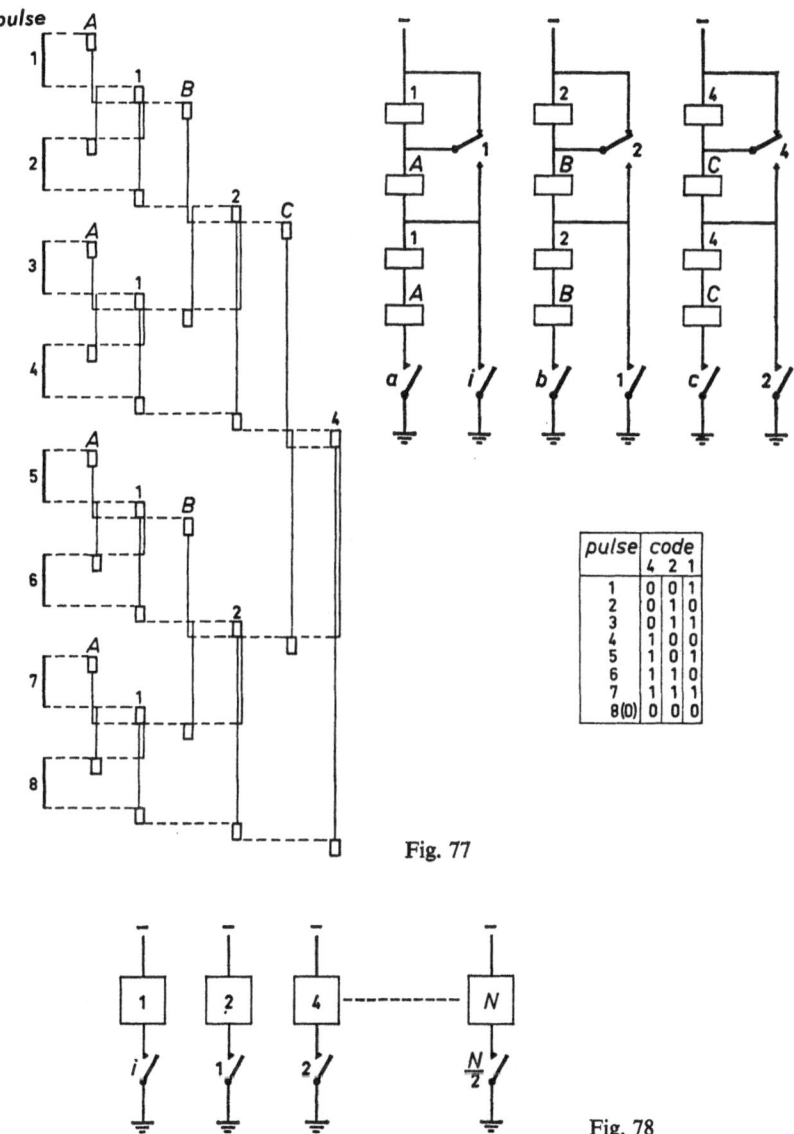

pulse	code		
	4	2	1
1	0	0	1
2	0	1	0
3	0	1	1
4	1	0	0
5	1	0	1
6	1	1	0
7	1	1	1
8(0)	0	0	0

Fig. 77

Fig. 78

retards them. This can be undesirable in high-speed counter circuits. The blocks in Fig. 78 can also be filled by counter elements where the relays are made to release faster by *opening the circuit* of the windings. Fig. 79 shows the circuit of such an element. The pulse contact unit is in the form of a change-over contact unit. As soon as this switches over at the start of the first pulse, relay *A* operates. A contact unit *a* prepares for the operation

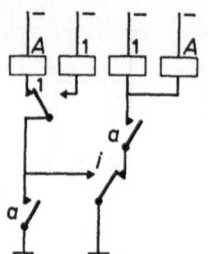

Fig. 79

of relay 1, but relay 1 does not actually operate until the end of the first pulse, when contact unit i returns to its original state. At the start of the second pulse, the circuit of relay A is immediately opened by the pulse contact unit, because contact 1 is switched over. Relay A thus releases quickly. Relay 1 remains energized during the second pulse, and then releases when its circuit is opened.

A relay can also be released by switching on a second winding in which the magnetic field is opposed to that in the first. The application of this principle for a counter element is shown in Fig. 80a and b.

In Fig. 80a, relay A operates at the start of the first pulse, and relay 1 at the end. At the start of the second pulse, a hold circuit is closed for relay 1. Relay A releases, however, because the current through the right-hand winding of A induces a magnetic field opposed to the original one. A disadvantage of this circuit is that the two contact units of relay A, which switch off the two windings, must break at precisely the same time. This can never be guaranteed, because after a relay has been used for a long time the precise moment at which its contact units open can vary somewhat. The circuit of Fig. 80b does not have this disadvantage, because at the start of the second

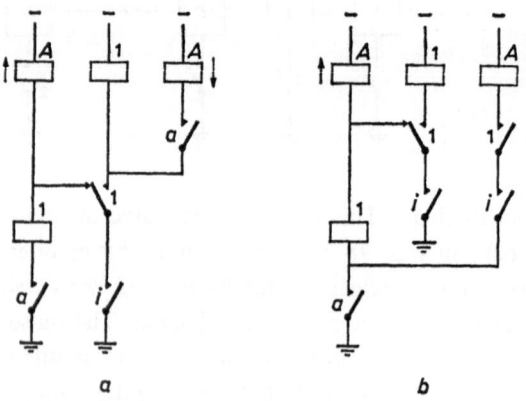

Fig. 80 a b

pulse both windings are switched off by the same contact unit *a*. An extra *i* contact unit is, however, needed in this circuit.

It is also possible to make a binary counter element with only one relay, the place of the second relay being taken by a capacitor. This principle is illustrated in Figures 81*a* and *b*.

At the start of the first pulse, the capacitor in Fig. 81*a* is fully charged. After the first pulse, contact unit *i* is returned to its original position and the capacitor is discharged through the top winding of relay 1. As soon as contact unit 1 is closed, current flows through both windings. The magnetic fields produced by these two windings are opposed, but that in the top winding predominates, so that the relay remains in the energized state. The capacitor voltage falls to a fraction of its original value. When contact unit *i* is switched over at the start of the second pulse, the capacitor is immediately completely discharged via contact unit 1 (now switched over). After the second pulse, the capacitor is charged via the bottom winding of relay 1. Moreover, during the charging of the capacitor the top winding of the relay is connected in parallel with the discharged capacitor in series with the resistance, so that now the bottom winding predominates, the field starts to reverse its direction and as the field goes through zero the relay releases.

In the circuit of Fig. 81*b*, which is based on the same principle, only one contact of relay 1 is needed. The capacitor is charged to the voltage provided by the voltage divider R_1–R_2; but since the resistance R_1 is many times higher than R_2, the capacitor still receives sufficient charge to switch relay 1 on after the first pulse. This circuit is not quite so quick, because more time is needed to discharge the capacitor via the high resistance R_1 at the start of the second pulse.

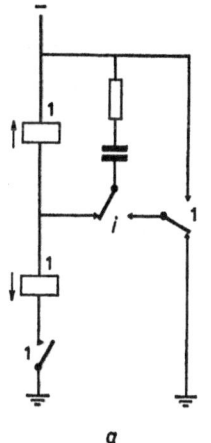 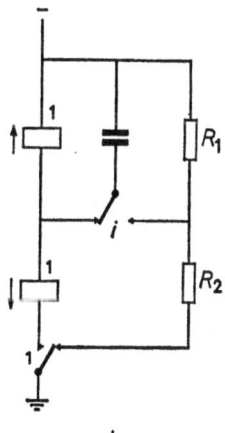

a   *b*   Fig. 81

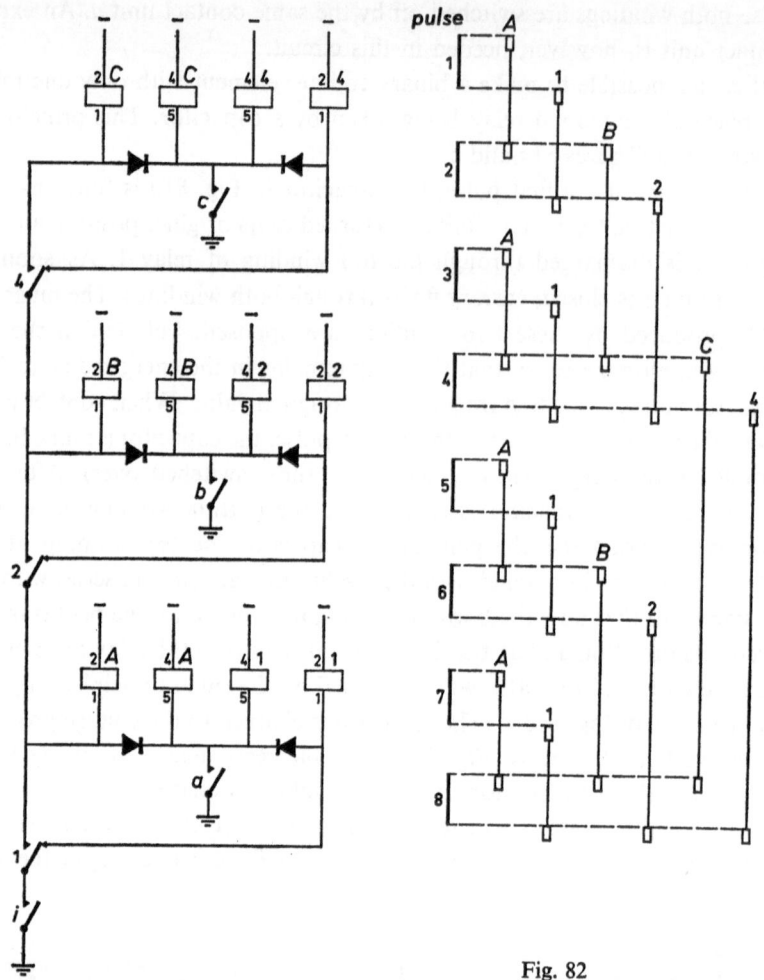

Fig. 82

Fig. 82 shows a binary counter circuit in which all elements are controlled by the pulse contact unit. This has the advantage that time shifts cannot occur.

In the counters we have been describing so far, time shifts do occur, as may be seen e.g. from the sequence diagram of Fig. 77, where after the 4th pulse relay 4 does not operate until first relay 1 and then relay 2 has released.

The circuit works as follows. When contact unit *i* is closed, both windings of relay 1 and winding 4-5 of relay *A* are switched on. The latter relay operates, but relay 1 does not, as its two windings are connected differentially. A hold circuit for relay *A* is closed by means of a contact unit *a*, as a result of which relay 1 also operates *after* the pulse, as then only the winding 4–5 carries current. When contact unit *i* is closed again (second pulse), current flows through both windings of relay *A*, because the change-over contact unit 1 is switched over. Relay *A* therefore releases. Relay 1 remains energized during the pulse via winding 4–5.

In the meantime, relay *B* has also operated at the start of the second pulse. Current then flows through both windings of relay 2. After the second pulse, relay 2 also operates. The complete operation of the circuit can be followed from the sequence diagram of Fig. 82.

5.7 Binary circuits for counting forwards and backwards

A counter can be operated by two pulse sources, to make it count both forwards and backwards.

A binary counter can count "backwards" by means of another switching sequence for the relays of the counter elements. Fig. 83 shows this for the counting of 4 pulses.

If we compare the circuits of Fig. 83 and Fig. 77, we see that the change-over contact unit 1, which shorts relay 1 in its non-energized state, is replaced by a change-over contact unit *a* which shorts relay *A* in its non-energized state. The hold contact unit *a* is also replaced by a hold contact unit 1. It may be seen from Fig. 83 that after the first, second, third and fourth pulses, positions 3, 2, 1 and 0 respectively are reached. The counter thus counts backwards. If the counter has to be able to count both forwards

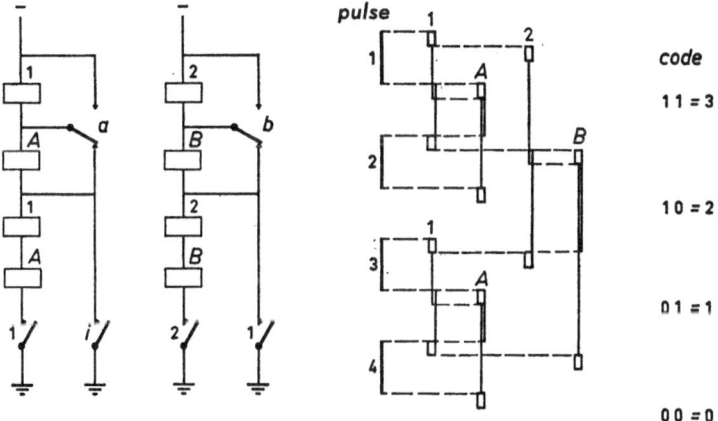

Fig. 83

and backwards, a change-over relay must be added. The function of this relay is to change the circuit of Fig. 77 into that of Fig. 83 and *vice versa*.

Fig. 84 gives the complete circuit for 4 pulses, the two pulse sources also being given for the sake of completeness.

If a pulse is received for counting forwards, relay I operates and the circuit behaves like that of Fig. 77. If however a pulse is received for counting backwards, first relay W and then relay I operates. The circuit then works like that of Fig. 83. Relay W takes care of the switching from forwards to backwards counting. Relay W must not release until relay I has released,

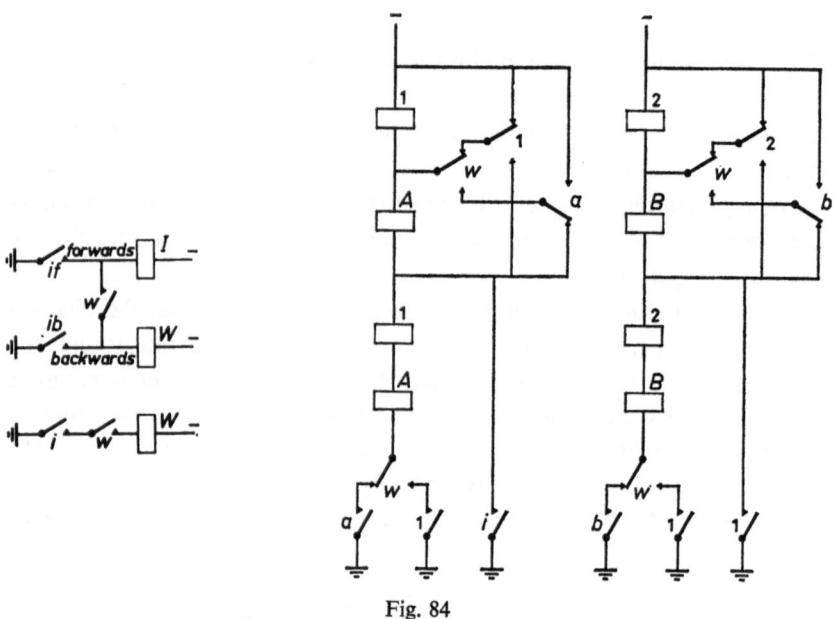

Fig. 84

to ensure that the pulse received is recorded completely. In order to achieve this, relay W is provided with a second winding which remains switched on until relay I has released.

5.8   Counter circuit based on the reflecting code

A counter circuit in which only one relay changes state at the start or end of each pulse cannot give rise to any undesirable intermediate states. A binary counter circuit does not have this property, but a counter based on the reflecting code does. A counter circuit based on this principle can be

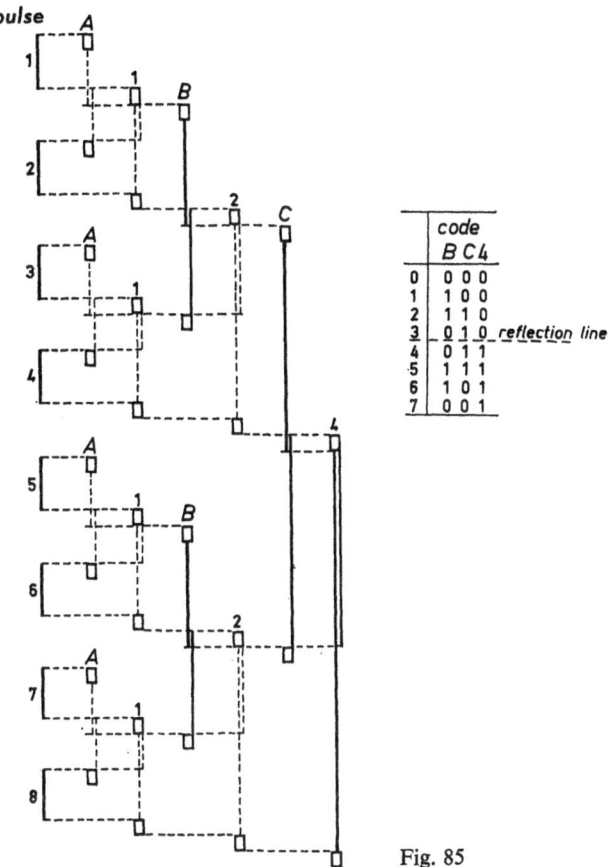

	code
	B C 4
0	0 0 0
1	1 0 0
2	1 1 0
3	0 1 0 _reflection_ line
4	0 1 1
5	1 1 1
6	1 0 1
7	0 0 1

Fig. 85

obtained from Fig. 77. Fig. 85 gives the sequence diagram of the binary counter of Fig. 77. It will be seen that the columns for relays *B*, *C* and 4 give the reflecting code.

5.9 Biquinary counter circuit

This counter circuit consists of five counter elements for counting up to five, two counter elements to indicate the first or second five, and one auxiliary relay. The circuit diagram is shown in Fig. 86.

Relay 1 operates at the start of the first pulse. After the first pulse, relay 1 remains energized, and relay *E* is also switched on. Pulses 2, 3 and 4 switch on relays 2, 3 and 4 in succession. Each time one counter relay operates, the one before releases *after* the pulse in question. During this pulse, it

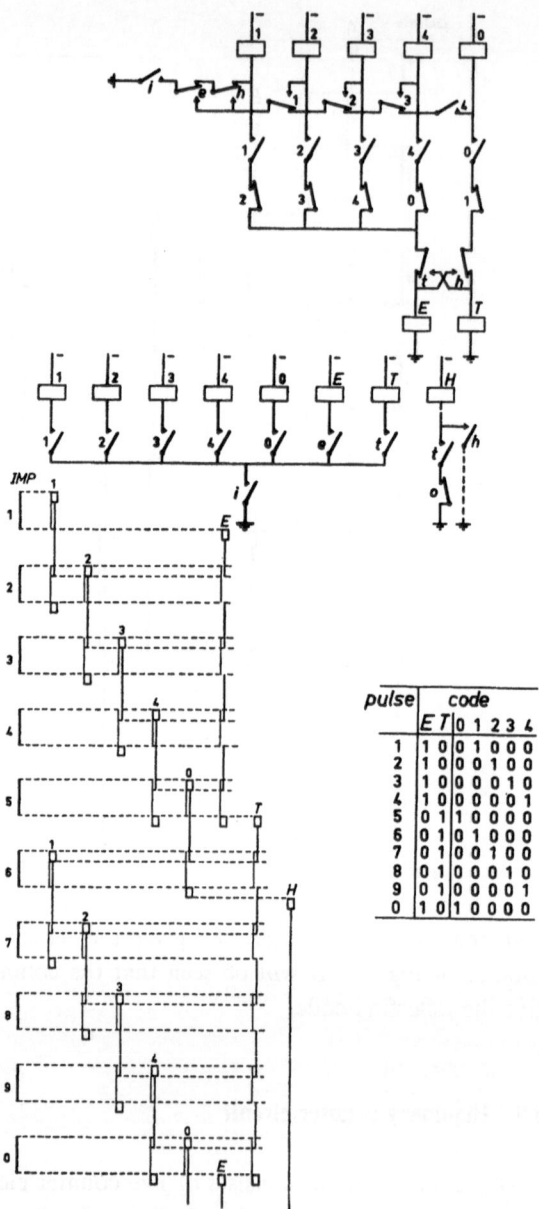

pulse	code					
	E T	0	1	2	3	4
1	1 0	0	1	0	0	0
2	1 0	0	0	1	0	0
3	1 0	0	0	0	1	0
4	1 0	0	0	0	0	1
5	0 1	1	0	0	0	0
6	0 1	0	1	0	0	0
7	0 1	0	0	1	0	0
8	0 1	0	0	0	1	0
9	0 1	0	0	0	0	1
0	1 0	1	0	0	0	0

Fig. 86

remains energized via a second winding. The fifth pulse causes relay O to operate, and the circuit of relay E is broken. This latter relay also has a second winding which keeps it energized during the fifth pulse, but after this pulse it releases. At the same moment, relay T operates in the hold circuit of relay O.

The sixth pulse operates relay 1 again, after which relay O releases. Relay T remains energized in the hold circuit of relay 1. When relay O releases, relay H operates, so that the seventh pulse can reach relay 2. The eighth and ninth pulses cause relay 3 and 4 respectively to operate, the preceeding relay releasing in each case. The tenth pulse causes relay O to operate again, and at the end of this pulse relay E operates again. Relay T then releases, at the same time as relay 4.

Use of the biquinary counter circuit has the advantage that one relay of the pair (E, T) and one relay of the five $(0, 1, 2, 3, 4)$ is switched on at any time. If by some fault in the circuit more or less than one relay from each group should be switched on, this can easily be found out by a detector circuit.

5.10 Cyclic 2-out-of-n codes, counting forwards and backwards

We have shown in Section 4.1 that the number of possible combinations in an m-out-of-n code is

$$\binom{n}{m} = \frac{n!}{m!(n-m)!}$$

We shall now consider the case where $m=2$, $n=7$, so $\binom{n}{m}=21$. For the realization of this code we shall consider relays designed so that, taking voltage, resistance and manufacturing tolerances into account, we may rely on the relay not operating at a current of i, but certainly operating at a current of $2i$. The principle of operation is shown in Fig. 87.

The resistance of relay A is relatively low compared to that of the resistance R, so that the current through the relay when 1 or 2 is closed is i, and the current when 1 and 2 are both closed is about $2i$. The operation of relay A depends on the coincidence of the two currents. This is difficult or impossible to achieve with the normal types of relays described in Chapter 1, because of the large spread in the operate value and release value (AT) of the relays.

A new type of relay which has been used more and more of recent years is the reed relay. This consists of a gas-filled glass tube containing two reeds

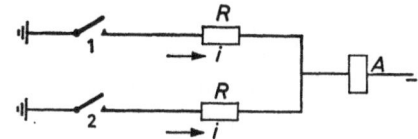

Fig. 87

of a magnetic material provided with a conductive surface coating. This tube is placed so that these reeds fall within the magnetic circuit of an electromagnet.

When the coil is switched on, the reeds are pulled against one another, thus making a contact. Since this relay can be produced completely automatically, there are practically no differences in the distance between the reeds, so that all relays of the same type operate at practically the same AT value. Moreover, the difference between the operate value and the non-operate value is much less than for the normal angular-armature types. Before discussing the circuit, which is given in Fig. 88, we will first work out the codes suitable for this purpose.

Now each bit occurs 6 times in a 2-out-of-7 code, namely together with each of the other six bits. There are also six possibilities for 2 out of 4 bits, namely 1–2 1–3 1–4 2–3 2–4 and 3–4. This fact is made use of to develop a cyclic 2-out-of-7 code, where the relays are switched on by the coincidence of two currents. For this purpose we make use of a matrix, e.g. like that shown in Table XXVI. Each row in this table contains four "ones" and

TABLE XXVI

	1	2	3	4	5	6	7
1	1	1	1	0	1	0	0
2	0	0	1	1	1	0	1
3	0	1	0	0	1	1	1
4	1	1	0	1	0	0	1
5	0	1	1	1	0	1	0
6	1	0	0	1	1	1	0
7	1	0	1	0	0	1	1

three "zeroes". The "ones" represent currents of value i. In the first row we thus have six possible cases of coincidence of 2 currents i, viz. 1–2 1–3 1–5 2–3 2–5 and 3–5. The "ones" in the other rows are so arranged that for each combination of 2 out of 7 (21 possibilities), coincidence occurs in only two different rows. For example, let us consider the combination 1–2. Coincidence for this combination occurs in rows 1 and 4. This thus ensures that combination 1–4 follows combination 1–2. We now look for the coincidence of bits 1 and 4 in the different rows, and find it in rows 4 and 6. Carrying on in this way, we get the complete code shown in Table XXVII. The arrangement of "ones" and "zeroes" in Table XXVII is one of 28 possible matrices of a particular type, which we shall now specify. If we call the arrangement 1 1 1 0 1 0 0 occurring in row 1 of table XXVII *pattern A*, then we find that each subsequent row also contains pattern A, shifted to the right 2 places each time. We therefore denote this matrix by $A2$, where A indicates the pattern and 2 the shift.

Now it is also possible to displace all the rows in the vertical direction, so that 1 1 1 0 1 0 0 comes in the second row, while the first row is now occupied by 1 0 1 0 0 1 1, which was previously in the seventh row. This procedure can be repeated seven times, so that $A2$ occurs in 7 different *forms*, denoted by $A2/1$, $A2/2$...$A2/7$. If in each of the seven cases we start with the combination 1–2 of the code to be developed, we find that we do

TABLE XXVII

	1	2	3	4	5	6	7	code		
1	1	1	1	0	1	0	0	1–2	1–5	1–3
2	0	0	1	1	1	0	1	1–4	1–6	1–7
3	0	1	0	0	1	1	1	4–6	6–7	4–7
4	1	1	0	1	0	0	1	5–6	3–7	2–4
5	0	1	1	1	0	1	0	3–6	2–7	4–5
6	1	0	0	1	1	1	0	5–7	3–4	2–6
7	1	0	1	0	0	1	1	2–3	2–5	3–5

indeed get seven different codes. Each of these codes offers 21 possibilities and is cyclic.

We shall now investigate how we can develop a code with the combinations in the order opposed to that in the above-mentioned code according to $A2/1$.

In Table XXVII the combination 1–2 follows from the combination 3–5 (transition from the last to the first combination) because 3 and 5 give coincidences in the first and second *rows*. We see however that the combination 1–2 gives the combination 3–5 (transition from the first combination to the last) if we look for coincidences in the *columns*, since the combination 1–2 gives coincidence in columns 3 and 5. It follows that the backwards-counting code can be found by looking for coincidences in the columns instead of in the rows. It is however simpler to rewrite the matrix, with the first row as the first column, the second row as the second column, and so on. This gives Table XXVIII.

The code can now be determined by coincidences in the rows, and runs in the reverse order to that produced by $A2/1$ (Table XXVII).

Closer inspection of Table XVIII shows that it has a different pattern, viz 1 1 1 0 0 1 0 (5th row), as opposed to pattern A (1 1 1 0 1 0 0). We

TABLE XXVIII

	1	2	3	4	5	6	7
1	1	0	0	1	0	1	1
2	1	0	1	1	1	0	0
3	1	1	0	0	1	0	1
4	0	1	0	1	1	1	0
5	1	1	1	0	0	1	0
6	0	0	1	0	1	1	1
7	0	1	1	1	0	0	1

will call this pattern B. The shift is 4, as may be seen e.g. by consideration of the three successive "ones" in rows 5 and 6. Moreover the matrix of Table XXVIII is the fifth of the seven possibilities, so that it may be denoted by $B4/5$.

It thus follows from the above that the codes $A2/1$ and $B4/5$ arise from one another, and possess opposite orders. Each of the codes produced from $A2/1...A2/7$ can be reversed by development from the corresponding matrices $B4/1...B4/7$, as follows:

$A2/1$ opposed to $B4/5$
$A2/2$ opposed to $B4/3$
$A2/3$ opposed to $B4/1$
$A2/4$ opposed to $B4/6$
$A2/5$ opposed to $B4/4$
$A2/6$ opposed to $B4/2$
$A2/7$ opposed to $B4/7$

It is natural to expect that $A4$ and $B2$ will also give 7 combinations each, which will pair off into opposite codes. These pairs are:

$A4/1$ opposed to $B2/2$
$A4/2$ opposed to $B2/5$
$A4/3$ opposed to $B2/1$
$A4/4$ opposed to $B2/4$
$A4/5$ opposed to $B2/7$
$A4/6$ opposed to $B2/3$
$A4/7$ opposed to $B2/6$

All the matrices and the codes obtained from them are given pair by pair in Tables XXIX to XLII.

TABLE XXIX

A2/1

	1	2	3	4	5	6	7			
1	1	1	1	0	1	0	0	1-2	1-5	1-3
2	0	0	1	1	1	0	1	1-4	1-6	1-7
3	0	1	0	0	1	1	1	4-6	6-7	4-7
4	1	1	0	1	0	0	1	5-6	3-7	2-4
5	0	1	1	1	0	1	0	3-6	2-7	4-5
6	1	0	0	1	1	1	0	5-7	3-4	2-6
7	1	0	1	0	0	1	1	2-3	2-5	3-5

1-2

B4/5

	1	2	3	4	5	6	7			
1	1	0	0	1	0	1	1	1-2	1-3	1-5
2	1	0	1	1	1	0	0	3-5	2-5	2-3
3	1	1	0	0	1	0	1	2-6	3-4	5-7
4	0	1	0	1	1	1	0	4-5	2-7	3-6
5	1	1	1	0	0	1	0	2-4	3-7	5-6
6	0	0	1	0	1	1	1	4-7	6-7	4-6
7	0	1	1	1	0	0	1	1-7	1-6	1-4

1-2

TABLE XXX

A2/2

	1	2	3	4	5	6	7			
1	1	0	1	0	0	1	1	1-2	2-6	1-6
2	1	1	1	0	1	0	0	2-5	4-6	1-7
3	0	0	1	1	1	0	1	2-4	6-7	1-5
4	0	1	0	0	1	1	1	5-6	1-4	2-7
5	1	1	0	1	0	0	1	4-7	5-7	4-5
6	0	1	1	1	0	1	0	3-5	3-4	3-7
7	1	0	0	1	1	1	0	2-3	3-6	1-3

1-2

B4/3

	1	2	3	4	5	6	7			
1	1	1	0	0	1	0	1	1-2	1-6	2-6
2	0	1	0	1	1	1	0	1-3	3-6	2-3
3	1	1	1	0	0	1	0	3-7	3-4	3-5
4	0	0	1	0	1	1	1	4-5	5-7	4-7
5	0	1	1	1	0	0	1	2-7	1-4	5-6
6	1	0	0	1	0	1	1	1-5	6-7	2-4
7	1	0	1	1	1	0	0	1-7	4-6	2-5

1-2

TABLE XXXI

A2/3

	1	2	3	4	5	6	7			
1	1	0	0	1	1	1	0	1-2	3-7	4-6
2	1	0	1	0	0	1	1	3-6	2-4	1-7
3	1	1	1	0	1	0	0	2-7	6-7	2-6
4	0	0	1	1	1	0	1	5-6	2-5	5-7
5	0	1	0	0	1	1	1	1-5	3-5	4-5
6	1	1	0	1	0	0	1	1-3	3-4	1-4
7	0	1	1	1	0	1	0	2-3	4-7	1-6

1-2

B4/1

	1	2	3	4	5	6	7			
1	1	1	1	0	0	1	0	1-2	4-6	3-7
2	0	0	1	0	1	1	1	1-6	4-7	2-3
3	0	1	1	1	0	0	1	1-4	3-4	1-3
4	1	0	0	1	0	1	1	4-5	3-5	1-5
5	1	0	1	1	1	0	0	5-7	2-5	5-6
6	1	1	0	0	1	0	1	2-6	6-7	2-7
7	0	1	0	1	1	1	0	1-7	2-4	3-6

1-2

TABLE XXXII

A2/4

	1	2	3	4	5	6	7			
1	0	1	1	1	0	1	0	1-2	1-4	2-4
2	1	0	0	1	1	1	0	4-7	2-7	1-7
3	1	0	1	0	0	1	1	5-7	6-7	3-7
4	1	1	1	0	1	0	0	5-6	3-6	3-5
5	0	0	1	1	1	0	1	2-6	1-3	4-5
6	0	1	0	0	1	1	1	1-6	3-4	2-5
7	1	1	0	1	0	0	1	2-3	1-5	4-6

1-2

B4/6

	1	2	3	4	5	6	7			
1	0	1	1	1	0	0	1	1-2	2-4	1-4
2	1	0	0	1	0	1	1	4-6	1-5	2-3
3	1	0	1	1	1	0	0	2-5	3-4	1-6
4	1	1	0	0	1	0	1	4-5	1-3	2-6
5	0	1	0	1	1	1	0	3-5	3-6	5-6
6	1	1	1	0	0	1	0	3-7	6-7	5-7
7	0	0	1	0	1	1	1	1-7	2-7	4-7

1-2

TABLE XXXIII

A2/5

	1	2	3	4	5	6	7			
1	1	1	0	1	0	0	1	1-2	2-5	2-7
2	0	1	1	1	0	1	0	1-5	5-7	1-7
3	1	0	0	1	1	1	0	3-5	6-7	1-4
4	1	0	1	0	0	1	1	5-6	4-7	1-3
5	1	1	1	0	1	0	0	3-7	1-6	4-5
6	0	0	1	1	1	0	1	4-6	3-4	3-6
7	0	1	0	0	1	1	1	2-3	2-6	2-4

1-2

B4/4

	1	2	3	4	5	6	7			
1	1	0	1	1	1	0	0	1-2	2-7	2-5
2	1	1	0	0	1	0	1	2-4	2-6	2-3
3	0	1	0	1	1	1	0	3-6	3-4	4-6
4	1	1	1	0	0	1	0	4-5	1-6	3-7
5	0	0	1	0	1	1	1	1-3	4-7	5-6
6	0	1	1	1	0	0	1	1-4	6-7	3-5
7	1	0	0	1	0	1	1	1-7	5-7	1-5

1-2

TABLE XXXIV

A2/6

	1	2	3	4	5	6	7			
1	0	1	0	0	1	1	1	1-2	3-6	5-7
2	1	1	0	1	0	0	1	2-6	3-5	1-7
3	0	1	1	1	0	1	0	1-3	6-7	2-5
4	1	0	0	1	1	1	0	5-6	1-5	1-6
5	1	0	1	0	0	1	1	1-4	4-6	4-5
6	1	1	1	0	1	0	0	2-4	3-4	4-7
7	0	0	1	1	1	0	1	2-3	3-7	2-7

1-2

B4/2

	1	2	3	4	5	6	7			
1	0	1	0	1	1	1	0	1-2	5-7	3-6
2	1	1	1	0	0	1	0	2-7	3-7	2-3
3	0	0	1	0	1	1	1	4-7	3-4	2-4
4	0	1	1	1	0	0	1	4-5	4-6	1-4
5	1	0	0	1	0	1	1	1-6	1-5	5-6
6	1	0	1	1	1	0	0	2-5	6-7	1-3
7	1	1	0	0	1	0	1	1-7	3-5	2-6

1-2

TABLE XXXV

A2/7

	1	2	3	4	5	6	7			
1	0	0	1	1	1	0	1	1-2	4-7	3-5
2	0	1	0	0	1	1	1	3-7	1-3	1-7
3	1	1	0	1	0	0	1	1-6	6-7	3-6
4	0	1	1	1	0	1	0	5-6	2-6	4-6
5	1	0	0	1	1	1	0	2-5	2-4	4-5
6	1	0	1	0	0	1	1	2-7	3-4	1-5
7	1	1	1	0	1	0	0	2-3	1-4	5-7

1-2

B4/7

	1	2	3	4	5	6	7			
1	0	0	1	0	1	1	1	1-2	3-5	4-7
2	0	1	1	1	0	0	1	5-7	1-4	2-3
3	1	0	0	1	0	1	1	1-5	3-4	2-7
4	1	0	1	1	1	0	0	4-5	2-4	2-5
5	1	1	0	0	1	0	1	4-6	2-6	5-6
6	0	1	0	1	1	1	0	3-6	6-7	1-6
7	1	1	1	0	0	1	0	1-7	1-3	3-7

1-2

TABLE XXXVI

A4/1

	1	2	3	4	5	6	7			
1	1	1	1	0	1	0	0	1-2	1-3	1-5
2	0	1	0	0	1	1	1	1-6	1-4	1-7
3	0	1	1	1	0	1	0	4-7	6-7	4-6
4	1	0	1	0	0	1	1	5-6	2-4	3-7
5	0	0	1	1	1	0	1	2-7	3-6	4-5
6	1	1	0	1	0	0	1	2-6	3-4	5-7
7	1	0	0	1	1	1	0	2-3	3-5	2-5

1-2

B2/2

	1	2	3	4	5	6	7			
1	1	0	0	1	0	1	1	1-2	1-5	1-3
2	1	1	1	0	0	1	0	2-5	3-5	2-3
3	1	0	1	1	1	0	0	5-7	3-4	2-6
4	0	0	1	0	1	1	1	4-5	3-6	2-7
5	1	1	0	0	1	0	1	3-7	2-4	5-6
6	0	1	1	1	0	0	1	4-6	6-7	4-7
7	0	1	0	1	1	1	0	1-7	1-4	1-6

1-2

TABLE XXXVII

A4/2

	1	2	3	4	5	6	7			
1	1	0	0	1	1	1	0	1-2	2-4	1-4
2	1	1	1	0	1	0	0	2-7	4-7	1-7
3	0	1	0	0	1	1	1	3-7	6-7	5-7
4	0	1	1	1	0	1	0	5-6	3-5	3-6
5	1	0	1	0	0	1	1	1-3	2-6	4-5
6	0	0	1	1	1	0	1	2-5	3-4	1-6
7	1	1	0	1	0	0	1	2-3	4-6	1-5
										1-2

B2/5

	1	2	3	4	5	6	7			
1	1	1	0	0	1	0	1	1-2	1-4	2-4
2	0	1	1	1	0	0	1	1-5	4-6	2-3
3	0	1	0	1	1	1	0	1-6	3-4	2-5
4	1	0	0	1	0	1	1	4-5	2-6	1-3
5	1	1	1	0	0	1	0	3-6	3-5	5-6
6	1	0	1	1	1	0	0	5-7	6-7	3-7
7	0	0	1	0	1	1	1	1-7	4-7	2-7
										1-2

TABLE XXXVIII

A4/3

	1	2	3	4	5	6	7			
1	1	1	0	1	0	0	1	1-2	3-5	4-7
2	1	0	0	1	1	1	0	1-3	3-7	1-7
3	1	1	1	0	1	0	0	3-6	6-7	1-6
4	0	1	0	0	1	1	1	5-6	4-6	2-6
5	0	1	1	1	0	1	0	2-4	2-5	4-5
6	1	0	1	0	0	1	1	1-5	3-4	2-7
7	0	0	1	1	1	0	1	2-3	5-7	1-4
										1-2

B2/1

	1	2	3	4	5	6	7			
1	1	1	1	0	0	1	0	1-2	4-7	3-5
2	1	0	1	1	1	0	0	1-4	5-7	2-3
3	0	0	1	0	1	1	1	2-7	3-4	1-5
4	1	1	0	0	1	0	1	4-5	2-5	2-4
5	0	1	1	1	0	0	1	2-6	4-6	5-6
6	0	1	0	1	1	1	0	1-6	6-7	3-6
7	1	0	0	1	0	1	1	1-7	3-7	1-3
										1-2

TABLE XXXIX

A4/4

	1	2	3	4	5	6	7			
1	0	0	1	1	1	0	1	1-2	4-6	3-7
2	1	1	0	1	0	0	1	2-4	3-6	1-7
3	1	0	0	1	1	1	0	2-6	6-7	2-7
4	1	1	1	0	1	0	0	5-6	5-7	2-5
5	0	1	0	0	1	1	1	3-5	1-5	4-5
6	0	1	1	1	0	1	0	1-4	3-4	1-3
7	1	0	1	0	0	1	1	2-3	1-6	4-7
										1-2

B2/4

	1	2	3	4	5	6	7			
1	0	1	1	1	0	0	1	1-2	3-7	4-6
2	0	1	0	1	1	1	0	4-7	1-6	2-3
3	1	0	0	1	0	1	1	1-3	3-4	1-4
4	1	1	1	0	0	1	0	4-5	1-5	3-5
5	1	0	1	1	1	0	0	2-5	5-7	5-6
6	0	0	1	0	1	1	1	2-7	6-7	2-6
7	1	1	0	0	1	0	1	1-7	3-6	2-4
										1-2

TABLE XL

A4/5

	1	2	3	4	5	6	7			
1	1	0	1	0	0	1	1	1-2	5-7	3-6
2	0	0	1	1	1	0	1	3-5	2-6	1-7
3	1	1	0	1	0	0	1	2-5	6-7	1-3
4	1	0	0	1	1	1	0	5-6	1-6	1-5
5	1	1	1	0	1	0	0	4-6	1-4	4-5
6	0	1	0	0	1	1	1	4-7	3-4	2-4
7	0	1	1	1	0	1	0	2-3	2-7	3-7
										1-2

B2/7

	1	2	3	4	5	6	7			
1	1	0	1	1	1	0	0	1-2	3-6	5-7
2	0	0	1	0	1	1	1	3-7	2-7	2-3
3	1	1	0	0	1	0	1	2-4	3-4	4-7
4	0	1	1	1	0	0	1	4-5	1-4	4-6
5	0	1	0	1	1	1	0	1-5	1-6	5-6
6	1	0	0	1	0	1	1	1-3	6-7	2-5
7	1	1	1	0	0	1	0	1-7	2-6	3-5
										1-2

TABLE XLI

A4/6									B2/3							
	1	2	3	4	5	6	7			1	2	3	4	5	6	7	
1	0	1	1	1	0	1	0	1-2 1-6 2-6	1	0	1	0	1	1	1	0	1-2 2-6 1-6
2	1	0	1	0	0	1	1	4-6 2-5 1-7	2	1	0	0	1	0	1	1	3-6 1-3 2-3
3	0	0	1	1	1	0	1	1-5 6-7 2-4	3	1	1	1	0	0	1	0	3-5 3-4 3-7
4	1	1	0	1	0	0	1	5-6 2-7 1-4	4	1	0	1	1	1	0	0	4-5 4-7 5-7
5	1	0	0	1	1	1	0	5-7 4-7 4-5	5	0	0	1	0	1	1	1	1-4 2-7 5-6
6	1	1	1	0	1	0	0	3-7 3-4 3-5	6	1	1	0	0	1	0	1	2-4 6-7 1-5
7	0	1	0	0	1	1	1	2-3 1-3 3-6	7	0	1	1	1	0	0	1	1-7 2-5 4-6
								1-2									1-2

TABLE XLII

A4/7									B2/6							
	1	2	3	4	5	6	7			1	2	3	4	5	6	7	
1	0	1	0	0	1	1	1	1-2 2-7 2-5	1	0	0	1	0	1	1	1	1-2 2-5 2-7
2	0	1	1	1	0	1	0	5-7 1-5 1-7	2	1	1	0	0	1	0	1	2-6 2-4 2-3
3	1	0	1	0	0	1	1	1-4 6-7 3-5	3	0	1	1	1	0	0	1	4-6 3-4 3-6
4	0	0	1	1	1	0	1	5-6 1-3 4-7	4	0	1	0	1	1	1	0	4-5 3-7 1-6
5	1	1	0	1	0	0	1	1-6 3-7 4-5	5	1	0	0	1	0	1	1	4-7 1-3 5-6
6	1	0	0	1	1	1	0	3-6 3-4 4-6	6	1	1	1	0	0	1	0	3-5 6-7 1-4
7	1	1	1	0	1	0	0	2-3 2-4 2-6	7	1	0	1	1	1	0	0	1-7 1-5 5-7
								1-2									1-2

A counter circuit working on the coincidence principle is shown in Fig. 88 and described below.

Initially relays 1 and 2 are operated, because relay S is released. When the "start" contact unit is closed, relay S operates, breaking the operate circuits of relays 1 and 2. These relays, however, remain energized via hold windings.

A circuit is now closed for relay V, as a result of which the first and second lines under the relays A to G are connected to earth. Relays A and D are then switched on via two resistances, and the other relays either through one resistance or not at all. Relays A and D are therefore the only ones to operate. This thus prepares for the combination 1-4, following on 1-2. As soon as relay I operates as the result of a count pulse, a hold circuit is closed for relays A and D.

The opening of the break contact unit i breaks the hold circuit for relays 1 and 2. After contacts 1 and 2 are broken, relay V releases, because contact unit i is also open.

After V has released, the switching-on circuits of relays A and D via the first and second lines are broken, but these relays remain energized via the second winding, because relay I is still energized.

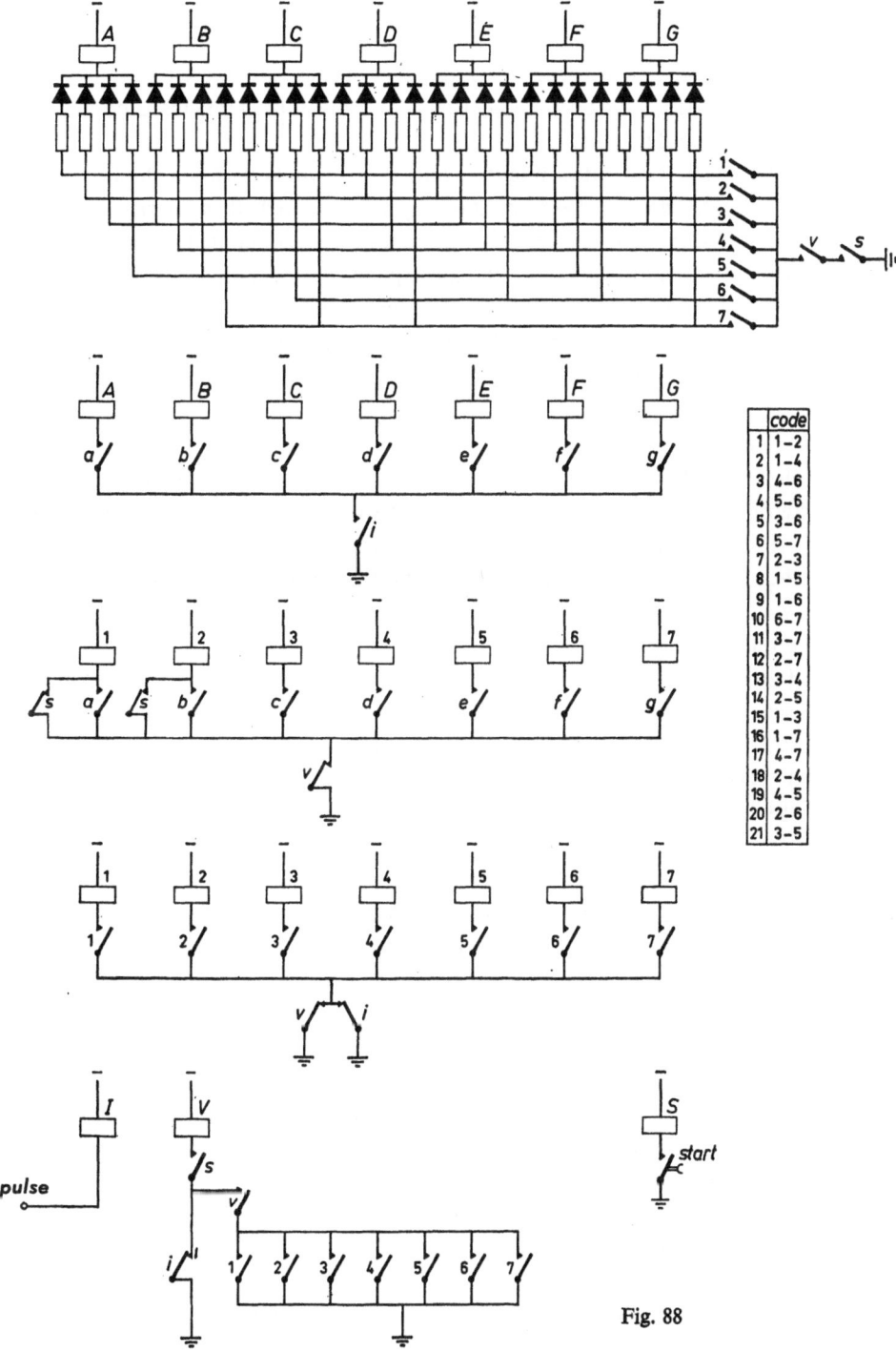

	code
1	1–2
2	1–4
3	4–6
4	5–6
5	3–6
6	5–7
7	2–3
8	1–5
9	1–6
10	6–7
11	3–7
12	2–7
13	3–4
14	2–5
15	1–3
16	1–7
17	4–7
18	2–4
19	4–5
20	2–6
21	3–5

Fig. 88

The release of relay V also causes relays 1 and 4 to operate via contact units a and d. After the count pulse relay I releases again, as a result of which the hold circuit for relays A and D is broken, and these relays release. At the same time, the energizing circuit for relay V is again closed by means of a break contact unit i.

After relay V operates, relays D and F are switched on via the first and fourth lines, because these are the only relays which then receive current through two resistances. This thus prepares for the combination 4–6, which follows combination 1–4. Each subsequent count pulse gives the next combination in the series, until after combination 3–5 relays A and B operate, thus preparing the initial combination 1–2 again. The diodes connected in series with relays A to G serve to decouple the circuits for these relays.

After the above, it is a simple matter to design a circuit which counts both forwards and backwards. This circuit is shown in Fig. 89; for the sake of simplicity, the relay S from Fig. 88 is omitted. We again start from the situation where relays 1 and 2 are switched on. Relay V is switched on, because relay I is switched off in the initial state. Relays A and D are switched on via contact units 1 and 2, and store the next position in the code (1 and 4). The relays CA and EA are also switched on via these same contact units 1 and 2; these relays store the *previous* position in the code (3 and 5), as may be seen from the code table of Fig. 89. If a pulse is applied as a result of which the code must go one step backwards, contact i_b is closed, causing first relay W and then relay I to operate. The operation of relay I causes relays 1 and 2 to release. Relay V then releases, because its hold circuit is opened.

After relay V has released, the counter relays 3 and 5 operate, because relay W is energized and these relays operate via contact units ca and ea.

Relay W must remain energized until relay V has operated again, to make sure that only the previous code position is assumed by the counter relays, and not the following one too.

Figures 88 and 89 are drawn for the codes $A2/1$ (forwards) and $B4/5$ (backwards). All the 14 code pairs mentioned in this chapter can be realized in a similar way, by connecting the resistances under the relays A to G and the relays $AA...GA$ to the 7 lines connected to contacts 1 to 7 in a different way.

A similar approach can be made to the development of 2-out-of-11 codes; here $(n-1)=10$, which is equal to the total number of combinations of 2 out of 5. The 2-out-of-11 codes can be developed with the aid of two patterns, each pattern with a shift of 3, 4, 5 or 9, as follows:

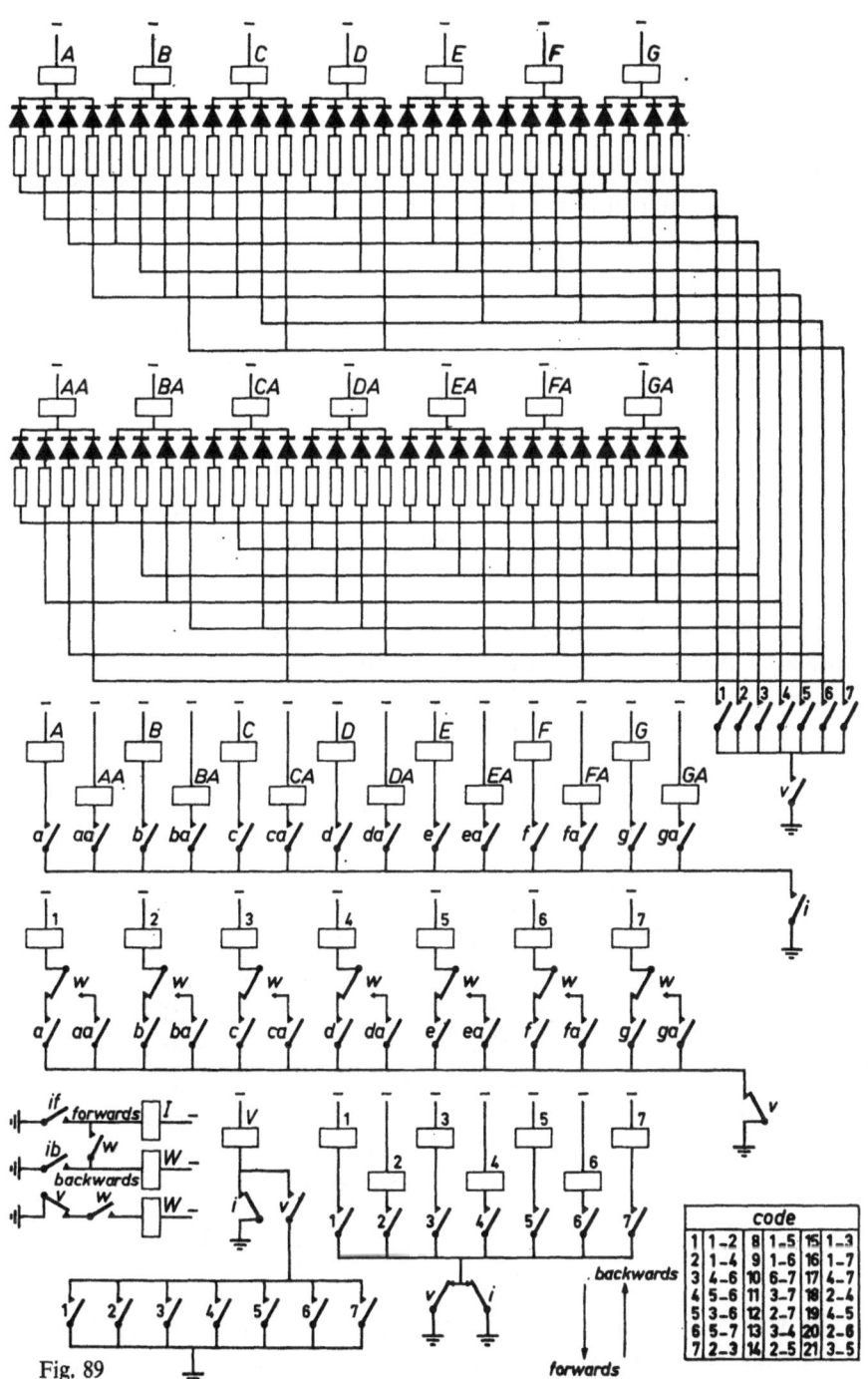

Fig. 89

code					
1	1–2	8	1–5	15	1–3
2	1–4	9	1–6	16	1–7
3	4–6	10	6–7	17	4–7
4	5–6	11	3–7	18	2–4
5	3–6	12	2–7	19	4–5
6	5–7	13	3–4	20	2–6
7	2–3	14	2–5	21	3–5

$A3/1$ 1 1 1 0 1 0 0 1 0 0 0 $B3/1$ 1 1 1 0 0 0 1 0 0 1 0
     0 0 0 1 1 1 0 1 0 0 1   0 1 0 1 1 1 0 0 0 1 0
           etc.   etc.

$A4/1$ 1 1 1 0 1 0 0 1 0 0 0 $B4/1$ 1 1 1 0 0 0 1 0 0 1 0
     1 0 0 0 1 1 1 0 1 0 0   0 0 1 0 1 1 1 0 0 0 1
           etc.   etc.

$A5/1$ 1 1 1 0 1 0 0 1 0 0 0 $B5/1$ 1 1 1 0 0 0 1 0 0 1 0
     0 1 0 0 0 1 1 1 0 1 0   1 0 0 1 0 1 1 1 0 0 0
           etc.   etc.

$A9/1$ 1 1 1 0 1 0 0 1 0 0 0 $B9/1$ 1 1 1 0 0 0 1 0 0 1 0
     1 0 1 0 0 1 0 0 0 1 1   1 0 0 0 1 0 0 1 0 1 1
           etc.   etc.

The row with three successive "ones" in positions 1, 2 and 3 can be displaced cyclically, so that each of the above 8 possibilities gives 11 variations, denoted by /1 to /11. The $8 \times 11 = 88$ code sequences derived from these 88 different arrangements can again be split up into 44 pairs, where the code order is reversed in the two members of each pair. As in the 2-out-of-7 code, each pair consists of an A pattern and a B pattern. The pairs are indicated below.

$A3/1–B4/8$	$A4/1–B3/5$	$A5/1–B9/2$	$A9/1–B5/2$
$A3/2–B4/5$	$A4/2–B3/1$	$A5/2–B9/8$	$A9/2–B5/4$
$A3/3–B4/2$	$A4/3–B3/8$	$A5/3–B9/3$	$A9/3–B5/6$
$A3/4–B4/10$	$A4/4–B3/4$	$A5/4–B9/9$	$A9/4–B5/8$
$A3/5–B4/7$	$A4/5–B3/11$	$A5/5–B9/4$	$A9/5–B5/10$
$A3/6–B4/4$	$A4/6–B3/7$	$A5/6–B9/10$	$A9/6–B5/1$
$A3/7–B4/1$	$A4/7–B3/3$	$A5/7–B9/5$	$A9/7–B5/3$
$A3/8–B4/9$	$A4/8–B3/10$	$A5/8–B9/11$	$A9/8–B5/5$
$A3/9–B4/6$	$A4/9–B3/6$	$A5/9–B9/6$	$A9/9–B5/7$
$A3/10–B4/3$	$A4/10–B3/2$	$A5/10–B9/1$	$A9/10–B5/9$
$A3/11–B4/11$	$A4/11–B3/9$	$A5/11–B9/7$	$A9/11–B5/11$

By way of example we give the development of codes $A4/6–B3/7$ from Tables XLIII and XLIV. All the code pairs given above can be derived in a similar way.

TABLE XLIII

A4/6

	1	2	3	4	5	6	7	8	9	10	11
1	0	0	1	1	1	0	1	0	0	1	0
2	0	0	1	0	0	0	1	1	1	0	1
3	1	1	0	1	0	0	1	0	0	0	1
4	0	0	0	1	1	1	0	1	0	0	1
5	1	0	0	1	0	0	0	1	1	1	0
6	1	1	1	0	1	0	0	1	0	0	0
7	1	0	0	0	1	1	1	0	1	0	0
8	0	1	0	0	1	0	0	0	1	1	1
9	0	1	1	1	0	1	0	0	1	0	0
10	0	1	0	0	0	1	1	1	0	1	0
11	1	0	1	0	0	1	0	0	0	1	1

TABLE XLIV

B3/7

	1	2	3	4	5	6	7	8	9	10	11
1	0	0	1	0	1	1	1	0	0	0	1
2	0	0	1	0	0	1	0	1	1	1	0
3	1	1	0	0	0	1	0	0	1	0	1
4	1	0	1	1	1	0	0	0	1	0	0
5	1	0	0	1	0	1	1	1	0	0	0
6	0	0	0	1	0	0	1	0	1	1	1
7	1	1	1	0	0	0	1	0	0	1	0
8	0	1	0	1	1	1	0	0	0	1	0
9	0	1	0	0	1	0	1	1	1	0	0
10	1	0	0	0	1	0	0	1	0	1	1
11	0	1	1	1	0	0	0	1	0	0	1

CODE

A4/6	B3/7					
↓	▲	1–2	2–5	3–5	3–8	1–8
		3–6	6–8	1–6	2–6	5–6
		9–11	4–10	7–11	9–10	4–7
		2–8	1–5	2–3	5–8	1–3
		6–10	6–7	6–9	4–6	6–11
		10–11	7–10	7–9	4–9	4–11
		8–11	1–10	2–7	5–9	3–4
		2–4	5–11	3–10	7–8	1–9
		3–9	4–8	1–11	2–10	5–7
		2–9	4–5	3–11	8–10	1–7
		8–9	1–4	2–11	5–10	3–7
						1–2

5.11 Counter circuits with capacitor memories

As has already been mentioned in Section 1 of this chapter, n relay pairs are needed for counting n pulses; but we have shown in Section 5.6 how one relay of each pair can be replaced by a capacitor.

We shall now describe another way in which relays can be saved by the use of capacitors. This method is in principle applicable to any code. Fig. 90 shows by way of example a 4-position binary ring counter. The counting code is also shown in Fig. 90.

Initially capacitor 1 is charged, and is used on receipt of the first pulse to energize relay 1. In position 1 this capacitor is not charged, but capacitor 2 is. When the second pulse is received, therefore, relay 1 releases, while relay 2 operates. In position 2 both capacitors are charged, so that on receipt of the third pulse both relay 1 and relay 2 operate. In position 3 neither

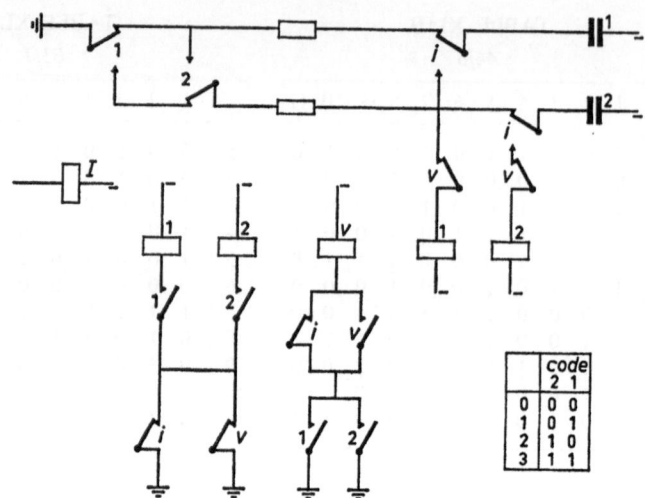

Fig. 90

capacitor is charged, so that on receipt of the fourth pulse the zero position is again reached. The function of relay V will be discussed in connection with the next counter circuit.

The charging circuits of capacitors 1 and 2 can be described by means of switching algebra:

$$f(1) = 1'2' + 1'2 = 1'$$
$$f(2) = 1\,2' + 1'2$$

Contact units 1 and 2 in series with the capacitors do indeed operate in accordance with these functions.

If a binary-decimal cyclic code has to be followed, we need a counter with 4 counter relays and 4 capacitors. The charging circuits for the capacitors can be determined from the code table (see Fig. 91). Capacitor 1 must be charged in all even positions in order to energize relay 1 on receipt of all odd-numbered pulses. Capacitor 1 can simply be operated by a break contact unit of relay 1, since this capacitor will then only be charged in even positions.

$$f(1) = 1' =$$

If the 4 relays were required to run through all 16 combinations, capacitor 2 would have to be charged in the following combinations:

$$f(2) = 12'4'8' + 1'24'8' + 12'48' + 1'248' + 12'4'8 + 1'24'8 + 12'48 +$$
$$1'248 = 12'(4'8' + 48' + 4'8 + 48) + 1'2(4'8' + 48' + 4'8 + 48).$$

which (for all 16 combinations) can be simplified to:

$$f(2)=12'+1'2$$

Since however we are dealing with the binary-decimal code, we must not go beyond the value 9. Relay 8 must thus ensure that capacitor 2 is not charged:

$$f(2)=8'(12'+1'2)=$$

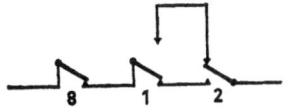

Capacitor 4 must be charged in positions 3, 4, 5 and 6:

$$f(4)=124'8'+1'2'48'+12'48'+1'248'$$

8' has no function in this expression, since in the binary-decimal code relay 8 never operates together with relays 1 and 2, or 4, or 1 and 4 or 2 and 4. The function may therefore be written:

$$124'+1'2'4 \ +12'4+1'24=$$
$$124'+4(1'2'+12' \ +1'2)=$$
$$124'+4(1'+2')=$$

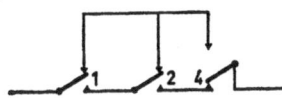

Finally, capacitor 8 must only be charged in positions 7 and 8:

$$f(8)=1248'+1'2'4'8$$

In the binary-decimal code combinations above 9 do not occur, so that in 1'2'4'8 the terms 2' and 4' have no function and may be omitted. In 1248' the term 8' has no function, so that:

$$f(8)=124+1'8=$$

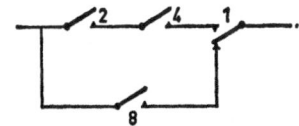

The counter circuit can now be realized by the same method as shown in Fig. 90. There are however more relays, and the contact network for charging the capacitors is adapted to the code. The full circuit is shown in Fig. 91.

This circuit operates as follows. Initially, all the relays are released, but capacitor 1 is charged. At the start of the first pulse, relay I operates, causing relay 1 to be switched on by the discharge of capacitor 1. Relay 1 remains energized via a hold winding.

At the end of the first pulse, relay V operates via the break contact i and the make contact 1. Capacitor 2 is now charged. At the start of the

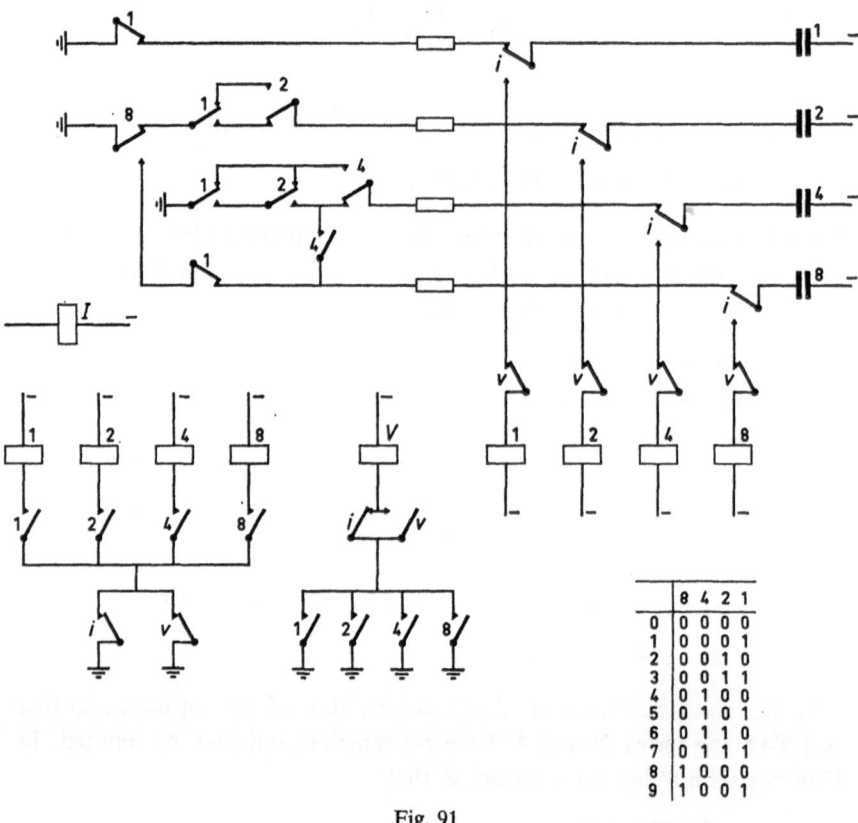

Fig. 91

second pulse relay 1 releases, because the hold circuit is switched off. Only now does relay V release, after which relay 2 operates as a result of the discharging of capacitor 2. Relay 2 is held in the energized state. At the end of the pulse, when relay I has released again, relay V operates again. Capacitors 1 and 2 are charged. The following pulses are counted in a similar way until we come to position 9. No capacitor is charged in this position, so that the 0 position is reached at the start of the 10th pulse. and at the end of the 10th pulse capacitor 1 is charged again.

Relay V has the function of making all counter relays release before the capacitors can be discharged. At the start of each count pulse, first all hold circuits of the counter relays are broken, and then relay V releases. Not until then can the new combination of relays be switched on via the capacitors.

The advantage of this method of counting is that the circuit for the counter relays is always the same. It is only the contact network for the charging of the capacitors which has to be designed to obtain the desired code.

We shall now show how the same circuit can be used for the 2-out-of-5 code. We shall make use of a counting code where as far as possible the relays are always switched on by a contact unit of the preceeding relay: relay 2 by a contact unit 1, relay 3 by a contact unit 2 and so on. Table XLV shows such a code.

TABLE XLV

	1	2	3	4	5
0	1	1	0	0	0
1	0	1	1	0	0
2	0	0	1	1	0
3	0	0	0	1	1
4	1	0	0	0	1
5	1	0	1	0	0
6	0	1	0	1	0
7	0	0	1	0	1
8	1	0	0	1	0
9	0	1	0	0	1

The contact network for each capacitor can now be calculated on the basis of this code. Capacitor 1 must be charged in positions 3, 4, 7 and 9, i.e.

$$f(1) = 1'2'3'45 + 12'3'4'5 + 1'2'34'5 + 1'23'4'5 =$$
$$5(1'2'3'4 + 12'3'4' + 1'2'34' + 1'23'4').$$

This shows that relay 5 must always be energized in order to charge capacitor 1. Relay 5 is *only* energized in positions 3, 4, 7 and 9, whence it follows that $f(1) = 5$, so that a make contact unit of 5 will suffice for this purpose.

$$f(1) = 5 =$$

Capacitor 2 is charged in positions 0, 5, 8 and 9, so that:

$$f(2) = 123'4'5' + 12'34'5' + 12'3'45' + 1'23'4'5$$

15' can be put outside brackets in the first, second and third terms:

$$f(2) = 15'(23'4' + 2'34' + 2'3'4) + 1'23'4'5$$

The break contact units can be left out as long as two relays are energized, so that:

$$f(2) = 15'(2+3+4) + 25$$

The term $(2+3+4)$ forms all possible combinations for a closed circuit together with 15' and may thus be omitted, so that we have finally:

$$f(2) = 15' + 25 =$$

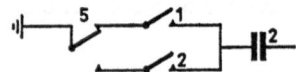

Capacitor 3 is charged in positions 0, 1, 4 and 6, so that:

$$f(3) = 123'4'5' + 1'234'5' + 12'3'4'5 + 1'23'45'$$

25' may be placed outside brackets in the 1st, 2nd and 4th terms:

$$f(3) = 25'(13'4' + 1'34' + 1'3'4) + 12'3'4'5$$

Here again we can simplify by leaving out the break contact units, giving:

$$f(3) = 25'(1+3+4) + 15$$

The terms between brackets can be omitted, because they form all possible combinations with 25':

$$f(3) = 25' + 15 =$$

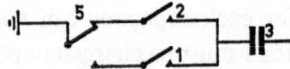

Capacitor 4 is charged in positions 1, 2, 5 and 7, which corresponds to the positions where relay 3 is energized, so that $f(4) = 3$.

$$f(4) = 3 =$$

Capacitor 5 is charged in positions 2, 3, 6 and 8, which corresponds to the positions where relay 4 is energized, so that $f(5) = 4$.

$$f(5) = 4 =$$

All the circuits for the charging of the 5 capacitors are shown in Fig. 92.

Finally, the entire counter circuit is shown in Fig. 93; it should be remarked in this connection that capacitors 1 and 2 must be charged in the initial state of all the counter relays. The counter circuit must thus be started. A relay S is included for this purpose, and is switched on the first time a pulse is received. This relay cannot release after that, because it is held during the pulses by a hold contact unit i, and after the pulses by contacts of the counter relays.

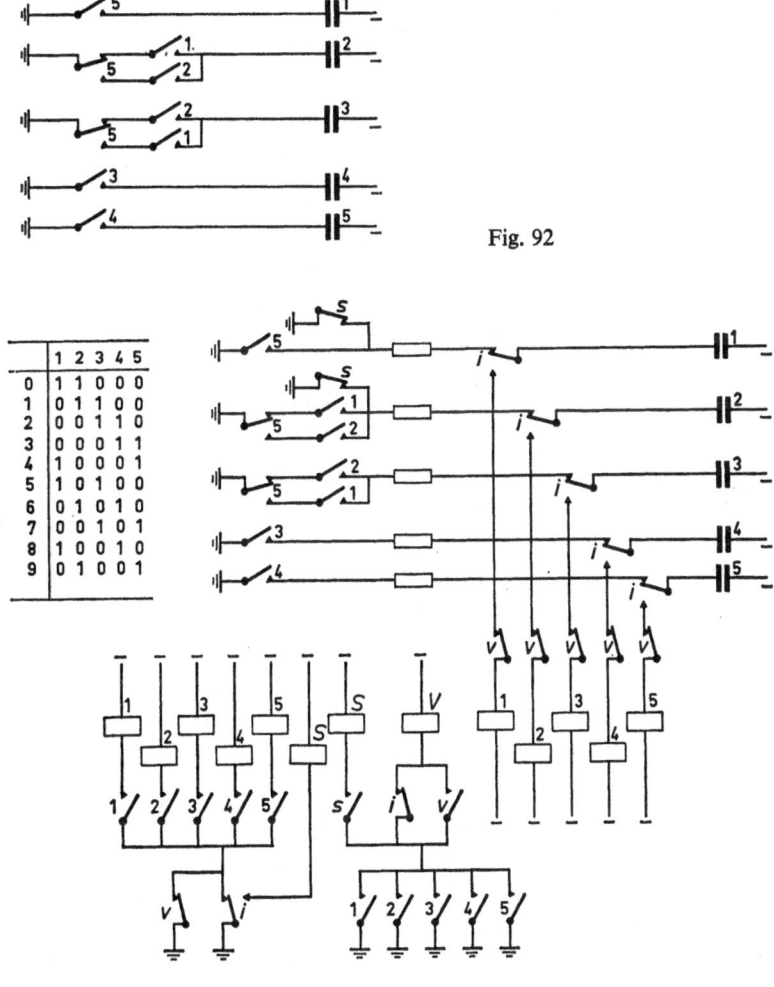

Fig. 92

	1	2	3	4	5
0	1	1	0	0	0
1	0	1	1	0	0
2	0	0	1	1	0
3	0	0	0	1	1
4	1	0	0	0	1
5	1	0	1	0	0
6	0	1	0	1	0
7	0	0	1	0	1
8	1	0	0	1	0
9	0	1	0	0	1

Fig. 93

5.12 Self-operating counter circuit with 26 outputs

In the previous sections we have given circuits which work with well known and much used codes. There are however an extremely great number of possible codes, so that the electrical engineer can himself construct a code which gives the desired number of outputs, taking into account the number of contacts available and the life time of the relays.

In this section we will give an example of such a design. The problem is to design a self-operating counter circuit (i.e. one without a pulse relay). The cycle time, i.e. the time in which the counter performs one cycle, must be about 2 s, i.e. about 77 ms per step. The counter is switched on for 4 hours a day. There are 60 days a year when the counter does not work at all. The life time of the counter is required to be 20 years, while the life time of a relay may be set at e.g. 100 million switching operations. If the binary code with 5 relays (32 possibilities) was used for the 26 combinations, relay 1 would be switched on and off 13 times per cycle. The total number of switching operations in 20 years would then be:

$$\frac{20 \times 300 \times 4 \times 3600 \times 13}{2} = 562 \times 10^6$$

Relay 1 would thus have a lifetime of

$$\frac{1}{5,62} \times 20 = \pm 3,5 \text{ years}$$

We must thus look for a code where as far as possible all relays perform the same number of switching operations. The number of switching operations per relay per cycle can be calculated as follows.

Number of cycles in 20 year's operation =

$$\frac{20 \times 300 \times 4 \times 3600}{2} = 43.2 \times 10^6$$

Each relay may therefore not be switched on and off more than twice per cycle, in order not to exceed the 100×10^6.

The number of change-over contact units needed for a contact tree with 26 outputs is $26 - 1 = 25$. Apart from these contact units, we also need a certain number for controlling the circuit. This control must be as simple as possible, which can be attained by composing the code as logically as possible: in other words, we try always to switch relay B on with relay A, relay D with relay C and so on. Fig. 94 shows a realization of this principle for 4 relays. Relay A operates in the zero position, and relays B, C and D follow

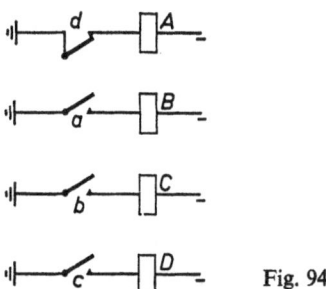

Fig. 94

in turn. Then relay A releases, and relays B, C and D after it, so that we are back at the zero position, and relay A can operate again. We thus get the following code:

A	B	C	D
0	0	0	0
1	0	0	0
1	1	0	0
1	1	1	0
1	1	1	1
0	1	1	1
0	0	1	1
0	0	0	1
0	0	0	0

Each relay is only switched on and off once per cycle. The control of this circuit is simple, and only needs one contact unit per relay. We therefore choose this principle for the desired 26-output circuit. Since we are allowed

TABLE XLVI

	A	B	C	D	E	F	G		A	B	C	D	E	F	G
1	0	0	0	0	0	0	0	14	0	0	0	0	0	0	1
2	1	0	0	0	0	0	0	15	1	0	0	0	0	0	1
3	1	1	0	0	0	0	0	16	1	1	0	0	0	0	1
4	1	1	1	0	0	0	0	17	1	1	1	0	0	0	1
5	1	1	1	1	0	0	0	18	1	1	1	1	0	0	1
6	1	1	1	1	1	0	0	19	1	1	1	1	1	0	1
7	1	1	1	1	1	1	0	20	0	1	1	1	1	0	1
8	1	1	1	1	1	1	1	21	0	0	1	1	1	0	1
9	0	1	1	1	1	1	1	22	0	0	0	1	1	0	1
10	0	0	1	1	1	1	1	23	0	0	0	0	1	0	1
11	0	0	0	1	1	1	1	24	0	0	0	0	1	0	0
12	0	0	0	0	1	1	1	25	0	0	0	0	1	1	0
13	0	0	0	0	0	1	1	26	0	0	0	0	0	1	0

to switch each relay on and off twice per cycle, use may be made of the code given on the previous page (see Table XLVI).

The circuit for relay A can be derived from the code. This relay must *operate* when relays E and F are not energized. This situation is found in positions 1 and 14. $f(A \text{ operate}) = e'f'$. Once relay A is energized, it must stay energized via a hold contact unit a until relay G has operated (position 8); but if relay F is not switched on, until relay E has operated (position 19). $f(A \text{ hold}) = ag' + e'f'$. Combining these two expressions, we obtain: $f(A) = e'f' + ag'$. The circuit for relay A is therefore as shown in Fig. 95. Relays B, C and D are each switched on when the preceeding relay (A, B and C respectively) has operated. Relay E is also switched on by a contact unit d, but in positions 24 and 25 (when relay D is released), relay E must remain energized until relay F has operated. $f(E) = d + ef'$. The circuit for this is shown in Fig. 96.

Relay F is switched on by a contact unit e, but only if relay G is not energized; and it must remain energized until relay E releases. $f(F) = eg' + ef$. The corresponding circuit is shown in Fig. 97.

Relay G operates as soon as relays D and F are energized, and remains energized until position 23 (D released, E energized, F released), so that $f(G) = fd + g(d + e' + f)$. The circuit for this is shown in Fig. 98.

Finally, the complete circuit is given in Fig. 99. The contacts of the relays used are also tabulated in this figure.

The relays are retarded by means of a copper sleeve round the core, to give the required operate and release times (about 77 ms).

This circuit can be used for the consecutive selection or switching on of a group of 26 devices.

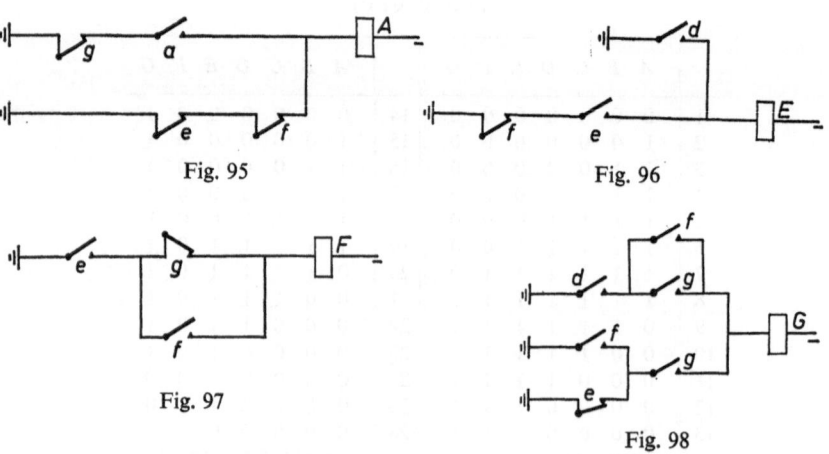

Fig. 95

Fig. 96

Fig. 97

Fig. 98

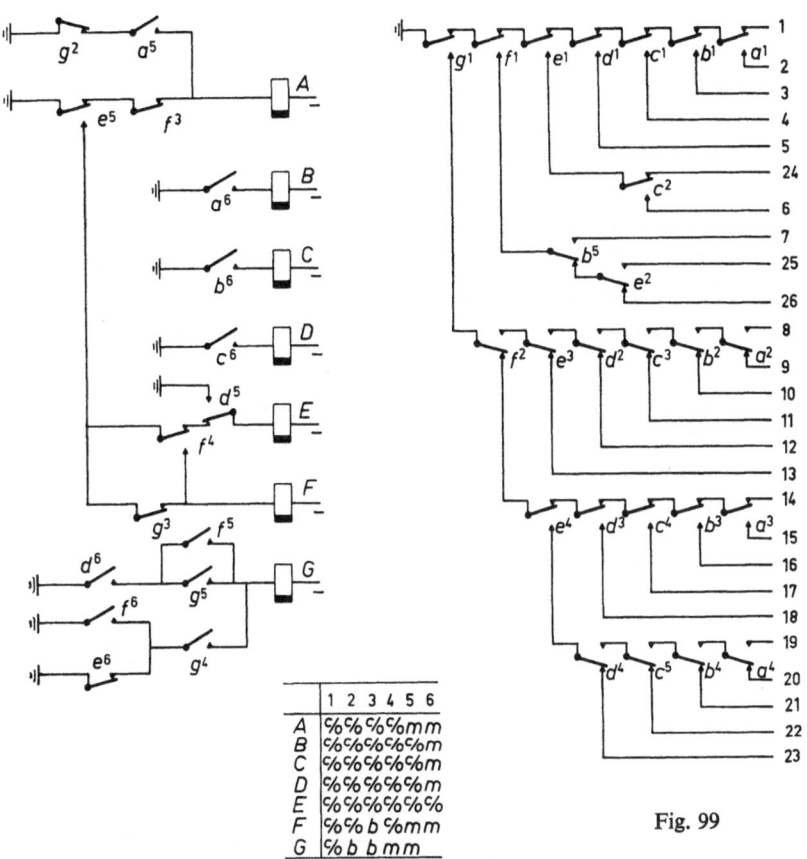

	1	2	3	4	5	6
A	%	%	%	%	m	m
B	%	%	%	%	%	m
C	%	%	%	%	%	m
D	%	%	%	%	%	m
E	%	%	%	%	%	%
F	%	%	b	%	m	m
G	%	b	b	m	m	

Fig. 99

5.13 Combination of two ring counters

Two cyclic counters, one with X positions and the other with Y, give XY possibilities when used together as long as X and Y are prime to one another.

Suppose that we combine a counter with 3 positions (A, B and C) and one of 4 positions (D, E, F and G); we thus get 12 positions in all, as may be seen from Table XLVII.

TABLE XLVII

		A	B	C	positions and combinations					
D		1	5	9	1	AD	5	BD	9	CD
E		10	2	6	2	BE	6	CE	10	AE
F		7	11	3	3	CF	7	AF	11	BF
G		4	8	12	4	AG	8	BG	12	CG

Output A of the first counter together with output D of the second form combination 1. Both counters are then moved on one place, giving outputs B and E which form together combination 2. The next step is to C and F, which give combination 3. After the next move, the first counter is back at A, but the second one is at G, which gives combination 4. In the next position the first counter is at B again, and the second at D, giving combination 5. All twelve combinations are formed one after the other in this way.

A circuit using a 3-position and a 4-position counter in this way is shown in Fig. 100. The two counters, with their 3 and 4 outputs respectively, switch on 12 relays, which register the successive combinations. The diodes decouple the windings of the relays.

The information supplied by the two counters can be registered in coded form. We take by way of example a 3-position and a 5-position counter; the 15 possibilities thus arising allow the binary coding of these positions. The circuit is shown in Fig. 101. The zero position of the binary code is

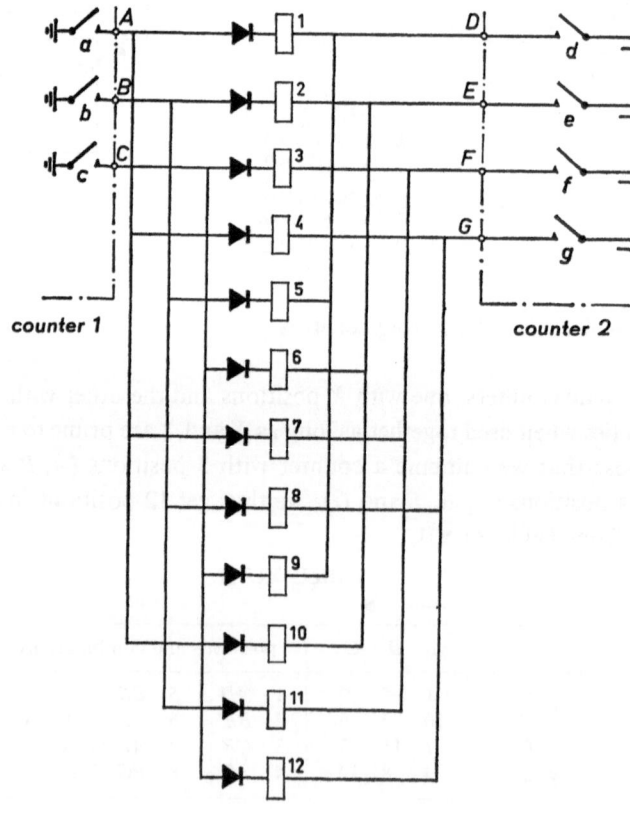

counter 1   counter 2

Fig. 100

TABLE XLVIII

	counter 1			counter 2					binary			
	A	B	C	D	E	F	G	H	8	4	2	1
1	1	0	0	1	0	0	0	0	0	0	0	1
2	0	1	0	0	1	0	0	0	0	0	1	0
3	0	0	1	0	0	1	0	0	0	0	1	1
4	1	0	0	0	0	0	1	0	0	1	0	0
5	0	1	0	0	0	0	0	1	0	1	0	1
6	0	0	1	1	0	0	0	0	0	1	1	0
7	1	0	0	0	1	0	0	0	0	1	1	1
8	0	1	0	0	0	1	0	0	1	0	0	0
9	0	0	1	0	0	0	1	0	1	0	0	1
10	1	0	0	0	0	0	0	1	1	0	1	0
11	0	1	0	1	0	0	0	0	1	0	1	1
12	0	0	1	0	1	0	0	0	1	1	0	0
13	1	0	0	0	0	1	0	0	1	1	0	1
14	0	1	0	0	0	0	1	0	1	1	1	0
15	0	0	1	0	0	0	0	1	1	1	1	1

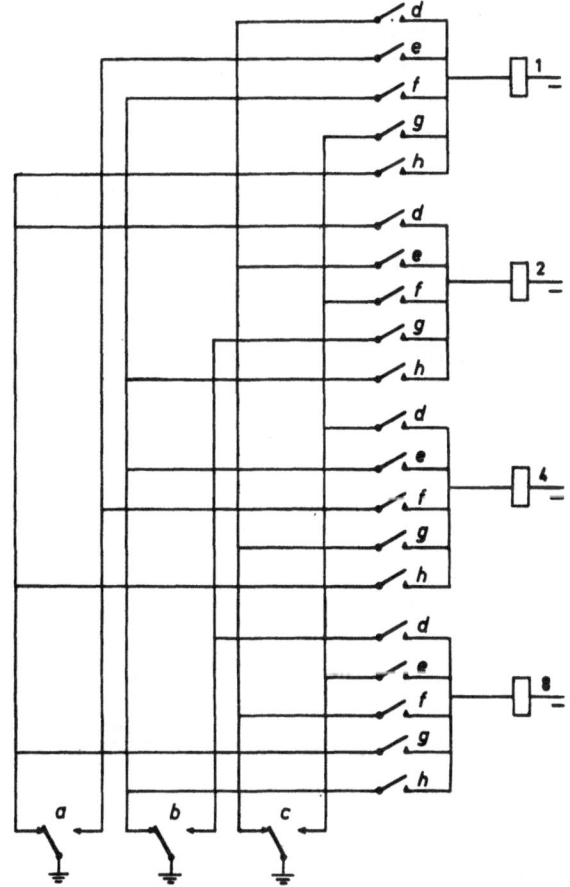

Fig. 101

not used. The positions of the two counters in the various combinations, and the resulting binary code, are shown in Table XLVIII.

The use of two counters has advantages if a large number of outputs are required. If for example 143 outputs are needed, one can use one 11-position counter and one 13-position. This leads to a considerable saving of relays.

5.14 Problems

1. Design a linear ring counter circuit with 4 positions, to switch red, green, yellow and white lamps on and off as follows.
   The red, green and yellow lamps burn in position 1.
   The red, yellow and white lamps burn in position 2.
   The white, green and red lamps burn in position 3.
   The green, white and yellow lamps burn in position 4.
   This circuit needs 8 contact springs.

2. How many relays are needed to count 14 pulses with
   a) a linear counter circuit
   b) a linear counter circuit with one auxiliary relay
   c) a binary counter circuit?

3. Draw a linear counter circuit for counting 4 pulses. Draw a circuit which is closed from the releasing of the pulse relay at the end of the first pulse to the operation of the pulse relay at the start of the 4th pulse, using not more than 10 contact springs.

4. Does the reflecting-code ring counter shown below work properly? Check this with the aid of a sequence diagram. Let the on/off ratio of contact

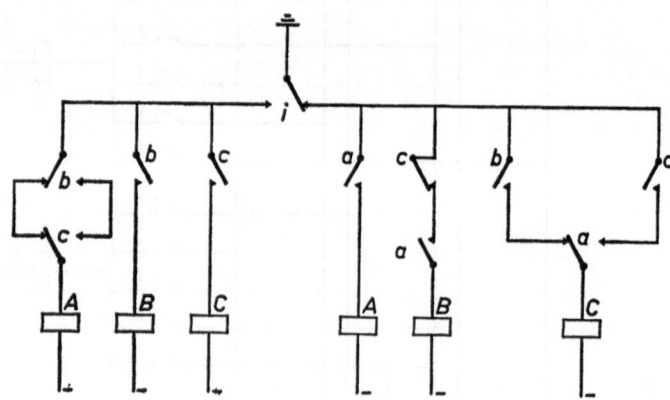

unit i be $80-80$ ms, the transit time of contact unit i be 5 ms, the operate time of all relays 25 ms and the release time 20 ms. Improve the circuit if necessary, and in this case give the proper sequence diagram.

5. A 21-position counter has to be designed on the principle described in section 11 of this chapter. The code is as follows:

1-2　4-5　1-7　3-5　1-6　2-5　1-5
2-3　5-6　1-3　4-6　2-7　3-6　2-6
3-4　6-7　2-4　5-7　1-4　4-7　3-7

Design the circuit for charging the 7 capacitors, using not more than 29 contact springs.

6. Design a counter circuit like that of fig. 93, using the following code. Use no more than 16 contact springs in the charging circuits for the capacitors.

	1	2	3	4	5
0	1	0	0	1	0
1	0	0	0	1	1
2	0	0	1	0	1
3	0	1	1	0	0
4	1	1	0	0	0
5	1	0	0	0	1
6	0	0	1	1	0
7	0	1	0	0	1
8	1	0	1	0	0
9	0	1	0	1	0

position of circuit	code						lamps
	1	2	3	4	5	6	
1	1	1	0	0	0	0	2 and 3
2	0	1	1	0	0	0	3 and 4
3	0	0	1	1	0	0	4 and 5
4	0	0	0	1	1	0	5 and 6
5	0	0	0	0	1	1	1 and 6
6	1	0	0	0	0	1	1 and 3
7	1	0	1	0	0	0	2 and 4
8	0	1	0	1	0	0	3 and 5
9	0	0	1	0	1	0	4 and 6
10	0	0	0	1	0	1	1 and 5
11	1	0	0	0	1	0	2 and 6
12	0	1	0	0	0	1	1 and 4
13	1	0	0	1	0	0	2 and 5
14	0	1	0	0	1	0	3 and 6
15	0	0	1	0	0	1	1 and 2

7. 6 relays form a counter circuit. In each position of this circuit, 2 out of six lamps are to be switched on. The code is given on the previous page together with the lamps to be switched on.

Use not more than 21 contact springs for switching on the lamps.

8. The code for a self-operating ring counter is given below.

   a) Design a circuit for this counter, using not more than 25 contact springs for the control.

   b) A red lamp must light up in position 4, and a yellow one in position 10. Use not more than 7 contact springs for this.

	A	B	C	D	E
0	0	0	0	0	0
1	1	0	0	0	0
2	1	1	0	0	0
3	1	1	1	0	0
4	0	1	1	0	0
5	0	0	1	0	0
6	0	0	1	1	0

	A	B	C	D	E
7	1	0	1	1	0
8	1	1	1	1	0
9	1	1	1	1	1
10	0	1	1	1	1
11	0	0	1	1	1
12	0	0	0	1	1
13	0	0	0	0	1

Chapter 6

DECODING CIRCUITS

In Chapters 4 and 5 we have discussed a large number of codes and the corresponding counter circuits by means of which information can be registered in coded form. In most cases, the coded information must again be presented in decimal form at a later stage, i.e. it must be decoded.

If the relays in which the information is stored contain a sufficient number of change-over contact units, these may be used to form "contact trees" for the decoding.

If however there are not enough change-over contacts per relay, diodes can be used for the decoding. We shall be returning to this later in this chapter.

6.1 General rules for the construction of a contact tree

Contact trees are networks with one input and as many outputs as the desired combinations of the relays used. Each relay combination corresponds to one output. The smallest tree is a change-over contact unit. The relay concerned can assume two states: energized and non-energized. This is indicated in Fig. 102. Inclusion of a second relay gives four outputs, as shown in Fig. 103 *a* and *b*. Binary counting is retained in this figure. Both ways of ordering the contacts give the desired outputs. The number of change-over contact units of a complete contact tree is always odd, being the sum of as many successive powers of 2 (starting from $2^0 = 1$) as there are

$$\begin{array}{cc} & 0 \\ & 1 \\ & 1 \end{array} \qquad \text{Fig. 102}$$

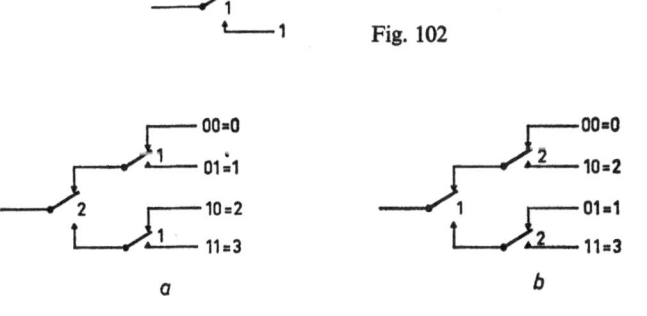

a   *b*   Fig. 103

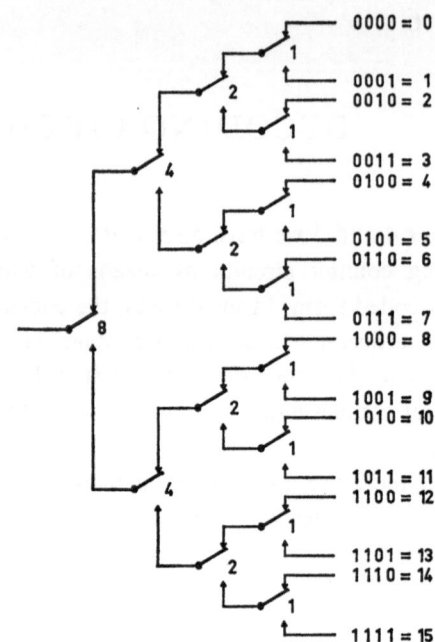

Fig. 104

relays. For example, in Fig. 103 we obtain 4 outputs with $1+2=3$ change-over contact units.

Fig. 104 shows a tree with 16 outputs, which according to the above rule requires $1+2+4+8=15$ change-over contact units. The outputs are indicated by binary numbers. The disadvantage of the circuit of Fig. 104 is the uneven distribution of the contact units between the four relays, viz 1, 2, 4 and 8. The various possible distributions are:

$$
\begin{array}{ccc}
1\ 2\ 4\ 8 & 1\ 3\ 3\ 8 & 1\ 4\ 4\ 6 \\
1\ 2\ 5\ 7 & 1\ 3\ 4\ 7 & 1\ 4\ 5\ 5 \\
1\ 2\ 6\ 6 & 1\ 3\ 5\ 6 &
\end{array}
$$

The preferred distribution can be found by dividing the contact tree into two parts (Fig. 105). If a contact unit a is used at the input of the tree, the other contact units must be distributed between B, C and D.

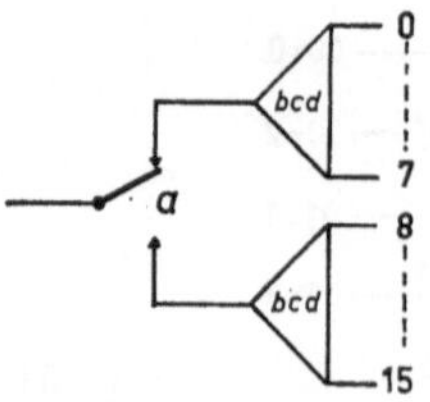

Fig. 105

$a+b+c+d = b\,c\,d + b\,c\,d$

1	1	2	4	8	= 124 + 124
2	1	2	5	7	= 124 + 133
3	1	2	6	6	= 124 + 142
4	1	2	6	6	= 133 + 133
5	1	3	3	8	= 124 + 214
6	1	3	4	7	= 133 + 214
7	1	3	5	6	= 142 + 214
8	1	4	4	6	= 133 + 313
9	1	4	5	5	= 124 + 331
10	1	4	5	5	= 142 + 313
11	1	4	5	5	= 214 + 241

The distributions tabulated above can each be split into two groups, each group serves 8 outputs and thus having 7 change-over contact units.

It may be seen that of the 11 distribution systems given, the greatest uniformity is attained with systems 9, 10 and 11.

6.2　Contact trees for a limited number of combinations out of *n*

A tree with 10 outputs can be used to turn a coded digit back into decimal form.

If we use a binary code, we can use part of the contact tree of Fig. 104 for this purpose. The combinations 1 0 1 0, 1 0 1 1, 1 1 0 0, 1 1 0 1, 1 1 1 0 and 1 1 1 1 do not occur, so that superfluous contacts can be omitted and the rule that the number of change-over contact units needed is one less than the number of outputs is still satisfied. The 9 contact units of such a tree are shown in Fig. 106. Here again, the contacts are distributed as evenly as possible between the various relays.

Another example in which only a relatively small number of the available possibilities are used is the circuit of Fig. 99 (Section 5.12). Of the $2^7 = 128$ possibilities, only 26 are used. The number of change-over contact units in the tree is then $26 - 1 = 25$.

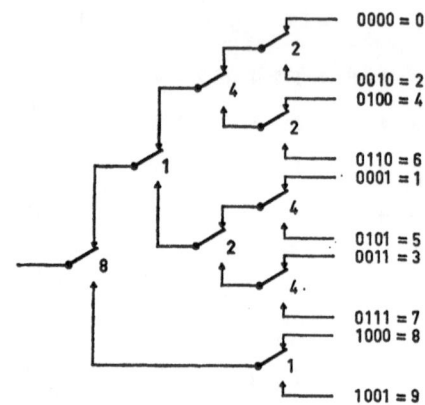

Fig. 106

6.3 Contact trees for the 2-out-of-5 code

If the 2-out-of-5 code is used, so that we can be sure that 2 out of the 5 relays are always switched on, we can also construct a contact tree with 9 change-over contact units. The circuit is shown in Fig. 107. The distribution is very even: relays 1, 2, 3 and 4 each have two contact units, while relay 5 has 1. The output $1+2$ is obtained from the check that relays 3, 4 and 5 are released, so that 1 and 2 are the only ones left on. In the cases

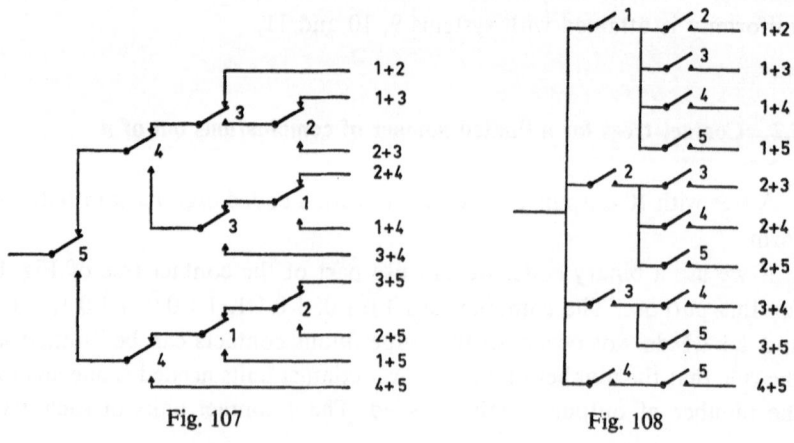

Fig. 107   Fig. 108

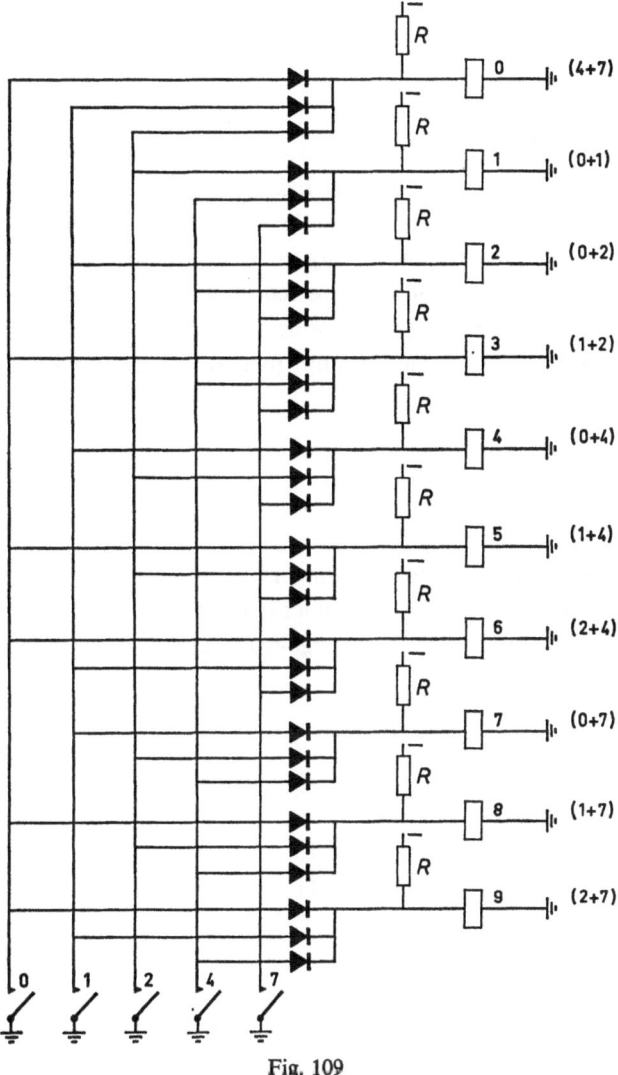

Fig. 109

of 1+3, 2+4 and 3+5, the check that 3 relays are released is used. One relay is already switched on, so that this relay forms a combination with the relay which is not in the circuit in question.

Another method of getting 10 outputs is shown in Fig. 108. The contact tree consists only of make contact units, and there are only two such units in series in each case. This advantage can be of importance for the switching of low voltages, as the contact resistance of a contact unit depends strongly on the voltage between the two contacts.

If e.g. a polarized relay is used, it may happen that only one make contact unit per relay is available for the decoding of information from a 2-out-of-5 code. It is still however possible to decode in this case with the aid of diodes, by *inhibiting* the outputs which are not indicated. Fig. 109 shows a circuit for decoding the 0 1 2 4 7 code. It is assumed in this case that two of the five contacts (0, 1, 2, 4, 7) are always closed. Each combination of two contact units forbids 9 outputs, but does not forbid the output corresponding to the combination in question. For example, relay 5 operates via the corresponding resistance R when the combination $1+4$ is fed in. The other 9 relays are shorted via one or two diodes. The three diodes of each output are connected to the three code contacts which are not related to the output in question.

6.4 Problems

1. Four relays can form the following 11 combinations. The other combinations do not occur. Draw a contact tree with at most 3 contact units per relay and not more than 10 change-over contact units in all.

	A	B	C	D		A	B	C	D
1	0	0	0	0	7	0	1	0	1
2	1	0	0	0	8	1	0	0	1
3	1	0	1	0	9	1	0	1	1
4	0	1	1	0	10	1	1	1	1
5	1	1	1	0	11	0	1	1	1
6	1	1	0	1					

2. Four relays each have 3 change-over contact units available for making a complete contact tree (16 outputs). How many auxiliary relays are needed for this purpose if 6 change-over contact units from each auxiliary relay can be used in the tree?
   Draw the circuit, with as few auxiliary relays as possible.

3. Deduce the code from the contact tree on page 163, and design another tree with the following distribution of the contact units:
   relay A 3 change-over contact units
   relay B 1 change-over contact unit
   relay C 5 change-over contact units
   relay D 6 change-over contact units.

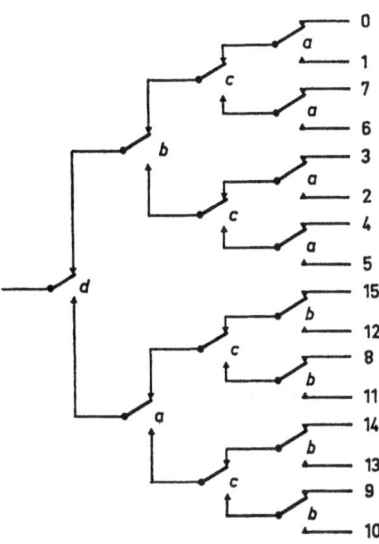

4. a) Design a contact tree for a 2-out-of 6 code, using relays A to F. The code is:

1 AB	6 AF	11 AE
2 BC	7 AC	12 BF
3 CD	8 BD	13 AD
4 DE	9 CE	14 BE
5 EF	10 DF	15 CF

Use not more than 14 change-over contact units and distribute them as follows among the relays:

A: 3 C: 2 E: 3
B: 1 D: 3 F: 2

b) Design a contact tree for the same code making use only of make contact units, of which the following are available:

5 on relay A 2 on relay D
3 on relay B 4 on relay E
1 on relay C 5 on relay F

5. One change-over contact unit on each of the relays 1, 2, 4 and 8 is available for the decoding of the binary-decimal code.
Realize the decoding to ten outputs, numbered 0 to 9, using not more than 30 diodes and 10 resistances.

6. Four relays A, B, C and D form a counter circuit with the following code.

	A	B	C	D			A	B	C	D
0	0	1	1	1		5	0	0	1	0
1	1	0	0	0		6	0	0	1	1
2	1	1	0	0		7	1	0	0	1
3	0	1	0	0		8	1	1	0	1
4	0	1	1	0		9	0	1	0	1

a) Design a contact tree giving the 10 code combinations as outputs (the other 6 possible combinations do not occur).

b) Ditto for the ten code combinations plus the zero position, i.e. 11 outputs in all (the other 5 combinations do not occur).

7. The code of problem 6 must be decoded to ten outputs (0 to 9), using only one change-over contact per relay. 30 diodes and 10 resistances may also be used. The outputs are to be used for energizing relays which are connected to earth.

8. Design a contact tree for the 1, 2, 4, 5 code, using not more than the following number of change-over contact units per relay:

relay 1 : 4   relay 4 : 2
relay 2 : 2   relay 5 : 1

Chapter 7

CHECKING CIRCUITS

These circuits are used to check whether a given number m out of n relays are switched on. The simplest example of such a circuit is clearly a make contact unit, which can be used to check whether a given relay is energized or not.

Another example is a circuit with 2 change-over contact units which, depending on how these are connected, can be used to indicate how many of two given relays are switched on.

In Fig. 110a relay G operates if one of the relays 1 and 2 is switched on.

In Fig. 110b the contacts are connected in another way, but the result is the same.

$$f(G) = 1'2 + 12' = (1' + 2')(1 + 2) = \text{odd number out of 2.}$$

In Figures 111a and 111b the contact units 1 and 2 are connected in another way, so that relay G is switched on if both relays are energized or non-energized.

$$f(G) = 1'2' + 12 = (1' + 2)(1 + 2') = \text{even number out of 2.}$$

When the number of relays is increased to n, a similar method can be used to investigate whether an even or an odd number of relays are switched on. This is discussed in more detail in this chapter.

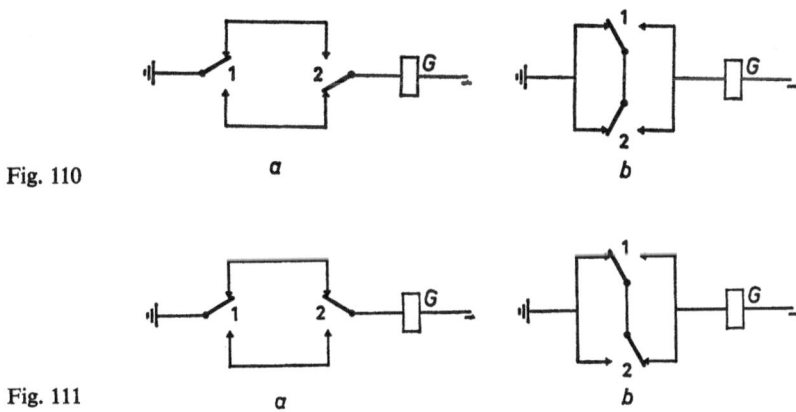

Fig. 110 a b

Fig. 111 a b

7.1 Check on *m* out of *n*

Suppose that we have to check an *m*-out-of-*n* code, where $n=4$ and $m=2$. This code has 6 combinations. However, a total of 16 combinations can be formed with 4 relays. In the code in question, only 6 of these are used, so the other 10 indicate that an error has been made. A contact network which indicates all values of *m* out of *n* is shown in Fig. 112. A checking circuit for 2 out of 4 can be derived from this (see Fig. 113). Combination of make and break contact units to change-over contact units gives the circuit of Fig. 114. This circuit is symmetrical. Relay *G* is only switched on by each of the 6 2-out-of-4 combinations.

Fig. 114 indicates how one can indicate when a *correct* combination (2-out-of-4) is present. If a *wrong* combination arises, the relay *G* of Fig. 114 will not operate. This is called negative indication of an error. A positive indication is given by making a relay (e.g. *F*) operate whenever one of the combinations of 0, 1, 3 or 4 out of 4 occurs. The circuit for doing this is the inverse of Fig. 114. As we have seen in Section 2.8, the inverse of a given circuit is obtained by replacing all make contact units by break contact units and *vice versa*, and all series connections by parallel ones and *vice versa*. An example of an inversion is shown in Fig. 115.

A simple method has been developed for finding the inverse of a given contact network, which we shall now illustrate with reference to the circuit of Fig. 114. First of all the circuit must be drawn in another way, as shown in Fig. 116.

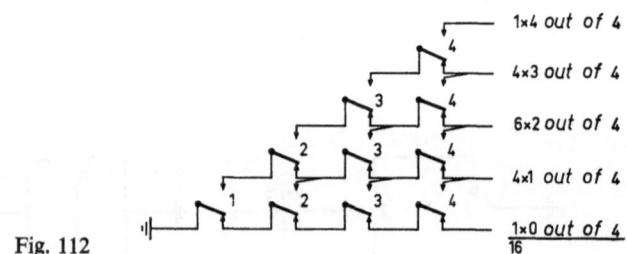

Fig. 112

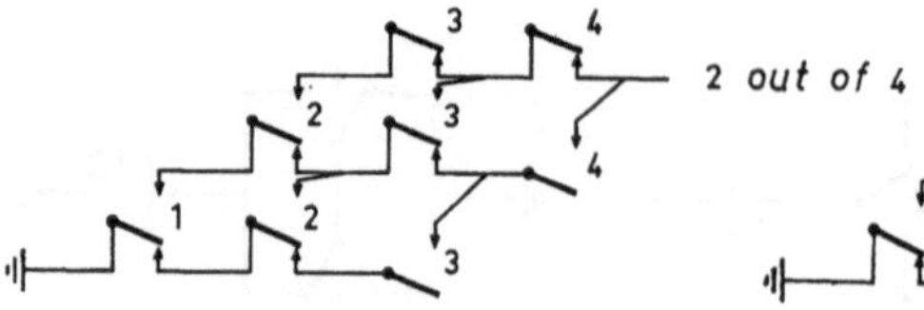

Fig. 113

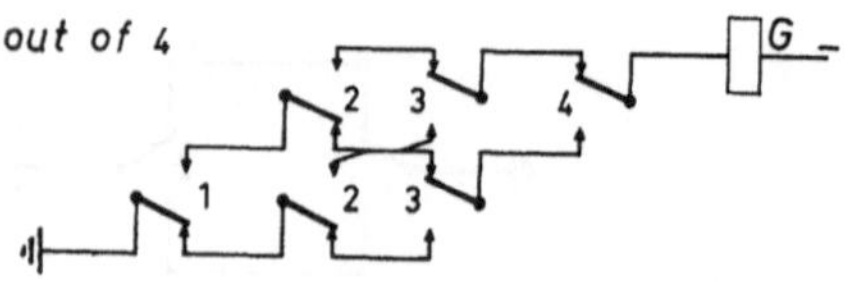

Fig. 114

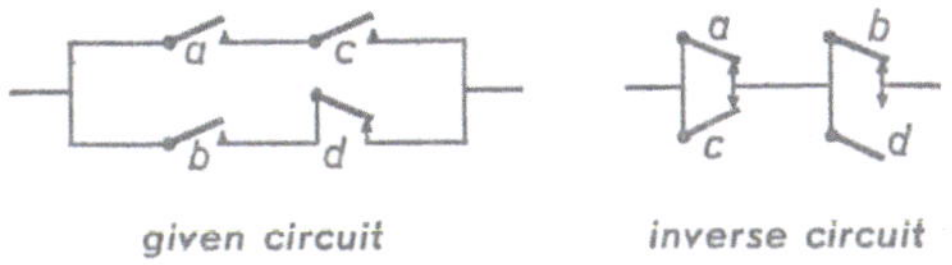

Fig. 115   *given circuit*   *inverse circuit*

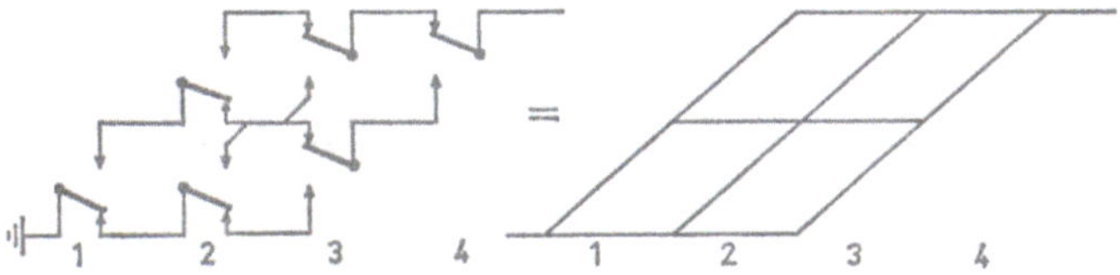

Fig. 116

It will be seen that in this new representation a make contact unit is represented by an oblique line, and a break contact unit by a horizontal line. In order to find the inverse circuit, we draw a new figure, in which each oblique line is crossed by a horizontal line, and each horizontal line by an oblique one (make contact unit replaced by break, and break by make). Fig. 117 shows both circuits drawn together. The input and output of the original circuit, with change-over contact units connected in *series*, are replaced by an input and output placed so that the change-over contact units switch in *parallel*. The new circuit is shown again in Fig. 118, and the contacts are then put in.

Finally, Fig. 119 gives the circuit obtained by combining make and break contact units to change-over contact units. This circuit energizes relay *F* every time a combination of 0, 1, 3 or 4 out of 4 occurs. Such checking circuits obey the rule that the number of contacts in a given circuit is equal to the number of contacts in the inverse circuit.

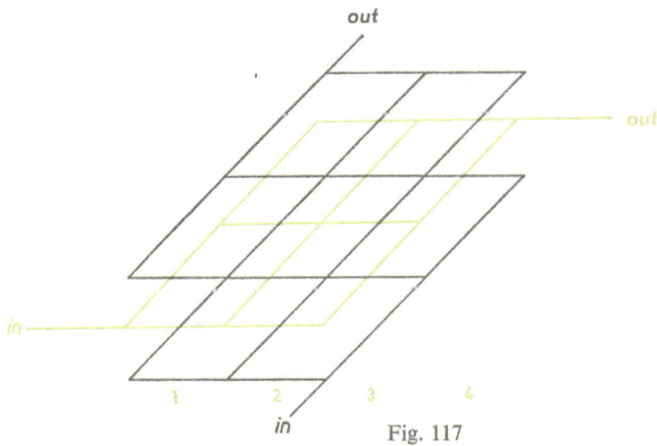

Fig. 117

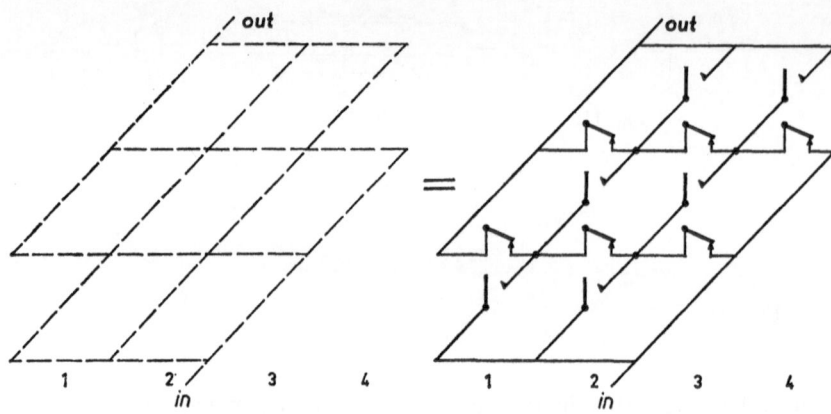

Fig. 118

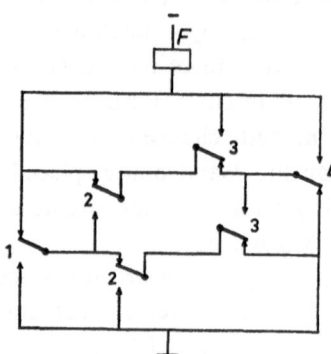

Fig. 119

7.2 Check on "odd" or "even" out of n

A check on the switching on of an even number of relays out of a total
of n relays could simply be obtained with the aid of a contact network like
that shown in Fig. 112. Connecting the outputs 0, 2, 4 etc. gives the desired
result directly. However, this method uses a relatively large number of
contact units. A simpler and cheaper solution can be obtained by working
with only 2 levels, of which the lower represents the even values, and the

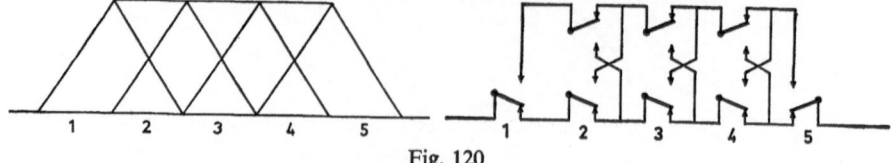

Fig. 120

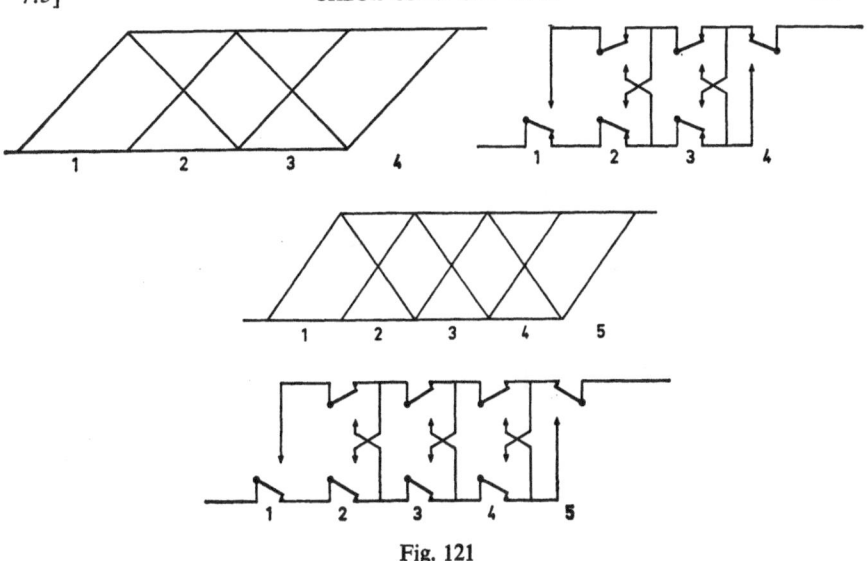

Fig. 121

upper the odd values. Fig. 120 shows this for even out of 5. A circuit for odd out of n can be obtained in a similar way, as may be seen from Fig. 121 for $n=4$ and $n=5$.

7.3 Check on m out of n, where the various values of m form an arithmetic progression

The check network of the previous section is conducting for *all* even values between 0 and n (Fig. 120), or for *all* odd values between 1 and n (Fig. 121). If however it is desired to construct a check network which is only conducting for a limited number of values of m, which together form an arithmetic progression, the following rules can be laid down.

Let the difference between successive values of m be r, and let the lowest value of m be a. The network can then be "folded back" from the $(r-1)$th layer $(r-1 \geqslant a)$.

Suppose that we want the circuit for 2 and 6 out of 7.

$a=2$ (also output layer)
$r =6-2=4$
$r-1=3$

The circuit is given in Fig. 122. Relay 5 is provided with two change-over contact units, a make contact unit and a break contact unit. Relay 6 needs a

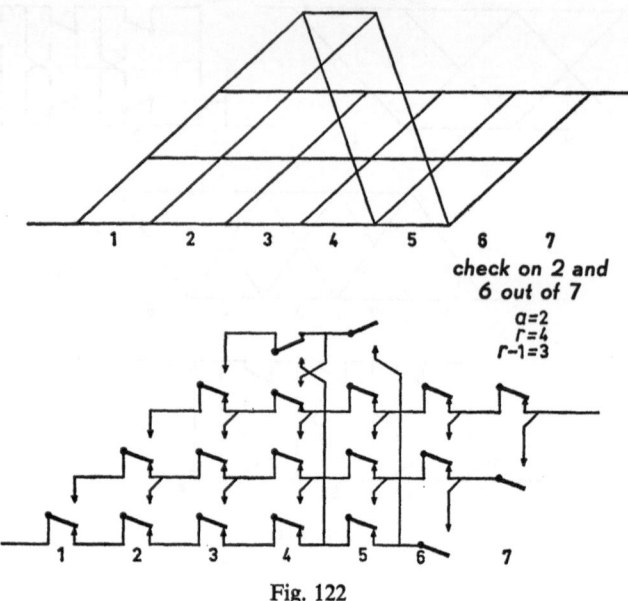

Fig. 122

change-over contact unit, a make contact unit and a break contact unit, while relay 7 has a make and a break contact unit. Contacts can be saved by combining make contact units with break contact units to give change-over contact units. The circuit simplified in this way is shown in Fig. 123.

Fig. 124 gives the circuit for 1 and 4 out of 5. We find:

$$r = 3 \qquad r - 1 = 2 \qquad a = 1$$

The switching diagram is shown in Fig. 124a, the corresponding circuit in Fig. 124b, and the simplified circuit obtained by combining make and break to change-over contact units in Fig. 124c.

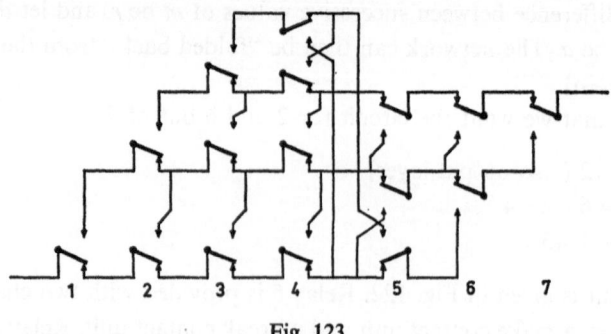

Fig. 123

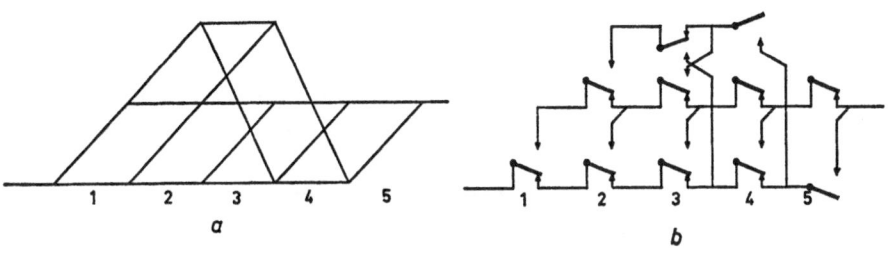

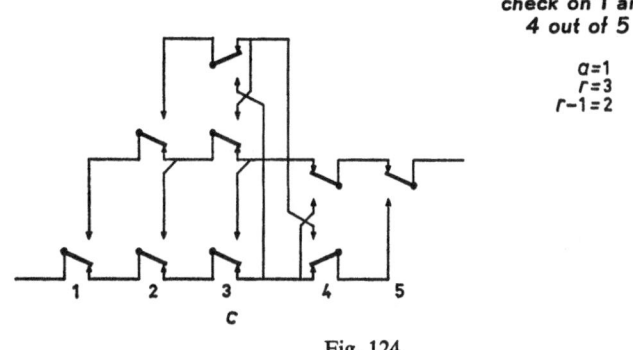

check on 1 and
4 out of 5

a=1
r=3
r−1=2

c

Fig. 124

These circuits can only be used if the difference between *n* and the highest value of *m* is less than the difference between successive values of *m*. For example, we cannot check for 1 and 3 out of 5 in this way, but have to use the complete circuit as shown in Fig. 125.

In this figure, the make and break contact units (of relay 5) have already been combined to a change-over contact unit. The circuit has been further simplified by connecting layers 0 and 1, and 1 and 3, after relay 4, so that relay 5 only needs one change-over contact unit.

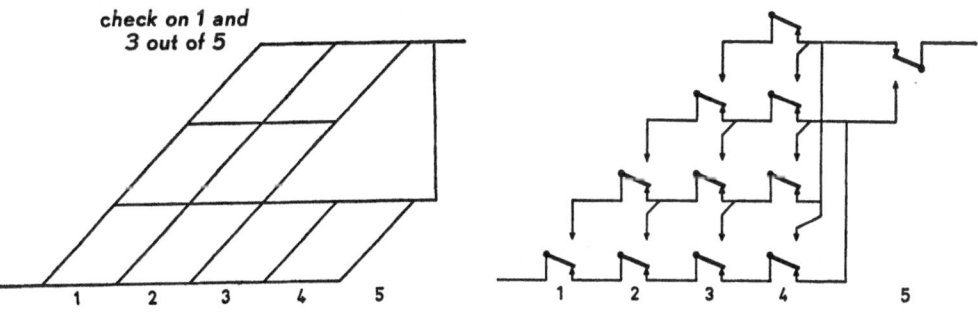

check on 1 and
3 out of 5

Fig. 125

7.4 Check on two successive relays out of n

Suppose that a circuit has to be closed when two successive relays out of n are energized, and the rest are not. In this case, each relay must obey the following rules.

a) It must "know" whether the previous relay is energized or not.
b) If the previous relay is not energized but the relay in question is, this information must be passed on to the next relay.
c) If the previous relay and the relay in question are both energized, this information must also be passed on to the next relay. These rules are illustrated in Fig. 126.

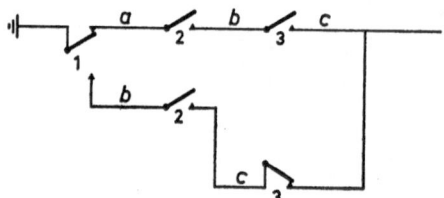

Fig. 126

Situations a, b and c can be indicated even more clearly for a group of six relays, as may be seen from Fig. 127. Relay 1 has only one input and two outputs (a and b). Relay 2 has two inputs and three outputs, viz:
a) relay 1 is not energized, nor is relay 2
b) relay 1 is not energized, but relay 2 is
c) relay 1 is energized, and so is relay 2.

The fourth combination (relay 1 energized but relay 2 not) has no consequences, as we have not asked for this. If, after relay 2, output c is reached, a check must still be made to ensure that all following relays are non-energized before a closed circuit is formed.

The above holds, *mutatur mutandis*, for all the other pairs of successive relays. Fig. 128 shows the circuit simplified by combining make and break to change-over contact units.

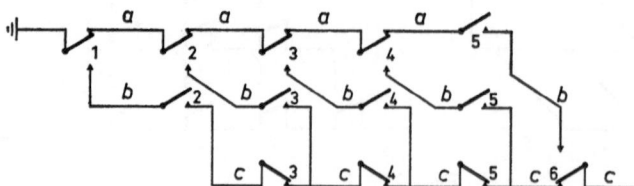

Fig. 127

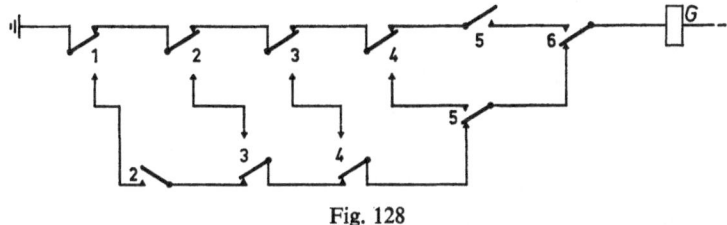

Fig. 128

7.5   Check on two code combinations which are complementary to one another

The circuit of Fig. 129 gives a closed circuit if relays 1 and 2 are *both* on or *both* off. The first situation gives:

$$f(G)=1'2'$$

The second situation gives:

$$f(G)=12$$

We may thus write

$$f(G)=1'2'+12$$

Addition of the code combinations gives $00+11=11$. The two combinations are thus complementary to one another, because their sum is equal

Fig. 129   Fig. 130

to the highest possible value represented by 2 bits. Fig. 130 gives a closed circuit if one of the two relays 1 and 2 is switched on.

$$f(G)=1'2+12'$$

The sum of the two code combinations is here again $01+10=11$.

Fig. 131 gives a similar circuit for 7 relays. This circuit also gives a closed circuit in 2 cases, viz for the code combinations 0101101 and 1010010, the sum of which is 1111111. This may be generalized to give the following rule:

> *If two groups of contact units connected in series are connected in parallel, and if the values of the contact units $1 \ldots n$ of the first group are the inverse of those of the contact units $1 \ldots n$ of the second group, then a circuit is only closed in 2 of the 2^n possible combinations.*

The circuit of Fig. 131 can be simplified as follows. The first part of Fig. 131 is reproduced in Fig. 132a. As we have seen in Section 3.4, we may interchange two contact units which are connected in series; if we do this in the top line of Fig. 132a, we obtain Fig. 132b. The connection indicated by the broken line in Fig. 132a may not be made, as this would completely wipe out the function of relay 1. In Fig. 132b, however, this may be done, as the make and break contact units of relay 1 are mutually exclusive. The same is true for relay 2. Fig. 132b is thus transformed into Fig. 133. Applying this method to all the contact units of Fig. 131, we obtain Fig. 134.

7.6 Check on all code combinations except those which are complementary to one another

This check network is the inverse of that discussed in the previous section: the network of Fig. 131 can be represented by:

$$f(G) = 1\ 2'3\ 4'5'\ 6\ 7' + 1'\ 2\ 3'\ 4\ 5\ 6'\ 7$$

and the inverse network by:

$$f(G) = (1' + 2 + 3' + 4 + 5 + 6' + 7)\ (1 + 2' + 3 + 4' + 5' + 6 + 7')$$

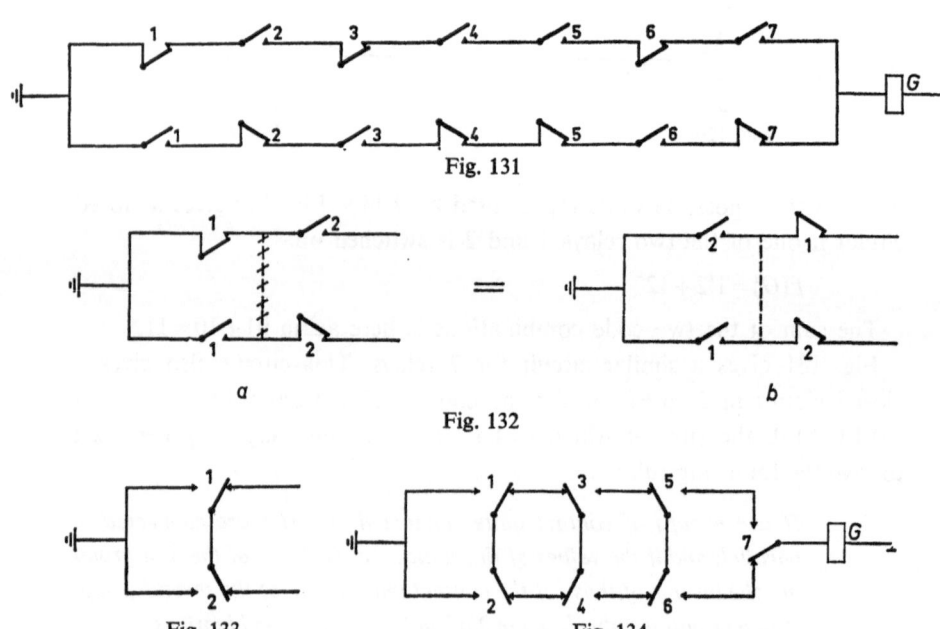

Fig. 131

Fig. 132

Fig. 133

Fig. 134

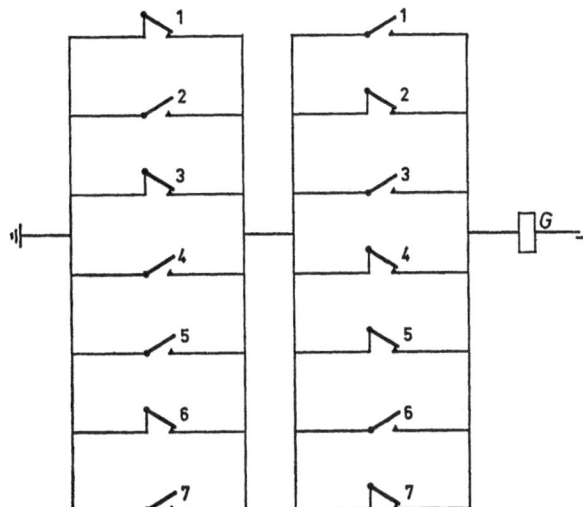

Fig. 135

This gives the circuit of Fig. 135. It may be seen from this figure that a closed circuit is always obtained, except when only relays 2, 4, 5 and 7 or only relays 1, 3 and 6 are energized. The make and break contact units of each relay are connected with one another, so they may be replaced by change-over contact units. The circuit simplified in this way is shown in Fig. 136.

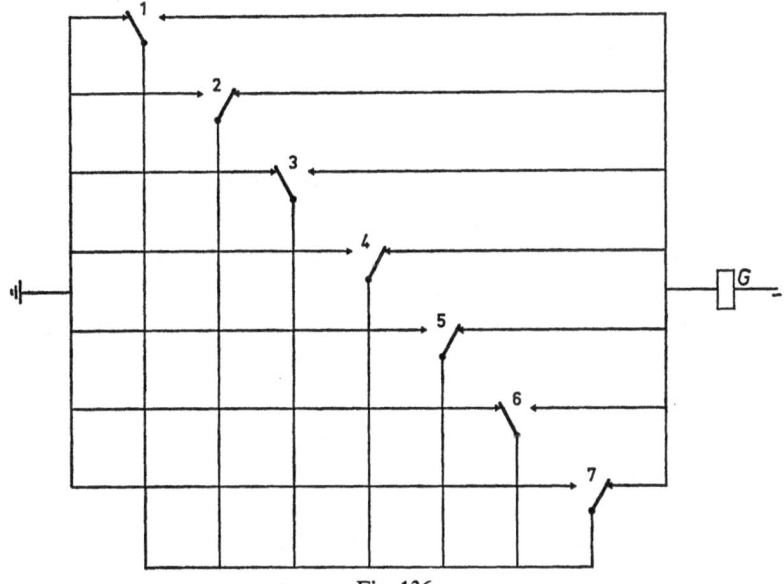

Fig. 136

7.7 Check on one or more particular code combinations

If with a given code one or more code combinations must be detected, an obvious solution is to use a contact pyramid, as shown in Fig. 137 for the checking of outputs 2 and 7.

Relay G can be connected to each of the ten outputs of the contact tree, so that any desired combination(s) of the code can be detected. However,

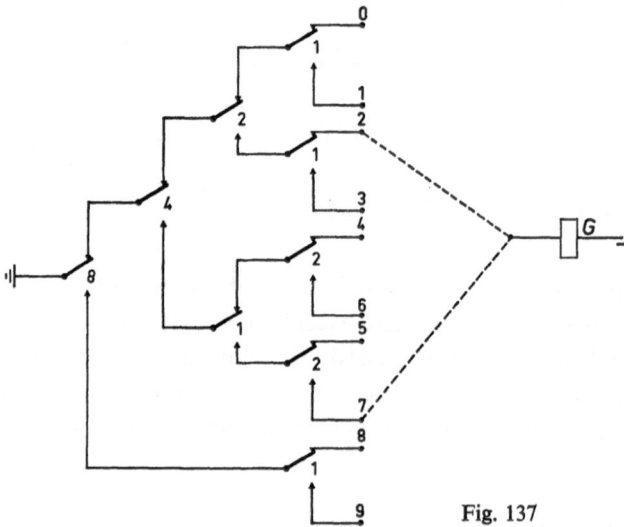

Fig. 137

the tree is built up of nine change-over contact units, only a few of which are in use at a time.

In almost all cases, a circuit using fewer contact units will suffice. For this purpose, 4 change-over contact units are arranged so that they can be connected in various ways, and the desired combinations can be obtained by connecting them in the appropriate way. Fig. 138 shows the basic circuit for the binary-decimal code.

There are 10 possible cases in which one combination is to be detected; the appropriate connections are shown in the following table.

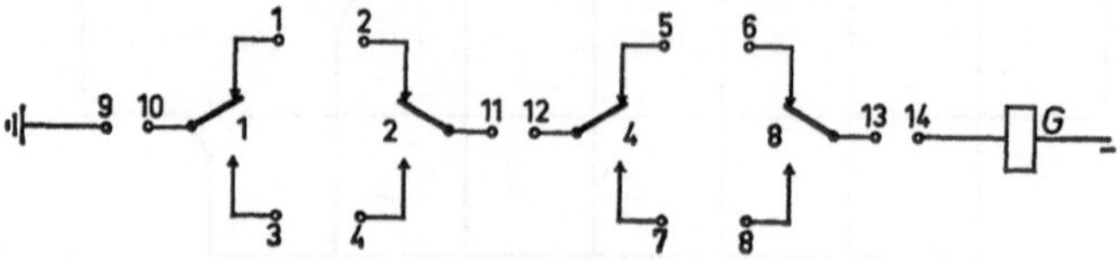

Fig. 138

digit	connect					digit	connect				
0	9–10	1–2	11–12	5–6	13–14	5	9–10	2–3	11–12	6–7	13–14
1	9–10	2–3	11–12	5–6	13–14	6	9–10	1–4	11–12	6–7	13–14
2	9–10	1–4	11–12	5–6	13–14	7	9–10	3–4	11–12	6–7	13–14
3	9–10	3–4	11–12	5–6	13–14	8	9–10	1–2	11–12	5–8	13–14
4	9–10	1–2	11–12	6–7	13–14	9	9–10	2–3	11–12	5–8	13–14

There are

$$\frac{10!}{2!\,8!} = 45$$

different combinations of two decimal digits. These too can all be detected with the connections shown in the table on page 178.

Some of the combinations shown in this table also satisfy the conditions of Section 7.5. For example, the combinations 6 and 9 contains two complementary numbers, viz.

$$1'\,2\,4\,8' \quad \text{and} \quad 1\,2'\,4'\,8$$

The circuit for detecting this is shown in Fig. 139.

There are the following numbers of possible combinations of 3, 4, 5, 6, 7, 8 or 9 out of 10:

$$3 \text{ out of } 10 = \frac{10!}{3!\,7!} = 120$$

$$4 \text{ out of } 10 = \frac{10!}{4!\,6!} = 210$$

$$5 \text{ out of } 10 = \frac{10!}{5!\,5!} = 252$$

$$6 \text{ out of } 10 = \frac{10!}{6!\,4!} = 210$$

$$7 \text{ out of } 10 = \frac{10!}{7!\,3!} = 120$$

$$8 \text{ out of } 10 = \frac{10!}{8!\,2!} = 45$$

$$9 \text{ out of } 10 = \frac{10!}{9!} = 10$$

comb.	connect					
0-1	9-11	2-12	5-6	13-14		
0-2	9-10	1-12	5-6	13-14		
0-3	9-10	1-2	3-4	11-12	5-6	13-14
0-4	9-10	1-2	6-11	13-14		
0-5	9-10	1-5	11-12	2-6		
0-6	9-11	2-5	4-7	10-12	3-7	13-14
0-7	9-10	1-2	3-7	11-12	1-6	13-14
0-8	9-10	1-2	11-12	5-14	4-5-6	13-14
0-9	9-10	1-6	3-8	11-13		
1-2	9-10	1-4	2-3	11-12	2-5	12-14
1-3	3-9	5-6	13-14	10-12	5-6	13-14
1-4	9-10	1-7	3-5	11-12		
1-5	9-10	3-11	2-6	13-14	2-6	13-14
1-6	9-10	1-4	3-5	11-12		
1-7	9-10	3-6	2-5	4-7	2-6-7	13-14
1-8	9-10	1-8	3-6	11-13	11-13	12-14
1-9	3-9	11-10	2-5	12-14	2-5	13-14
2-3	4-9	11-12	5-6	13-14		
2-4	9-10	1-6	11-13	2-7	4-5	12-14
2-5	9-10	1-4	3-7	2-5-6	11-12	13-14
2-6	9-10	1-6	11-13	4-14	4-6	13-14
2-7	9-10	1-5	3-7	11-12	4-6	13-14
2-8	9-10	1-5	11-12	2-8		

comb.	connect					
2-9	9-12	1-2-5	10-11	4-6	3-8	13-14
3-4	9-12	4-5	1-7	10-11	2-3-6	13-14
3-5	9-10	3-6	11-13	4-5	2-7	12-14
3-6	9-11	4-6	10-13	1-7	3-5	12-14
3-7	9-10	3-4	6-11	13-14		
3-8	9-12	2-3-5	1-8	10-11	4-6	13-14
3-9	9-10	3-5	6-7	2-8	4-6	13-14
4-5	2-9	11-12	6-7	13-14		
4-6	9-10	1-12	6-7	13-14	6-7	13-14
4-7	9-10	1-2	3-4	11-12	6-7	13-14
4-8	9-10	1-2	11-12	5-8	3-8	13-14
4-9	9-10	1-2	10-12	6-7	6-7	13-14
5-6	9-11	1-2-5	2-3	11-12		
5-7	9-10	1-4	6-7	13-14	5-8	13-14
5-8	9-11	3-12	10-12	3-6	6-7	13-14
5-9	9-10	1-2-7	11-12	5-8		
6-7	4-9	2-3	6-7	13-14	4-6	13-14
6-8	9-10	11-12	11-12	5-8	1-4	3-6-14
6-9	2-7-9	1-2-7	11-12	10-13	10-13	3-8-14
7-8	4-5-9	11-12	5-8	6-7		
7-9	9-10	2-3-7	1-2	4-6	5-8	13-14
8-9	2-9	11-12	11-12	5-8		

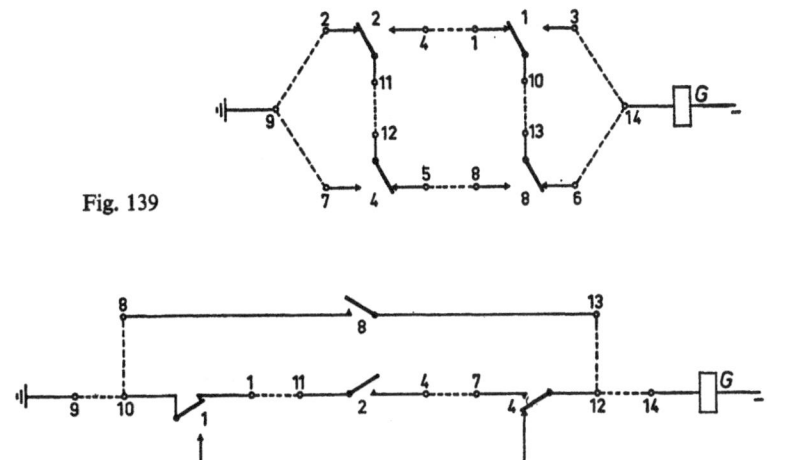

Fig. 139

Fig. 140

Many of these can also be detected with the basic circuit of Fig. 138.

By way of example, Fig. 140 shows the circuit for the detection of the combination of 1, 3, 6, 8 and 9.

7.8 Problems

1. 5 relays can be energized in all possible combinations. A lamp must be made to light up for the following combinations. This is possible with 26 contact springs.

A	B	C	D	E		A	B	C	D	E
1	1	0	0	1		0	0	1	1	1
0	1	0	1	1		1	0	1	0	1
1	0	0	1	1		1	1	0	1	0
0	1	1	1	0		0	1	1	0	1
1	1	1	0	0		1	0	1	1	0

2. A contact network using six relays must be conducting in all positions of the 6 relays exept the following. Draw the contact network, using not more than 34 contact springs.

A	B	C	D	E	F		A	B	C	D	E	F
1	0	1	0	0	0		1	0	0	0	1	0
0	1	0	1	0	0		0	1	0	0	0	1
0	0	1	0	1	0		1	1	0	0	0	0
0	0	0	1	0	1		0	1	1	0	0	0
1	0	0	1	0	0		0	0	1	1	0	0
0	1	0	0	1	0		0	0	0	1	1	0
0	0	1	0	0	1		0	0	0	0	1	1
							1	0	0	0	0	1

3. Draw a network which is closed when 0, 1, 3 and 4 relays out of 4 are energized (18 contact springs).

4. Design a contact network which lights up a lamp when 3 out of 6 relays are energized, except when the 3 relays which are energized are A, C and E or B, D and F. This is possible with a total of only 30 contact springs.

5. A check circuit must be able to transmit 15 *similar* combinations out of a total of 64 possible ones.
   a) How many relays are needed to present these 64 combinations?
   b) Is it possible to use a particular code for transmitting the 15 required combinations, and if so which?
   c) Draw the check network (in principle).

6. Five relays close a circuit in the following combinations: $ab'cde'$ and $a'bc'd'e$.
   Design a network which closes a circuit in the other 30 combinations, using not more than one change-over contact per relay.

7. Design a network which closes a circuit whenever an odd number of relays out of 5 are switched on, using not more than 24 contact springs.

8. Of 7 relays, only the combinations 0, 2, 3, 5 and 6 out of 7 are permitted. Design a check circuit which switches on a lamp if a forbidden combination arises, using not more than 15 change-over contact units.

9. Draw the following circuit with 18 contact springs:

$$ab'c'd' + a'b'cd' + abcd' + ab'cd + a'bc'd' + abc'd + a'b'c'd + a'bcd$$

Chapter 8

SOME COMPUTING CIRCUITS WITH RELAYS

8.1 General concepts

As has already been mentioned in Section 4.8, the digits of decimal numbers can be separately represented in binary form. The highest binary value occurring in this presentation is 1 0 0 1 (9). The values 1 0 1 0 to 1 1 1 1 are thus not used.

If however the digits 8 and 6 are added, the result is 14 ("4, carry 1"). The addition in the binary-decimal notation must give the same result:

$$
\begin{array}{r r}
8 & 1\ 0\ 0\ 0 \\
+\ 6 & 0\ 1\ 1\ 0 \\
\hline
14 & 1\ 1\ 1\ 0
\end{array}
$$

We must be able to deduce from this result that a carry must be made to the tens, while the result must also be corrected so that 0 1 0 0 (4) is left in the units. Both the carry and the correction are taken care of by adding 0110 (6) to the result

$$
\begin{array}{r}
1\ 1\ 1\ 0 \\
+0\ 1\ 1\ 0 \\
\hline
\end{array}
$$

carry   ←1 0 1 0 0 = 4

This can be explained as follows. Let us suppose that the result of the addition of two digits is 10. The binary notation for this is 1 0 1 0. The result for the units must however be 0 = 0 0 0 0, since after the "1" has been carried, there should be nothing left in the units. We can get 0 0 0 0 in the units by adding on to the result until we just get a five-figure binary number (1 0 0 0 0 = 16): add: 10 and 6

$$
\begin{array}{r}
1\ 0\ 1\ 0 \\
0\ 1\ 1\ 0 \\
\hline
\end{array}
$$

carry   ←1 0 0 0 0

Another example: the sum of 8 and 9.

$$
\begin{array}{r}
1\ 0\ 0\ 0 \\
1\ 0\ 0\ 1 \\
\hline
1\ 0\ 0\ 0\ 1
\end{array}
$$

correction 0 1 1 0

carry   ←1 0 1 1 1(7)

The *subtraction* of two numbers is not so simple, as here we sometimes have to "borrow 1" from the next higher place. This also happens in the decimal notation.

$$
\begin{array}{r}
3\ 7\ 2\ 5 \\
-2\ 4\ 5\ 6 \\
\hline
\end{array}
$$

The 6 in the units cannot be subtracted from 5 without giving a negative result, but it can be subtracted from 15. The digit 2 in the tens is reduced to 1 in connection with this.

Both in the decimal and in the binary notation, we can replace subtraction by *addition* in order to get round the difficulty of "borrowing". We then proceed as follows:

$$
\begin{array}{r}
3\ 7\ 2\ 5 \\
-2\ 4\ 5\ 6 \\
\hline
1\ 2\ 6\ 9
\end{array}
\qquad
\begin{array}{r}
3\ 7\ 2\ 5 \\
+7\ 5\ 4\ 3 \\
\hline
1\ 1\ 2\ 6\ 8 \\
\longrightarrow\ 1 \\
\hline
1\ 2\ 6\ 9
\end{array}
$$

The "9's complement" of 2456 is 7543 (9999−2456=7543). The addition gives the result 11268. This result is 9999 too high. We can correct for this by subtracting 10 000 from the result and adding 1, which can be done by taking a 1 from the ten-thousands and adding it on to the units. This is known as an "around-end carry".

Addition of the complementary value (the "1's complement") is also used for the subtraction of binary numbers.

Maximum value of a 6-figure binary number = 1 1 1 1 1 1

given the number = 1 0 1 0 0 1

the complementary value = 0 1 0 1 1 0

It may be seen from this that the complementary value can easily be obtained by replacing all "1"s by "0", and all "0"s by "1".

Suppose that the following subtraction has to be carried out:

$$
\begin{array}{r}
1\ 1\ 0\ 1\ 1\ 0\ 1\ 0 \\
-1\ 0\ 1\ 1\ 0\ 0\ 1\ 1
\end{array}
\qquad
\begin{array}{r}
1\ 1\ 0\ 1\ 1\ 0\ 1\ 0 \\
\text{complement} = +0\ 1\ 0\ 0\ 1\ 1\ 0\ 0 \\
\hline
1\ 0\ 0\ 1\ 0\ 0\ 1\ 1\ 0 \\
\text{!}\hspace{5.5em}\longrightarrow 1 \\
\hline
1\ 0\ 0\ 1\ 1\ 1
\end{array}
$$

The desired result is obtained by adding the complement and carrying a "1" from the highest place to the lowest.

This method is also used for the subtraction of binary-decimal digits. Since however in this case each decimal digit is represented separately in the binary notation, we must also determine in each case when we have to "borrow 1" from the next higher place. We shall be returning to this point below.

8.2 Addition

In the addition of numbers in the binary-decimal notation we distinguish between two forms of carry, the bit carry and the digit carry. The addition of 2 bits gives four possibilities:

	1	2	3	4
A	0	1	0	1
B	0	0	1	1
	0	1	1	10

Only in case 4 is a bit carry involved. Fig. 141 shows a simple circuit for carrying out the addition. If relay A or B is energized, relay R is switched on. If relay A and B are both energized, or both not energized, relay R is not switched on.

Fig. 142 shows this circuit, extended by inclusion of relays C and D. In all cases where an odd number of the relays A, B, C and D are energized

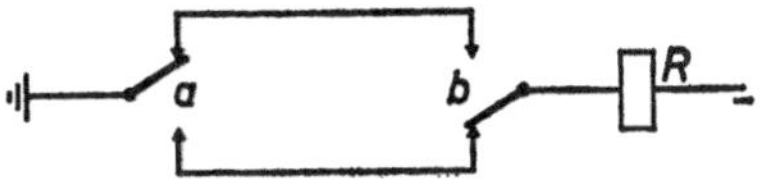

Fig. 141

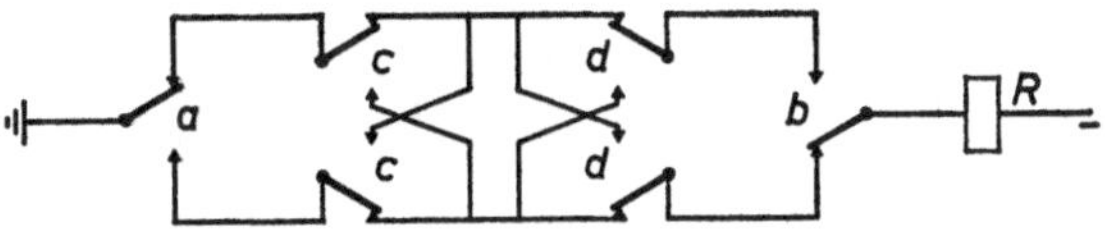

Fig. 142

relay R is switched on. If an even number of relays are energized, relay R is not switched on.

A bit carry is involved whenever two or more "ones" have to be added. In this connection, we must also take possible "carries" from the previous place into account. Consider the following addition:

$$A \ 0 \ 1 \ 1$$
$$B \ 0 \ 0 \ 1$$

Addition of the units gives 0, carry 1, since both A and B are given the value 1 (see Fig. 141). We represent the carry by $T1$; we now find:

	(4)	(2)	(1)
A	0	1	1
B	0	0	1
T	$T2$	$T1$	
	1	0	0

The addition of the "twos" again gives the result 0, carry 1 (the 1 originally present + the carry $T1$). Finally, the addition of the "fours" gives the result 1 (from the second carry $T2$).

From the above we may deduce that if no carry is received, a carry is produced if A and B are both 1. If however a carry is received, a carry is produced even if A or B is 1. The circuit of Fig. 143 follows. Relay T_2 operates if $a2$ and $b2$ are closed. If a carry has been received, so that relay T_1 is energized, relay T_2 operates if contact unit a_2 or b_2 is closed.

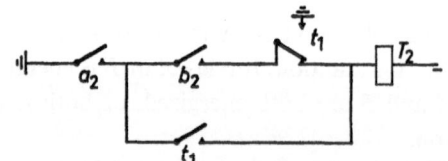

Fig. 143

If a carry is produced, this can quite simply be introduced into an adder circuit. In order to see this, let us have a look at the 8 possible combinations of a and b with or without carry.

	1	2	3	4	5	6	7	8
A	0	1	0	1	0	1	0	1
B	0	0	1	1	0	0	1	1
T	0	0	0	0	1	1	1	1
	0	1	1	0	1	0	0	1

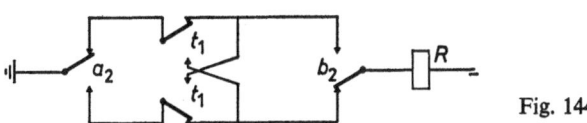

Fig. 144

The result of cases 1 to 4, where no carry is received, is the reverse of that for cases 5 to 8, where a carry *is* received. This is fully realized by the circuit of Fig. 144.

In cases 1, 4, 6 and 7 the result is 0 and relay R does not operate. In cases 2, 3, 5 and 8, relay R does operate.

By way of example, let us consider the addition of the numbers 654 and 295. If we represent these numbers with the aid of relays, we obtain:

hundreds				tens					units				
−	$A4$	$A2$	−(6)	−	$A4$	−	$A1$	(5)	−	$A4$	−	−(4)	
−	−	$B2$	−(2)	$B8$	−	−	$B1$	(9)	−	$B4$	−	$B1$(5)	
$T4$	$T2$	−		−	−	$T1$			$T4$	−	−		
1	0	0	0 (8)	1	1	1	0	(14)	1	0	0	1	(9)

This result must still be corrected: a "one" must be carried from the tens to the hundreds, and the tens must be given their correct binary form $(1\ 1\ 1\ 0 = 14 = \text{carry} + 0\ 1\ 0\ 0)$. A correction term must therefore be added on to the result.

hundreds				tens				units			
1	0	0	0	1	1	1	0	1	0	0	1
		TH			$C4$	$C2$			−	−	
				$TC4$	$TC2$				−	−	
1	0	0	1 (9)	0	1	0	0 (4)	1	0	0	1 (9)

Here $C2$ and $C4$ take care of the correction by adding on the value 6, $TC2$ and $TC4$ take care of the resulting bit carries, and TH of the digit carry to the hundreds.

The circuits for realizing the first operation is shown in Fig. 145. The result of this operation is represented by the relays $R_1 - R_8$, which are provided for each digit.

Relay R_1 is only switched on via the contact units a_1 and b_1, since no carry is received for the addition of the lowest bits. With the relay R_2, the possible carry resulting from the addition of the lowest bits must be taken into consideration. The contact units t_1 are therefore added in series with the contact units a_2 and b_2. Similar carry contact units are provided in the energizing circuits of relays R_4 and R_8.

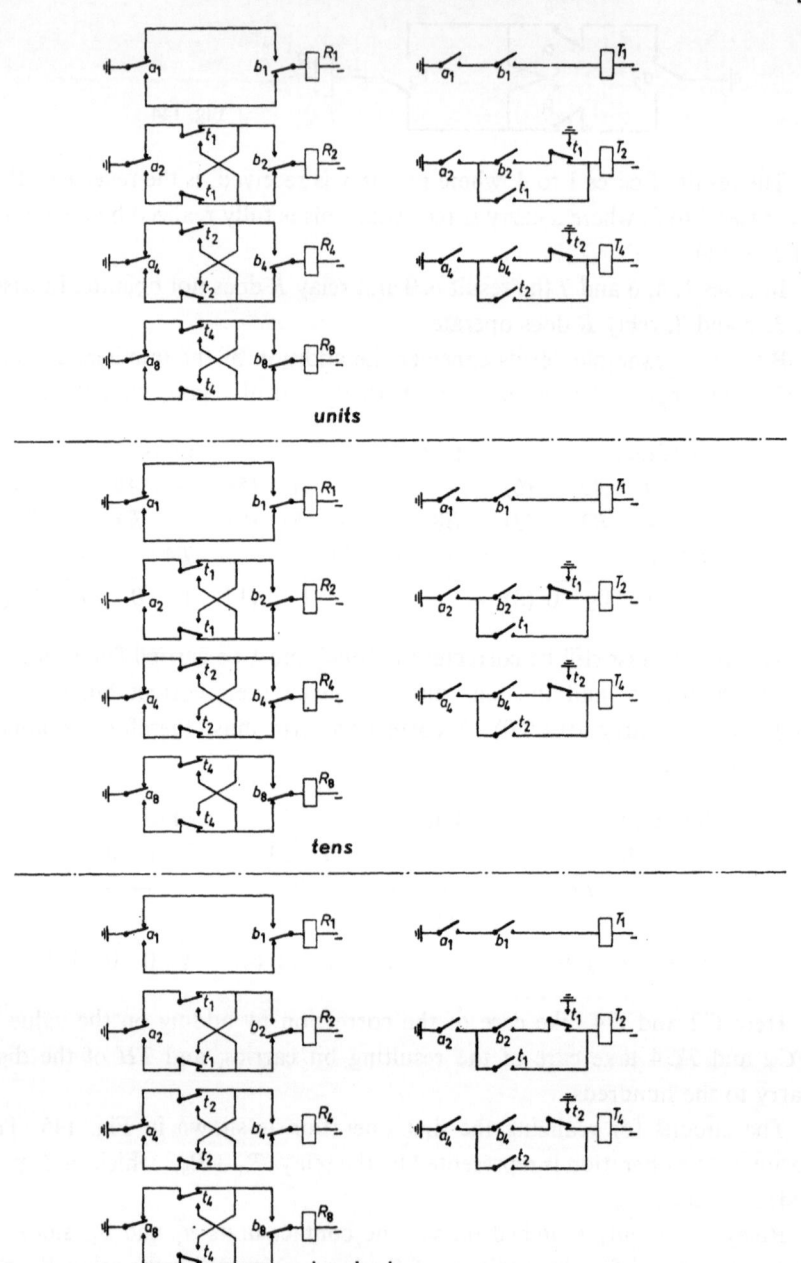

Fig. 145

The switching on of the carry relay T_1 depends only on the positions of the contact units a_1 and b_1. The switching on of relay T_2, however, depends not only on the positions of a_2 and b_2 but also on the position of the contact units t_1, since whether T_2 will be switched on or not also depends on whether a carry is received or not. The switching on of relay T_4 is subject to similar conditions. Relay T_8 is not shown, because it plays no part in this operation. The result of the first phase of the addition can be determined with reference to Fig. 145.

Let us consider the positions of the various relays at the end of the first phase of the addition of 587 and 399.

hundreds				tens				units			
– $A4$ – $A1$ (5)				$A8$ – – – (8)				– $A4$ $A2$ $A1$ (7)			
– – $B2$ $B1$ (3)				$B8$ – – $B1$ (9)				$B8$ – – $B1$ (9)			
$T4$ $T2$ $T1$				– – –				$T4$ $T2$ $T1$			
$R8$ – – – (8)				1 – – – $R1$ (1)				1 – – – – (0)			

This preliminary addition does not give the sum of the numbers in question. All digits of the result must be corrected, as follows:

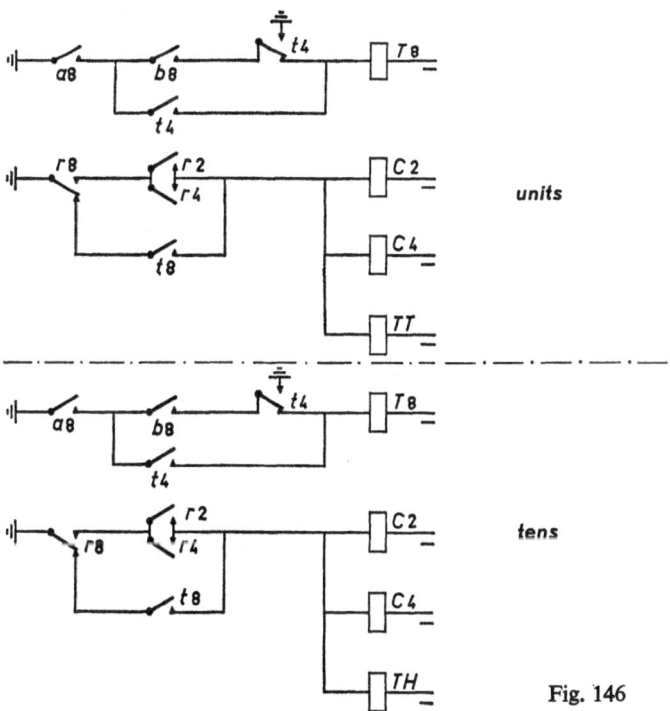

Fig. 146

a) the units give the result 16 ($\underline{1}$). We must add 6 to this, and arrange for a digit carry to the tens.

b) the tens must be corrected in a similar way, giving a carry to the hundreds, while at the same time a carry is received from the units.

c) the hundreds receive a carry from the tens.

Fig. 146 takes care of the various corrections.

The carry relays T_8, which were omitted in Fig. 145, come into play here. There are two groups of possibilities:

a) the sum of two digits a and b is less than 16, but greater than 9. In this case relay R_8 is switched on, and also relay R_2 and/or R_4. A circuit is then closed for the relays $C_2 + C_4$ (adding on 6), and TT (carry to tens) or TH (carry to hundreds).

b) The sum of two digits is greater than or equal to 16. In this case relay R_8 is not switched on, but relay T_8 is. This relay indicates that the value 16 has been reached or exceeded. In this case too, the relays C_2, C_4 and TT or TH are switched on.

The correction term is now added on to the result of the preliminary addition, during which process bit carries can also be produced (relay TC):

	hundreds				tens			
	$R8$	–	–	–	–	–	–	$R1$
+6						$C4$	$C2$	
digit carry			TH					TT
bit carry	–	–	–		$TC4$	$TC2$	$TC1$	
	$V8$	–	–	$V1$ (9)	$V8$	–	–	– (8)
	units							
	–	–	–	–				
+6		$C4$	$C2$					
digit carry								
bit carry	–	–						
	–	$V4$	$V2$	– (6)				

The circuit for this addition is given in Fig. 147. The final result is represented by the relays V.

In the energizing circuits for the relays $V_1 - V_8$ for the units, no digit carry is received, while relays V_1 and V_2 also receive no bit carry. Relay V_1 thus follows only contact unit r_1. Relay V_2 depends on the preliminary result (r_2), and on the possibility that 6 must be added to correct the result ($C_2 + C_4$).

Relay V_4 also depends on the same factors as V_2, but in addition it can receive a bit carry if relays R_2 and C_2, and hence also TC_2, are switched on.

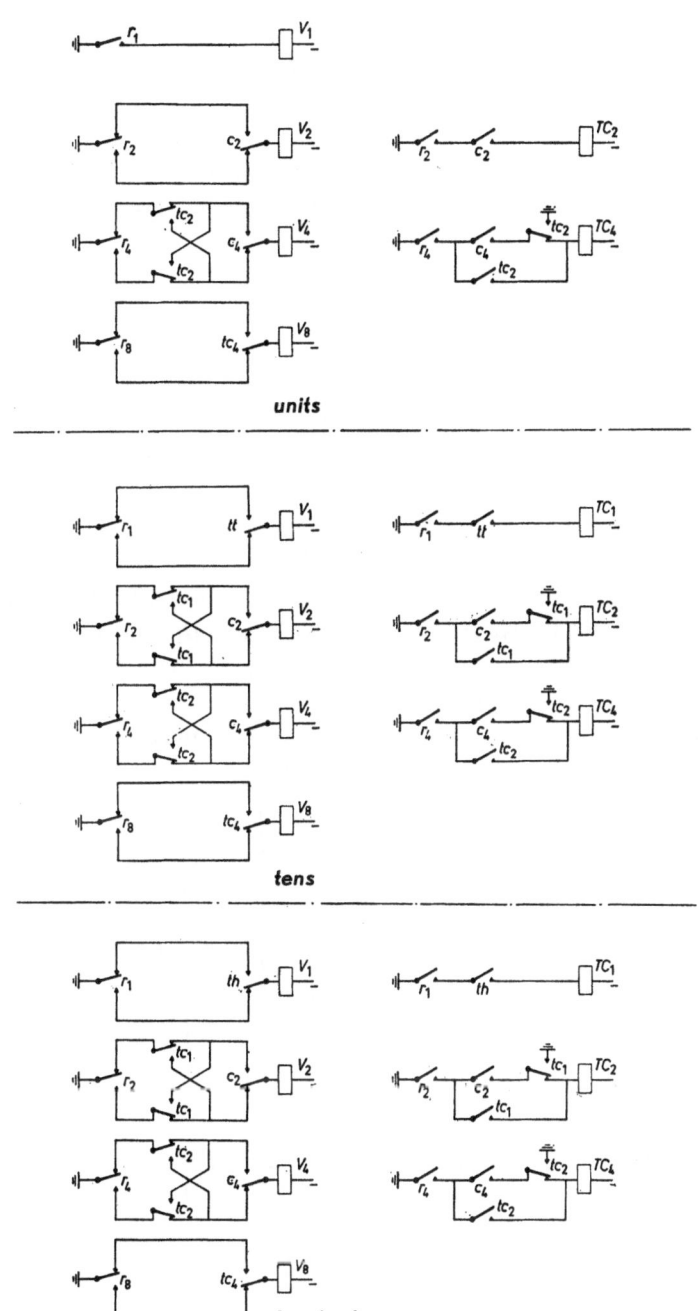

units

tens

hundreds

Fig. 147

Finally, relay V_8 again depends only on the position of contact unit r_8 and on a possible bit carry, received if 2 or 3 out of the relays R_4, C_4 and TC_2 are energized.

The V relays for the tens work in exactly the same way, except that there may be a digit carry from the units, indicated by relay TT.

The same holds for the V relays for the hundreds, except that here the possible carry is indicated by the relay TH.

8.3   Subtraction

As has already been mentioned in Section 8.1, a subtraction is transformed into the addition of the complement of the subtractor, with an around-end carry.

Fig. 148 shows a circuit which can be used for both the addition and the subtraction of two digits. We shall consider the addition $5+3=8$ and the subtraction $5-3=2$ with this circuit. First the addition. In this case, relay S is not energized. Relays A_1 and A_4 (digit 5) and B_1 and B_2 (digit 3) are switched on. Because contact units b_1 and a_1 are both closed, relay T_1 operates, as a result of which relay T_2 operates via contact units t_1 and b_2. Relay T_4 then also operates, because contact units t_2 and a_4 are closed.

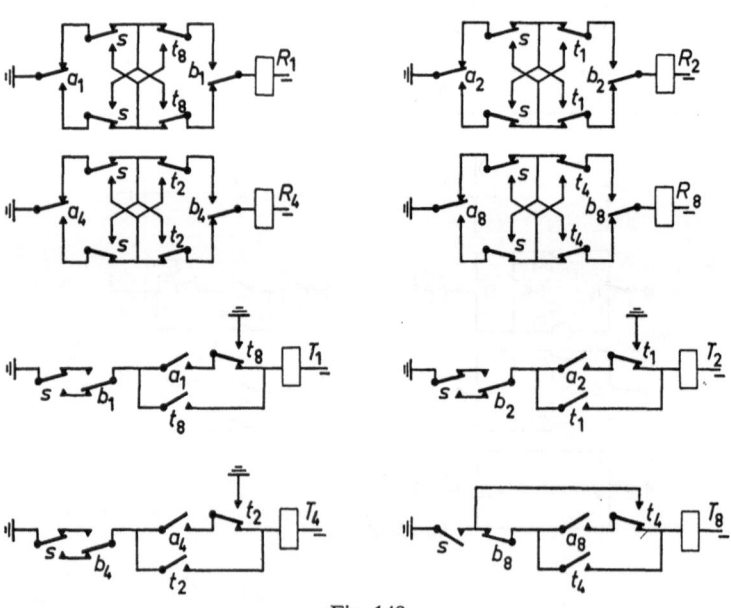

Fig. 148

Relay T_8 cannot operate during addition, because contact unit s is not closed. The relays finally switched on are thus A_1, A_4, B_1, B_2, T_1, T_2 and T_4. Consequently, a circuit is closed for relay $R8$ only, indicating the result of $5+3=8$.

For the subtraction of these digits, relay S is switched on. Here again relays A_1, A_4, B_1 and B_2 are energized. Relay T_1 cannot operate now, because the function of contact unit b_1 has been reversed by a contact unit s. Relay T_2 cannot operate either, because relays T_1 and A_2 are not energized. Relay T_4 can however operate via contact units a_4, b_4 and s. A circuit is also closed for relay T_8, via contact units t_4, b_4 and s. The operation of T_8 causes relay T_1 to be switched on after all, via the contact units t_8 and a_1. Relay T_2 stays non-energized, however, since even though contact unit $t1$ has switched over, contact units a_2 and b_2 prevent a circuit from being closed for T_2. The relays switched on are thus A_1, A_4, B_1, B_2, T_1, T_4 and T_8. As a result of this, only relay R_2 can operate. This gives the result of $5-3=2$.

Closer inspection of Fig. 148 shows that for subtraction the change-over contact units s and $t8$ in the circuit of relay R_1 are superfluous, since these contact units are always in the same position. An around-end carry is always needed in a subtraction, which means that relay S must always be energized. Under these circumstances, the relay T_8 can be switched on directly by a make contact unit s in the circuit for subtraction. These simplifications are introduced in Fig. 149, which shows the circuit for subtracting two three-figure numbers.

When two three-figure numbers have to be subtracted, further complications arise owing to the fact that one may have to "borrow 1" from the next highest digit. In order to illustrate this, let us consider the subtraction of 399 from 587. The first phase of this subtraction can be carried out with the circuit of Fig. 149. The positions of the relays are given below:

	hundreds				tens				units			
minuend	$-$	$A4$	$-$	$A1$	$A8$	$-$	$-$	$-$	$-$	$A4$	$A2$	$A1$
subtractor	$-$	$-$	$B2$	$B1$	$B8$	$-$	$-$	$B1$	$B8$	$-$	$-$	$B1$
operation	S	S	S	S	S	S	S	S	S	S	S	S
bit carry	$T4$	$-$	$T1$	$T8$	$-$	$-$	$-$	$T8$	$T4$	$T2$	$T1$	$T8$

This may be expressed as follows in binary notation:

0	1	0	1		1	0	0	0		0	1	1	1
1	1	0	0		0	1	1	0		0	1	1	0
1	0	1	1		0	0	0	1		1	1	1	1
0	0	1	0		1	1	1	1		1	1	1	0

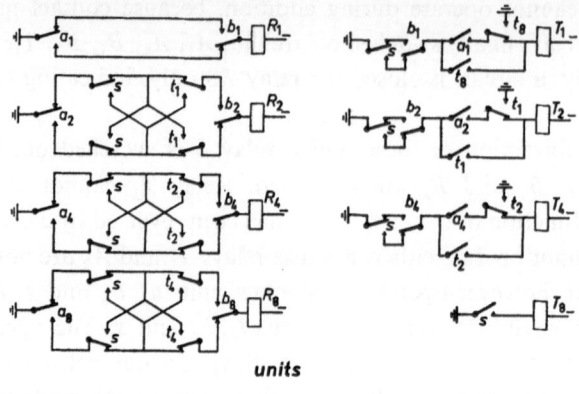

units

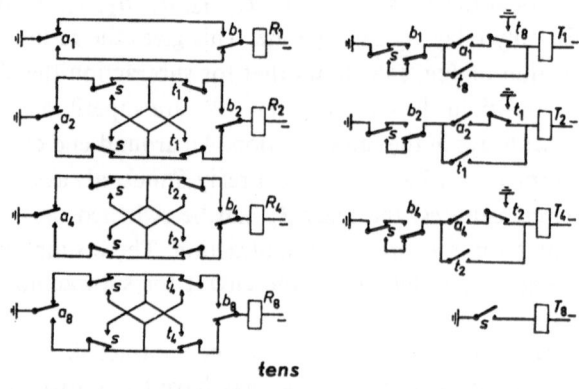

tens

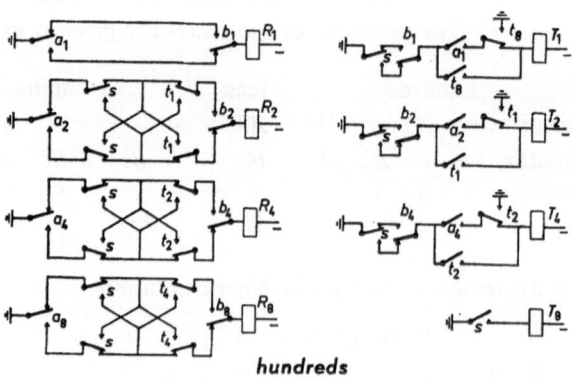

hundreds

Fig. 149

It may be remarked in this connection that the relays S and B look after the second row of "ones" or "zeroes" *together*. If relay B is energized, relay S inverts the value 1 to 0. If relay B is not energized, relay S inverts the value 0 to 1. As described above, an even number of "ones" gives the result 0 and an odd number of "ones" gives the result 1. The result of the first operation, expressed in the binary notation, is given in this circuit by the relays $R1-8$ for all digits. The result is thus:

$$-\quad -\quad R2\quad -\qquad R8\quad R4\quad R2\quad R1\qquad R8\quad R4\quad R2\quad -$$

The second operation takes care of the correction resulting from the need to "borrow 1" from a higher place if the subtraction of two digits would give a negative result. In the example given here, this is the case in both the units and the tens, so that we must borrow 1 from both the tens and the hundreds. If the tens are reduced by 1, the value of the units is increased by 10. In order to correct the result, therefore, we must add 10 on to the units.

1 must be subtracted from the tens because of the borrow by the units, and 10 must be added on because of the borrow from the hundreds. Finally, 1 must be subtracted from the hundreds because of the borrow by the tens. Whether or not it is necessary to borrow 1 is determined by the circuit of Fig. 150. In every conceivable case that the digit of the minuend (relays A_1-A_8) is less than the digit of the subtractor (relays B_1-B_8), relays D_2 and D_8 operate, indicating the addition of ten. Relay LT indicates that 1 has been borrowed from the tens, while relay LH carries out the same function for the hundreds. In the example given of the subtraction 587–399, the relays D_2 and D_8 do indeed operate for the units and the tens, as do the relays LT and LH.

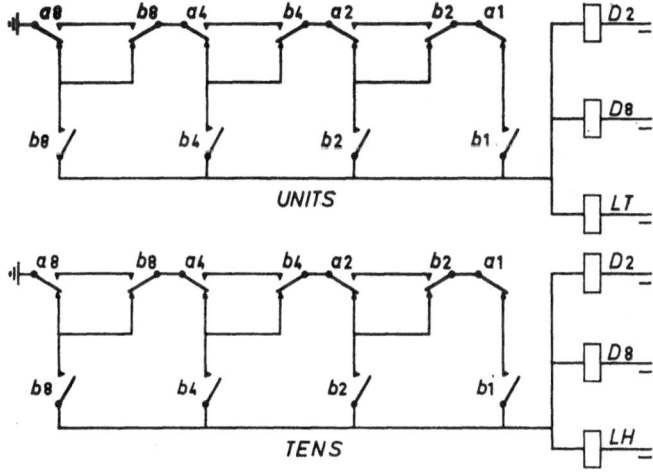

Fig. 150

The subtraction of 1 (0 0 0 1) from a digit can easily be realized by the addition of 1 1 1 1. In the case considered here we actually add 1 1 1 0 (complement of 0001); but since there is always an around-end carry, it is simpler to omit this around-end carry and add 1 1 1 1.

Fig. 151 shows the second operation, the determination of the total correction. The positions of the relays D_2, D_8, LT, LH and the bit-carry relays are given below. The result is indicated by means of the relays $C_1 - C_8$:

	hundreds				tens				units			
add 10	–	–	–	–	D8	–	D2	–	D8	–	D2	–
subtract 1	LH	LH	LH	LH	LT	LT	LT	LT	–	–	–	–
carry	–	–	–	–	T4	T2	–	–	–	–	–	–
total correction	C8	C4	C2	C1	C8	–	–	C1	C8	–	C2	–

In the third operation the correction is added on to the result of the first operation. The circuit for this is given in Fig. 152. The result is indicated by relays $E_1 - E_8$ for all digits. The positions of the relays are given below.

	hundreds				tens				units			
1st operation	–	–	R2	–	R8	R4	R2	R1	R8	R4	R2	–
2nd operation	C8	C4	C2	C1	C8	–	–	C1	C8		C2	
carry	T4	T2	–		T4	T2	T1		T4	T2		
final result	–	–	–	E1	E8	–	–	–	E8	–	–	–
				1		8				8		

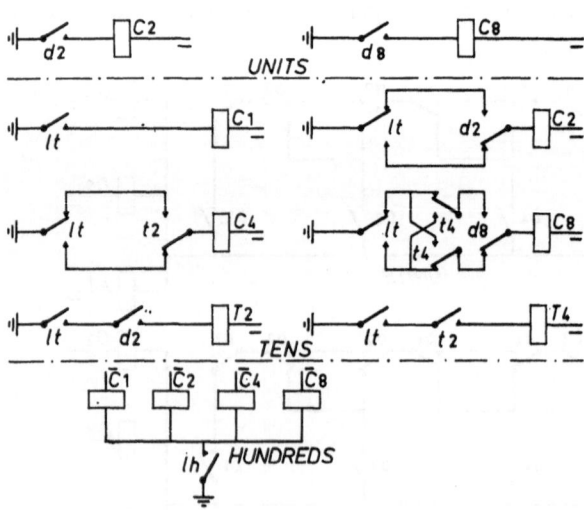

Fig. 151

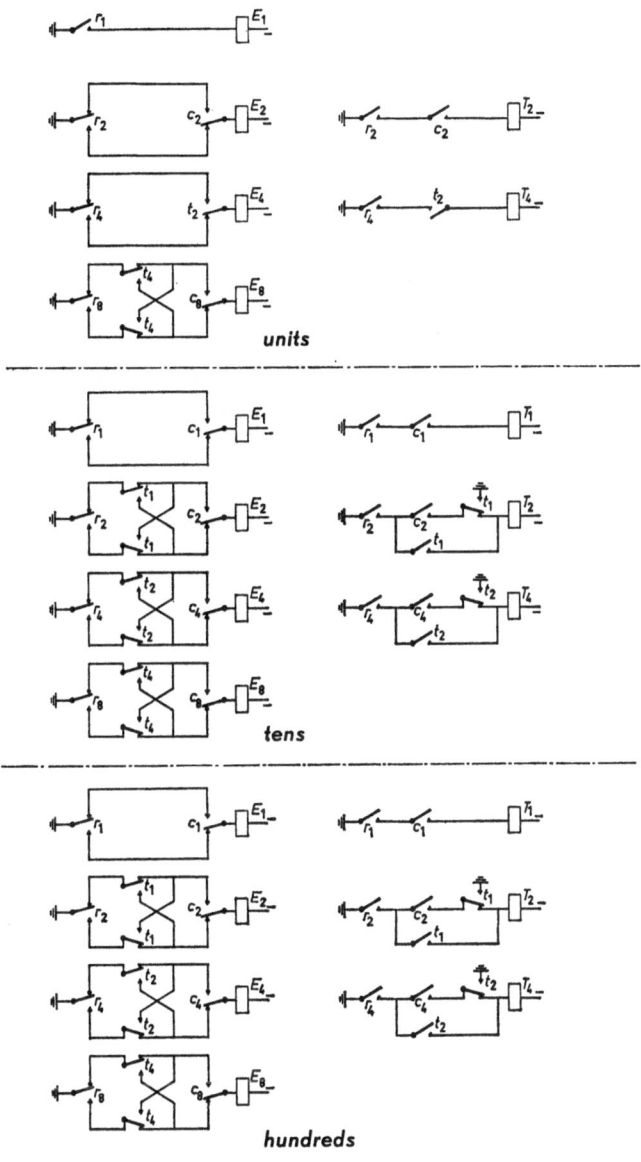

Fig. 152

It follows from our discussion of the correction that relays C_1 and C_4 are never needed for the units, as we never borrow 1 from the units. These relays can thus be omitted, which simplifies the circuits for the relays E_1 and E_4. The carry relay T_1 for the units in Fig. 152 is also superfluous, since relay C_1 has been omitted, so that T_1 can never operate. As a result of this, only two contact units are needed in the circuit for relay E_2. Relay T_8 is not needed for any of the digits, because the correction always involves an addition, so that the around-end carry can be left out.

We have only given the *computing* operations above, without considering how the type of operation (addition or subtraction) is determined, and how the three operations are initiated in turn.

8.4 Problems

1. The binary number 1 1 0 1 0 0 must be subtracted from 1 1 0 0 1 1 1. By how much does the result of this operation exceed 1 0 1 1 1 0?

2. How much is 1 1 0 1 0 1 1 − 1 1 0 1 1 0 + 1 1 0 1 − 1 1 0 1 0 1?

3. Carry out the following binary-decimal additions:

   a) 327 + 498
   b) 587 + 312
   c) 627 + 293.

4. Carry out the following binary-decimal subtractions:

   a) 837 − 354
   b) 926 − 287
   c) 435 − 312.

Chapter 9

LOCKING CIRCUITS

9.1 Purpose and design of locking circuits

It quite often happens in switching techniques that a number of devices must work with one central device. The latter may only be connected to one of the former at a time. An example of this is found in some telephone systems, where the number dialed by the person who is calling is stored in a recording circuit. The making of the connection takes relatively little time, so that *one* common device can perform this function for a *number* of recording circuits (the operation of recording circuits is dealt with in Chapter 11).

In this application of locking circuits, there is thus a *probability* that one of the peripheral devices will have to *wait* for a connection with the common device. It is then important that this probability should be distributed as evenly as possible over the various devices of the group, so that the principle "first come, first served" is observed as far as possible.

Another example of locking circuits is found in the distribution of calls over a number of devices. An example of this is the distribution of connecting devices among telephone subscribers who are calling up. If for example 100 subscribers must share 10 devices for making a connection, it is *desirable,* but not necessary, that these 10 devices should be uniformly loaded and thus called on in turn. If however the distribution of the calls over the 10-connecting devices is less "fair", this does not matter to the callers. It is only the consideration that a uniform loading of the 10 devices will increase the life of the telephone exchange which makes it attractive to distribute the calls at least relatively "fairly".

It thus appears from the above that the design of a locking circuit must take into consideration the function that this circuit must perform, viz:

a) if there is a "queue", the circuit must keep to the right order as far as possible;

b) if the circuit is used for distributing calls, the "fairness" of the distribution is not so important.

Let us suppose that several of the equivalent devices A–E of Fig. 153 want to be connected to device R. Only one connection may be made at a time, so that the other devices must wait. After the connection has served its purpose, one of the waiting devices may be connected with R, and so on. The device V placed between $A–E$ and R is a connecting circuit, which realizes the desired connection.

The connecting circuit contains equipment to arrange for the locking function, and equipment which makes the connection. The locking circuit VG need not be a part of V, but may be an independent device which gives V certain instructions to make connections. This is illustrated in Fig. 154. Locking circuits can be constructed with electromechanical or with electronic components.

In many cases it is possible to split the device V up into sections which can be placed in the devices $A–E$. This is often advantageous, because if e.g. two further devices F and G have to be added to the group, V is then automatically expanded to deal with the larger group. When studying the equipment it is then however difficult to determine which components look after the locking and which look after the coupling. A circuit based on the principle of Fig. 154, using a simple rotary selector as locking circuit, is shown in Fig. 155.

The rotary selector is started when one or more devices connect the start wire with earth by closing the contact unit s, which causes relay ST to operate. The selector rotates as a result of the energizing of the rotary magnet D, until the selector wiper makes contact via one of the contact units s of one of the devices $A–E$, as a result of which relay T can operate. A contact unit of relay T stops the selector by breaking the circuit for D, while a make contact unit of T will energize one of the relays $A–E$ via another

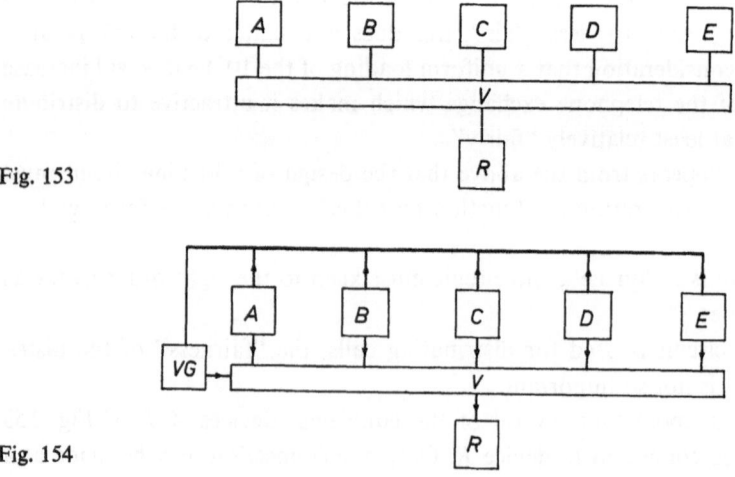

Fig. 153

Fig. 154

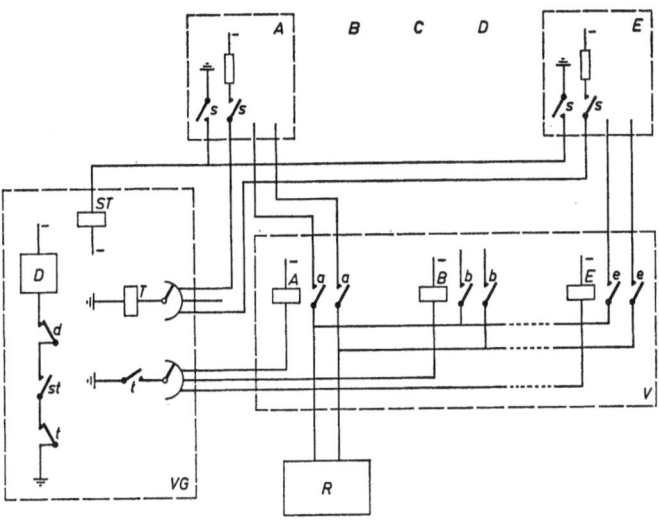

Fig. 155

arm of the selector. This relay will now establish the connection between R and the device in question (e.g. A). When the connection between R and A has fulfilled its purpose, contact units s, which held relays ST and T, are switched over and relay T can release. Relay ST will remain energized if one of the devices B–E still connects ST to earth. Relay A releases as a result of the releasing of relay T. The selector can now continue rotating until the next device is selected, and so on.

9.2 Locking circuits with relays

In the above example we used a rotary selector as a locking circuit. A locking circuit with relays must carry out the same function. There must thus be a line via which the various devices can announce that they want to be connected with the central device R. Let us call this line the "call line". There must also be a line by means of which the locking circuit indicates which device may make the connection with the central device R. Let us call this line the "select line". The call line and the select line are normally formed by a chain of relay contact units. Fig. 156 shows such a locking circuit for the devices A–E.

The relays A–E also act as auxiliary relays for the select line of Fig. 156. It will be clear that if device A has already "called up", devices B–E can no longer do this. Device A is thus selected. If however device E calls up,

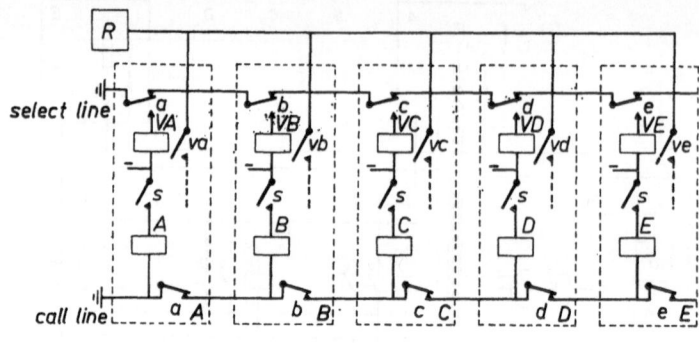

Fig. 156

devices A–D can still do so, and the select line to device E is then broken and replaced by e.g. one to the device B. This makes for irregular operation, and it is more difficult for the last device in the line to be connected than for the ones before it.

By giving the select line the "opposite" direction to the call line, we can ensure that a relay once energized stays energized until the connection has served its purpose, which prevents a connection from being broken off by a device which is earlier in the line. However, such a locking circuit is still not quite fair since a device which is earlier in the call line has a better probability to be selected. In the circuit of Fig. 157, during the time that e.g. device C is connected to the common device, devices D and E cannot call up, but devices A and B can. After C, first B and then A is selected. After A is finished with the common device, all the devices again get the chance to operate their call relay; but relay E will only be able to operate if it is faster than one of the relays A to D, relay D can only operate if it is faster than one of the relays A to C, and so on.

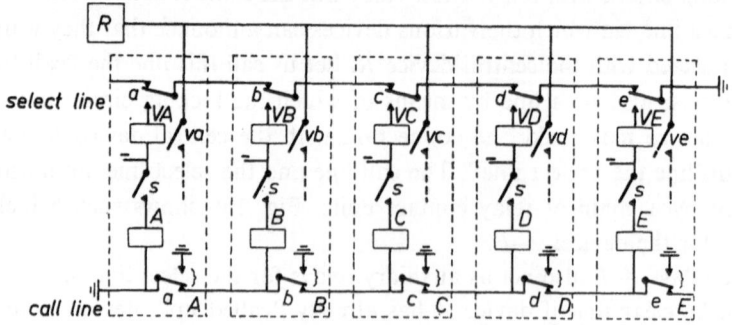

Fig. 157

This too is naturally a great disadvantage. It could happen that relay E never gets the chance to operate, because it is slower. Once a relay has operated, it is held by its own make-before-break contact unit. Relay E only gets the chance to operate for a very short time, while relay A has the chance for much longer, and will thus be used much more.

Fig. 158 shows a locking circuit in which many of the disadvantages mentioned above are no longer found. This will be clear from study of the sequence diagram of Fig. 158. The time during which the relays A to E can operate is increased by the time of one of the relays VA to VE. The relays VA to VE are used in the connecting circuit between devices A to E and the common device R. By making use of contact units va to ve in the call line, we no longer need to ensure that the call relay is held by one of its own contact units, since only one of the relays VA to VE can be energized so that no contact unit in the call line will open before the call relay in question releases.

In the examples considered, both the call line and the select line are series circuits. If many contact units are connected in series, there is a considerable chance of a poor contact, because the voltage drop across each contact unit is low. Important series circuits of this kind are therefore often executed in duplicate. Moreover, a differential relay is then connected in series with each series circuit to provide a check on the operation of both circuits; it is assumed that the two branches will not both break down at the same time. The operation of the differential relay is retarded, so that a difference in the closing of the contact units in the two circuits will not make it operate.

Fig. 159 shows an example of the select line of Fig. 158, executed in duplicate with a check relay. As soon as insulation occurs in one of the make or break contact units of one of these two lines, relay X operates and closes an alarm circuit. The locking circuit however continues to work in the normal way, as each branch is capable of maintaining the function of the circuit alone. Relay X does not indicate precisely where the defect is: this must be found by testing; but at least one knows that there is a defect somewhere.

9.3 Gate-lockout circuits

The locking circuits with relays described so far have the disadvantage that the devices near the beginning of the call line have more opportunity of calling up (and are thus used more) than the devices nearer the end. The latter devices must then also wait longer to make use of the central device.

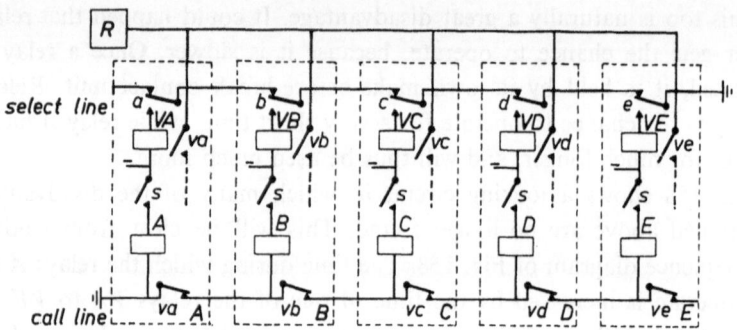

device *A* is finished

devices *A–E* can call up
(*C* and *E* do in fact call up)

device *E* is connected to *R*
device *E* is finished

device *C* is connected to *R*
(devices D and E cannot call up)
device *B* now calls up

device *C* is finished
device *B* is connected to *R*
(*C, D* and *E* cannot call up)

device *B* is finished

all devices can call up again

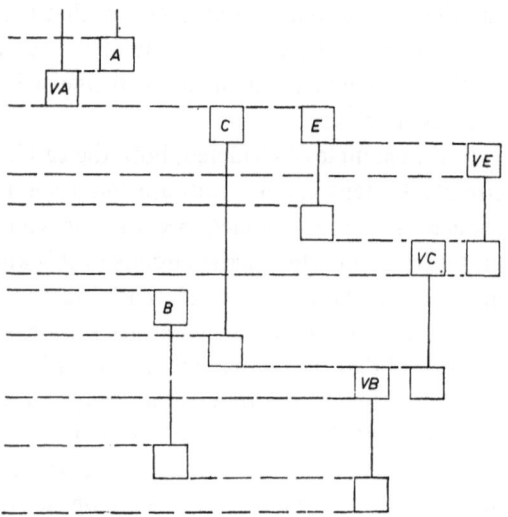

Fig. 158

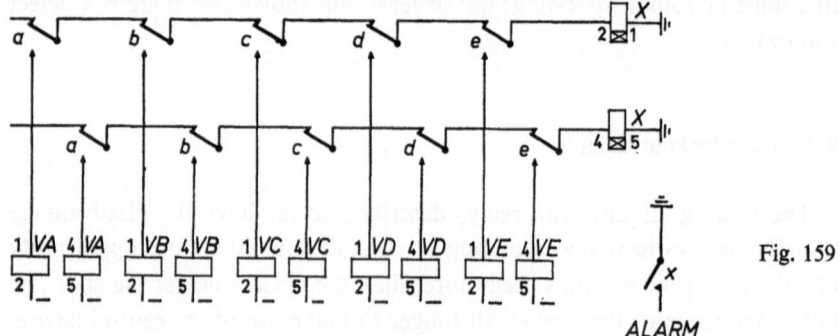

Fig. 159

In gate-lockout circuits, this "unfairness" is done away with. If one device calls up, all other devices are also given the opportunity to call up for a certain (short) period of time. The call line is then broken, so that no new calls are possible. The devices which have called up are then dealt with in numerical order. When all the devices which called up have been dealt with, the waiting devices are again given the opportunity of calling up. This manner of operation may be compared with that of a canal lock, where a number of ships can take their place before the lock gates are closed. Ships cannot now enter the lock until these ships have left via the second set of lock gates, and these have been closed again after them. Locking circuits like that of Fig. 160 are therefore called gate-lockout circuits.

In Fig. 160 the relays A to N are the call relays, and relays VA to VN are the coupling relays. A device which calls up closes the corresponding contact unit s, as a result of which the call relay operates if relay SL is not energized. When one or more of the relays A to N operate, relay SL is switched on. The operation of this relay is retarded, so that all the call relays could possibly operate during the operate time of SL. Once SL is energized, however, all the call-relay circuits are broken by the break contact units sl, except where the call relays have already operated: these call relays are held by their own make contact units. The closing of the make contact unit sl allows the relays VN to VA to be energized in succession, if the corresponding call relay is energized. When the last device has been dealt with, and the call relay which was energized for the longest time has released, relay SL will release again. The closing of the break contact units sl now allows the call relays to operate again. For the sake of simplicity, the select line in Fig. 160 is drawn as a single line, i.e. without check relay. The call line is no longer a series circuit, but has become a separate circuit for each device.

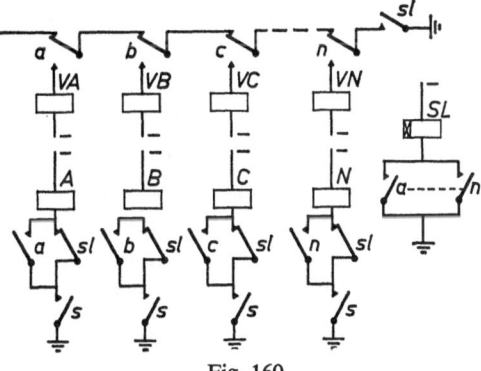

Fig. 160

Many variations of the circuit of Fig. 160 are possible. In Fig. 161, the relay *SL* is energized in the quiescent state. When one of the relays *A* to *N* operates, relay *SL* releases (retarded). During the release time, the other call relays still have the opportunity of operating, giving the lock effect again. The various devices are then selected in turn, until the last call relay releases and relay *SL* can operate again, etc. One disadvantage of the circuits of Fig. 160 and 161 is the large number of contact units *sl* needed.

Another type of gate-lockout circuit is shown in Fig. 162. In this circuit the various devices call up by closing the corresponding contact unit *s*, thus connecting the call relay to minus. If the locking circuit is not occupied at that moment (relay *SL* not energized), the device in question can operate

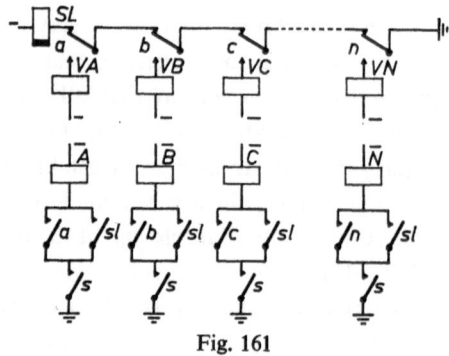

Fig. 161

via the closed break contact unit of *SL*. The call relay closes a hold circuit, so that it does not release when the contact unit *sl* opens. The contact units *a* to *n*, connected in parallel, energize the relay *SL*. Break contact units of this relay open the call line, so that no more call relays can operate. The devices are now dealt with in order, from *N* to *A*. When device *N* is finished, it opens its contact unit *s*, so that relay *N* releases. The following sneak circuit is however maintained as long as a call relay is energized: earth via contact unit *n*, *N*4–5, *N*2–1, e.g. *B*1–2, contact unit *s* of device *B* to minus. This circuit can hold relay *N* unless the winding 4–5 has just as many turns as winding 1–2, so that the relay is fully differential. This however can easily be seen to. We now have a gate-lockout circuit which only needs a few contact units *sl*. The relay *X* checks the lines for the coupling relays *VA* to *VN*. Further, relay *SL* is made with two windings, and for extra security a differential relay can also be included in series with the two windings of this relay.

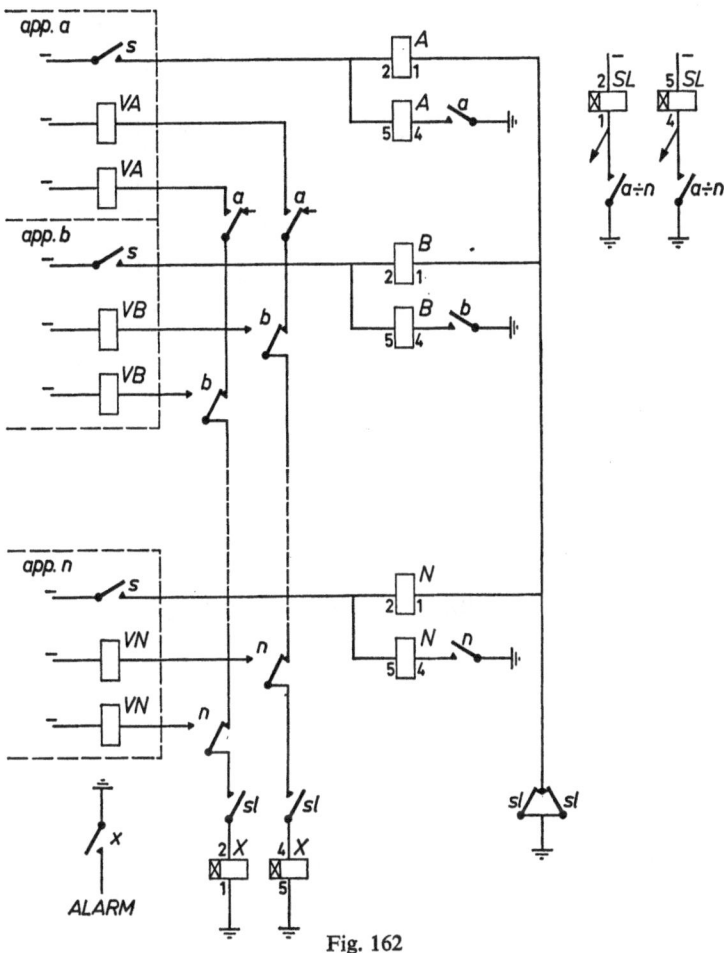

Fig. 162

With this gate-lockout circuit we have thus succeeded in making the call time of all the different devices the same. Because of the fixed position of the earth on the select line, there is however preference between the different devices as far as selection is concerned: in Fig. 162, device N (if it calls up together with other devices) will always be selected first. If however we arrange for the earth not to be at a fixed spot on the select line, but to move about, this preference for a particular device is also done away with. Fig. 163 shows a circuit in which earth is alternately at the beginning and the end of the call line, which is achieved with the aid of a binary counter element.

In Fig. 164 the earth is connected via a step by step selector switch, which moves on one step each time the locking circuit is filled.

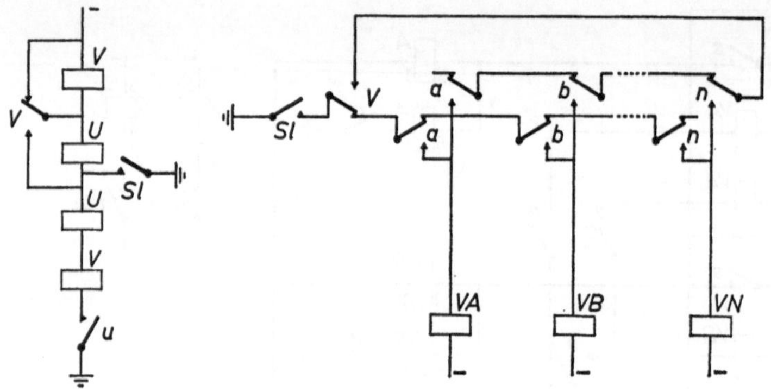

Fig. 163

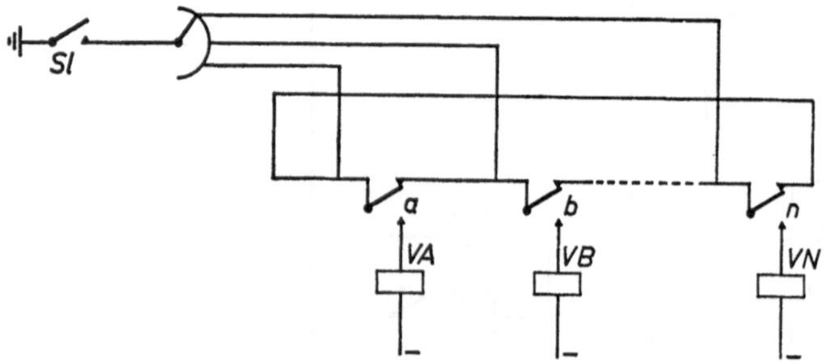

Fig. 164

9.4   Multiple locking circuits

If many devices have to work with one common device, long contact chains are needed if the solutions described above are used. This is a disadvantage of these circuits, as is the large number of call relays needed in this case. In order to prevent this, the chain can be divided into parts, each of which forms a separate locking circuit. These locking circuits however all make use of the same groups of relays, which makes for a considerable saving of relays.

Fig. 165 shows a multiple locking circuit for 100 devices, which are divided into 10 groups of 10. If one device of a given group of 10 calls up, the corresponding relay *TT* operates via make contact unit *tsl*. The operation of relay *TT* causes relay *TSL* to release (retarded), and one of the relays 1*T* to 10*T* to operate. This circuit also exhibits the above-mentioned lock effect, since the releasing of relay *TSL* will prevent any more of the relays 1*TT* to 10*TT* from operating. Depending on whether one or more devices from a given group call up at the same time, the operation of one of the relays *T* will cause the operation of one or more relays *EH*. The relays *EH* will

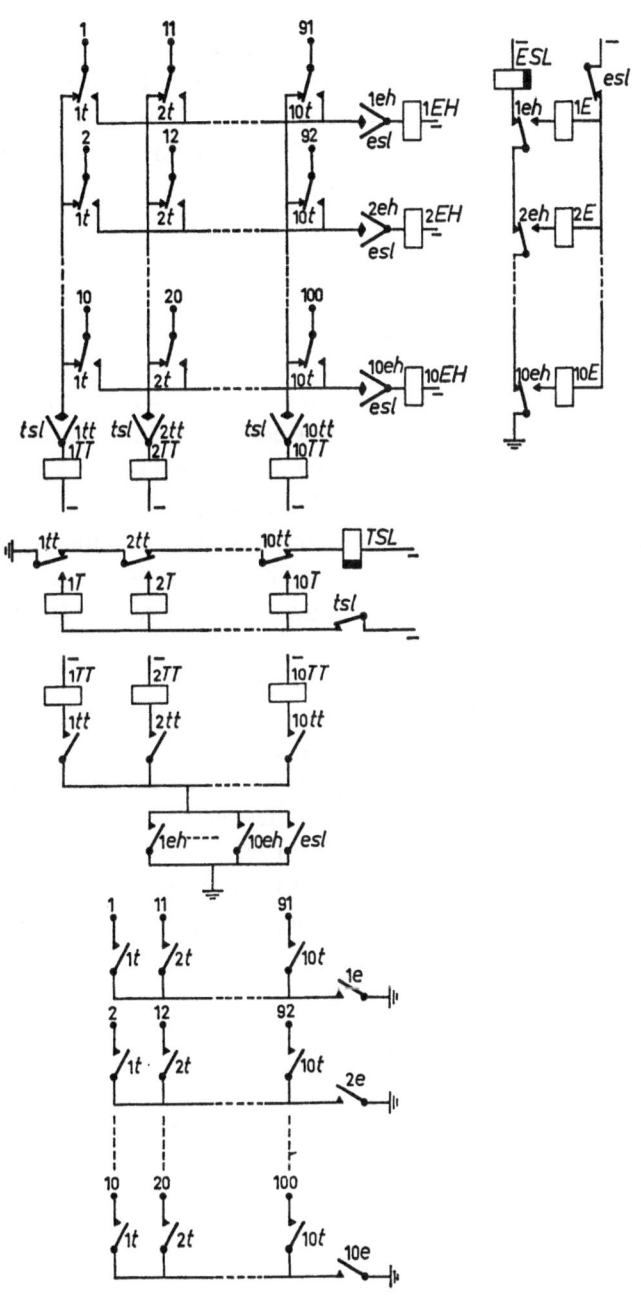

Fig. 165

(after the retarded release of *ESL*) cause one of the relays 1*E* to 10*E* to operate via a break contact unit *esl*, according to the normal lock effect. The operation of the relay *T* will cause the relay *TT* to tend to release. To prevent this, a hold circuit is formed by make contact units *eh* and *esl* in parallel.

There is now only one of the relays *T* and one of the relays *E* energized. The "select line" is here in the form of a matrix network, which allows the appropriate one of the 100 devices to be selected by these two relays. We thus obtain in this way a circuit without too many contact units in series, and where the number of relays is not proportional to the number of devices.

9.5 Locking by potential changes

In Section 1.7 we discussed various kinds of selectors. Those selectors which are not operated by pulses must be stopped immediately the rotary selector wiper reaches a marked contact. Moreover a second selector (which may be rotating at the same moment) must be prevented from stopping at the same contact. In order to determine whether a selector has reached a given contact, and if so to annul the probability of the testing of a second device as quickly as possible, use is made of a test circuit or locking circuit. We shall be returning to this below.

If despite these measures two devices should test at the same point, we speak of a "double test". Various measures can be taken to make the chance of a double test as small as possible. If as in Fig. 166 the test time is a s, and the first device begins testing at time t_1, then the test will be over at time $t_1 + a$. A second device, starting to test at time t_2, will finish at time $t_2 + a$. There is now a certain time b in which both devices test. If it is possible to reduce the test time a, then the probability that t_2 will fall within the time interval t_1 to $t_1 + a$ is smaller, and hence the probability of

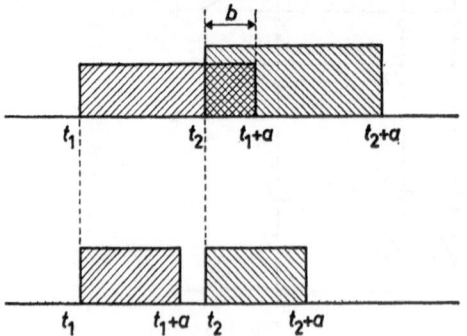

Fig. 166

Fig. 167

a double test is also smaller. The reduction of the test time is thus a means of reducing the probability of a double test (see Fig. 167).

Testing is normally done to earth or to a voltage point. However, one also sometimes tests for the *absence* of e.g. earth. This is called testing with a negative criterion but in telephony is referred to as "earth testing". The danger of such a test method is that a poor contact or a broken wire can cause a double test.

An example of an earth testing is shown in Fig. 168. A selector wiper rotates over a contact bank, where the busied contacts are connected to earth. The selector wiper must touch two contacts simultaneously when rotating, as otherwise it will find no earth between two contacts. One then speaks of a "bridging" selector wiper.

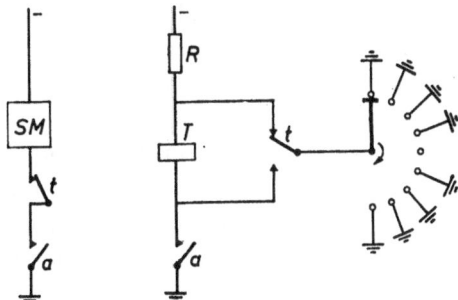

Fig. 168

The selector starts rotating as soon as relay *A* operates, while relay *T* is also prepared for operation. As soon as the selector wiper passes the contact which is not connected to earth, the short-circuit via the relay *T* is broken, and this relay operates. The operation of relay *T* switches off the selector magnet, and at the same time the selector contact where one is testing is connected to earth via contact units *t* and *a*. This prevents a second selector from testing at this point. If however two selectors arrive at the free output simultaneously and the test relays *T* operate practically simultaneously, then a double test is produced. The probability of a double test can be reduced by making the operate time of the relays *T* small. For this reason, high-speed relays are often used for testing.

With high-speed selectors, it is also necessary to make the test time very short in order to allow the selector to stop in time at the marked contact. In this case too we make use of a high-speed relay with one change-over contact unit. Fig. 169 shows a test circuit making use of a high-speed relay.

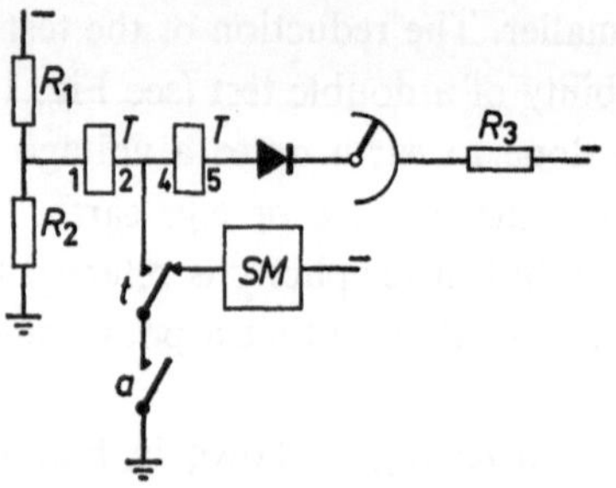

Fig. 169

The resistance of winding 1–2 of relay T is high compared to that of T4–5. The winding is connected to a tapping of a potentiometer consisting of the resistances R_1 and R_2. As soon as the marked contact is found, relay T will operate and connect the two T windings to earth via contact unit t. This causes winding T1–2 to be magnetized in the opposite direction. The winding T4–5 (low-ohmic) receives a higher energization. The voltage of the selector wiper at the moment when contact unit t is switched over is determined by the low-ohmic winding T4–5, the diode and the resistance R_3, the total being such that relay T has enough AT to hold. When a double test occurs, the voltage drop produced by *two* currents in the resistance R_3 is so great that the winding T4–5 does not have enough AT to hold the relay T (since the AT of winding T1–2 are opposed to those of winding T4–5). When relay T releases, the selector rotates further. This circuit has certain drawbacks. The full battery voltage cannot be used for the test, and a high resistance of the test relay and thus also a large number of turns makes the operate time longer. A test relay with only one winding with a low resistance and relatively few turns is much faster. A circuit using such a relay is shown in Fig. 170. Here again use is made of a potentiometer. Directly after the test at the marked contact the potential of point a becomes lower than that of point c, which makes it impossible for another test circuit to test.

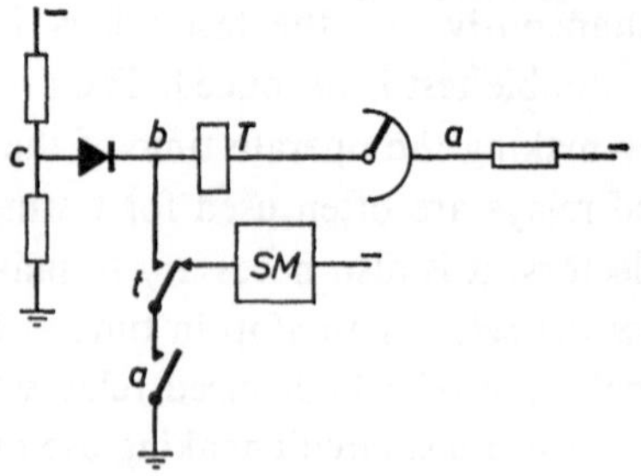

Fig. 170

The diode prevents a coupling between points b and c after the test relay has operated. If faster test methods have to be used, one makes use of an electronic circuit.

9.6 Combination of a locking circuit and a check circuit

It sometimes happens that the combination of a locking circuit and a check circuit can perform a useful function in a control circuit.

By way of example we shall consider here the remote control of high-tension transformers. Five transformers are available for the energy supply, of which only three may be switched on at a given moment. The two others act as a stand-by. The locking circuit indicates which transformers are to be switched on, and the check circuit prevents more than 3 transformers from being switched on. If a transformer is not switched on despite an instruction to this effect, an alarm signal is given indicating the transformer at fault. The selection of the transformers to be switched on must be fully automatic, and they must be switched on in cyclic succession.

Fig. 171 shows a circuit for carrying out this task.

In the quiescent state of the circuit relay G is switched on, as a result of which a circuit is also closed for relay V. The relays K_1 to K_5 are switched on via the break contact units of relay S. As a result of this relay S also operates. Relays K_1 to K_5 remain switched on via hold contact units. Relay G releases as a result of the opening of contact unit s. After relay G has released, a circuit is closed via contact unit k_1 and one wire of the cable to transformer station 1, where relay T_1 operates. As a result of this, transformer 1 is switched on, and as soon as this has been accomplished the contact unit a_1 is closed at station 1. This thus provides a confirmation signal, which switches on the relay B_1 via one wire of the cable. A contact unit of relay B_1 is used to close a hold circuit for relay T_1. The operation of relay B_1 breaks the hold circuit of relay K_1, which releases (retarded), after which a circuit is closed for relay T_2 at station 2. The switching on of the transformer and the giving of the confirmation signal are carried out as described for station 1. After the third transformer has been switched on in the same way, the relays B_1, B_2 and B_3 are all energized, as a result of which the relay G operates via the check circuit formed by the contacts of relays B_1 to B_5. It is now impossible to switch on another station. Until relay G operates, the pilot light LF burns, as a sign that less than three transformers are switched on.

Relay V, which was switched on during the quiescent state of the circuit

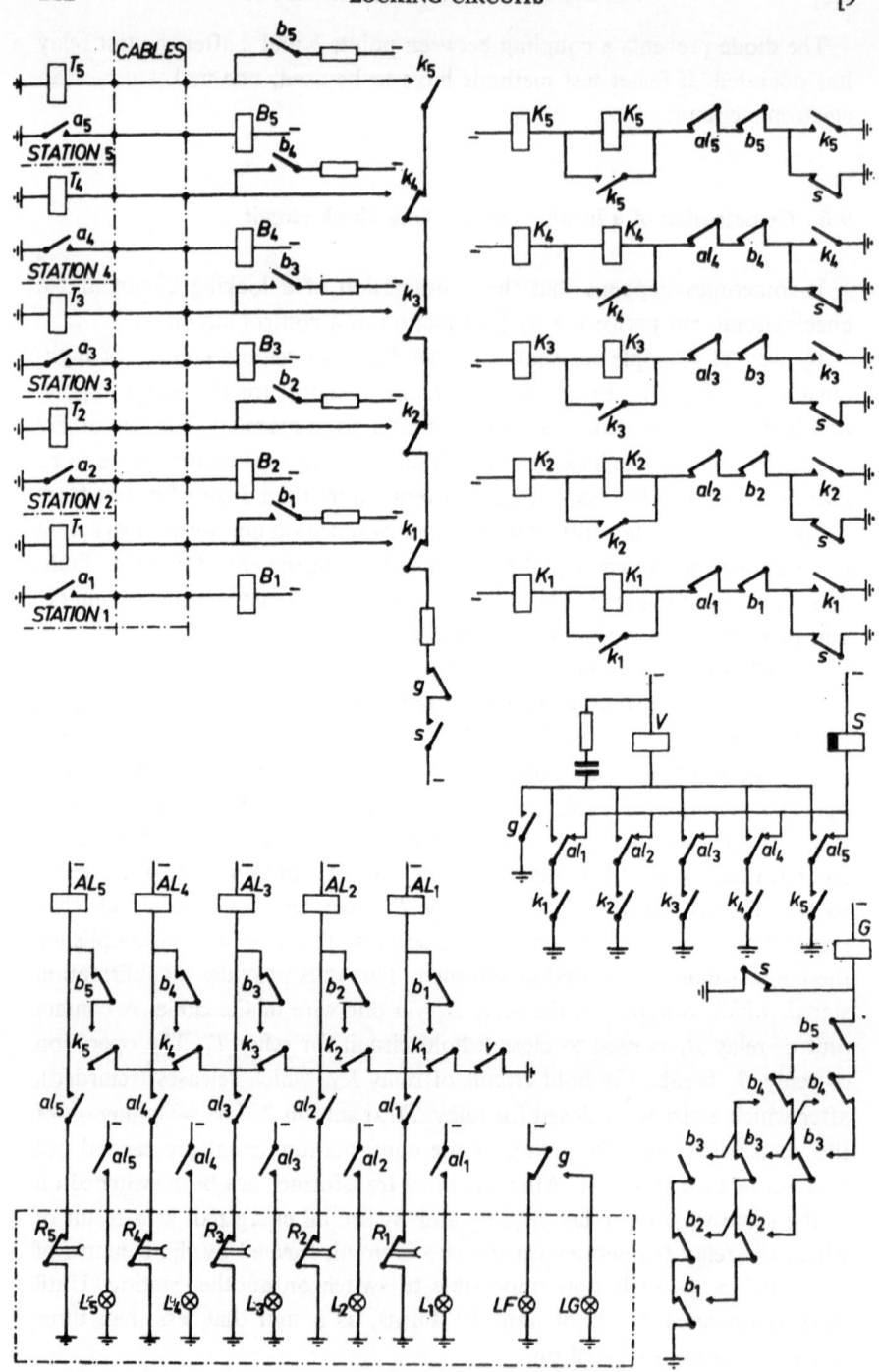

Fig. 171

and was switched off by the releasing of relay G, does not release during the selection of the three stations, because a capacitor is connected in parallel with it, giving a release time of e.g. 5 s. Since the switching on of the stations takes much less time than this, relay V can be switched on again before it releases, by means of a make contact unit of relay G. The normal operating state has now been reached.

If one of the selected transformers falls out of operation, the corresponding contact unit a opens, as a result of which relay B releases and the check circuit indicates that less than three transformers are switched on. This indication is caused by the releasing of relay G, which also initiates the selection of the fourth transformer. As soon as confirmation of this has been received by operation of relay B_4, relay G is switched on again via the check circuit.

The relays K_1 to K_5 operate fast and release slowly. The latter is necessary to ensure that contact unit g can open the select line of contact units k_1–k_5 (which are connected in series) before yet another transformer is switched on as a result of the releasing of a relay K.

The releasing of relays K_1 to K_5 is retarded by short-circuiting one of the windings of the relay in question by means of a contact unit k.

If relay G is not switched on, indicating that not enough transformers are on, and as a result of this the relay T of one of the stations is switched on, it is expected that this will have the effect of switching on the corresponding transformer. If no confirmation signal (operation of relay B) is received, as a result of a fault in the line or something wrong in the circuit for switching the transformer on, then relay V releases after about 5 s, switching on an alarm relay AL which indicates the faulty station. This relay is held via a hold contact unit al and a push-button R. The station in question is indicated by the corresponding lamp L. Meanwhile, the operation of relay AL causes the corresponding relay K to be switched off, so that the selection circuit is switched one stage further and the next station is selected. Before relay K releases (retarded), relay V is also switched on again. After relay G has operated again, relay V is held via a contact unit g.

If the last relay K releases as a result of the switching on of the last transformer, relay S also releases, as a result of which the relays K of the stations which are not in use operate and these stations are again ready as stand-by.

In the above an alarm is given if an attempt to switch a transformer on fails. During the operation of a transformer the corresponding contact unit a remains closed. If the transformer falls out of operation, it is true that the next transformer is switched on, but no indication is given of this. If the breakdown of a transformer is not already indicated in any other way, the circuit of Fig. 171 can be supplemented by a suitable alarm circuit for this purpose. This circuit is shown in Fig. 172. This extra alarm signal can be obtained by the addition of five relays. As soon as relay B operates as a sign that a transformer has been switched on, the corresponding relay U is immediately switched on. This relay is held by a hold contact unit and an push-button D. Now as soon as the breakdown of a transformer causes the corresponding relay B to release, the corresponding alarm lamp UL is switched on.

In the examples given in Fig. 171 and 172 the locking circuit and the check circuit cooperate to the full. In practice, however, the switching on of the

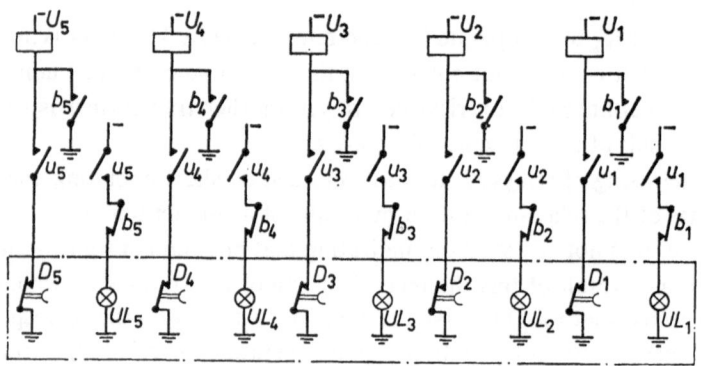

Fig. 172

transformers as realized in this circuit is still not reliable enough, because there is insufficient check on the intactness of the line.

Let us suppose for example that the wire connecting contact unit a and relay B is broken; then when the transformer in question is switched on, no confirmation signal will be given. As a result of this a fourth transformer will be switched on, which is against the rules. For this reason, a quiescent-current check is nearly always carried out on the various wires in the cable. This check can be realized by means of an extra check relay. Fig. 173 shows the check circuit for station 1. As long as the transformer in question is not being used, relays D_1 and E_1 are connected to the wires of the cable. These relays have a high resistance (e.g. 20 000 ohm), so that relay T_1 cannot operate in series with relay D_1. If the line is broken, relay D_1 and/or relay E_1 releases, as a result of which the alarm relay AL_1 is switched on via an inter-rupter, which closes once in half a second.

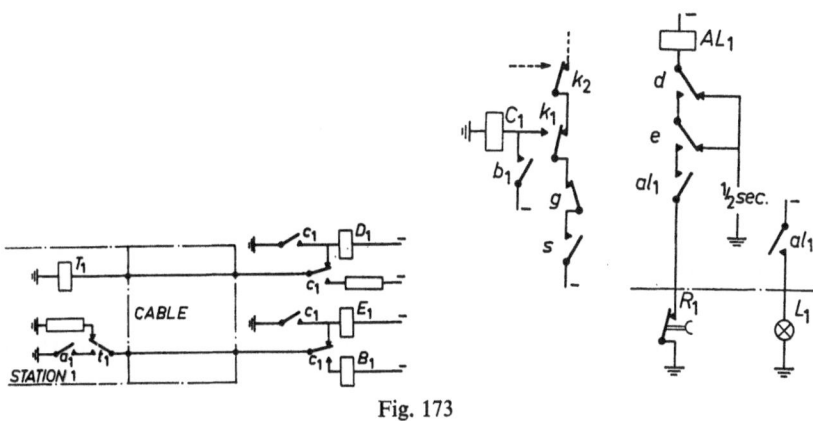

Fig. 173

Relay K_1 is switched off via a contact unit of AL_1, so that station 1 can no longer be selected. Thanks to the regular switching on and off of relay AL_1, the lamp L_1 flashes, thus indicating that something has gone wrong in the line to station 1. This thus allows a clear distinction to be made between the case that the instruction to switch on has no effect (see Fig. 171) and a break in the line.

If station 1 has to be switched on, relay C_1 is switched on via the select line. As a result of this, the check relays D and E are kept switched on and a low resistance is connected to the cable wire to T_1, so that this relay operates. In the station the high resistance, via which relay B cannot operate, is switched off by a contact unit of relay T_1. Only if the transformer is correctly switched on by relay T_1 and contact unit a is closed is a circuit formed for relay B_1. As soon as relay B_1 has operated, relay C is switched on by a contact unit b_1. For the rest, the operation of this circuit is identical with that of Fig. 171.

9.7 Problems

1. At four places A, B, C and D a transmitter is installed which can transmit information by means of punched tape to a central receiver.

   By each transmitter there is a non-locking button T by means of which a call sign may be given. When the tape has been inserted in the transmitter, a make contact unit z is closed. The transmitter is started by switching on an electromagnet ZM as soon as the receiver is available.

   a) What type of locking circuit comes into consideration in this case?

   b) Draw a complete circuit.

2. Three devices A, B or C can be connected to a device D by closing a contact unit a, b and c respectively.

In order to prevent a double test, a cyclic counter for 3 positions ensures that at any given moment only one of the devices A, B and C can be connected to D.

The transfer of information between A, B or C and D takes place via two wires.

Design the locking circuit and explain its operation.

3. Design a locking circuit for connecting the devices A, B or C to a device D under the following conditions:

   a) If A, B and C request to be connected with D simultaneously, the connections $A-D$, $B-D$ and $C-D$ are made in that order.

   b) If while one of A, B and C is connected to D the two others simultaneously request to be connected, the order given in a above must be adhered to.

   c) How would you design the circuit if the order is not important and A has priority over B and C, and B over C?

4. A relay SL occurs in Fig. 162.

   a) Why is the operation of this relay retarded?

   b) Why are two windings applied?

   c) How could you supplement this circuit to check whether both windings of relay SL are in good order?

5. Why is the operation of relay X in Fig. 162 retarded?

6. In the circuit of Fig. 162 both windings of relays A, B, ..., N have the same number of turns.
   Why?

7. In the circuit shown below, the voltage is 48V, the current through relay T_1 before it operates is 40 mA, and after it operates 60 mA (selector K_1 connected to point C).

   a) What is the value of the resistances R_1 and R_2?

   b) What is the voltage of point C with respect to earth after relay T_1 has operated?

c) If relay T_1 has operated and selector K_2 is set to the same point, what is then the voltage difference between points M and C?

d) What current then flows through relay T_2?

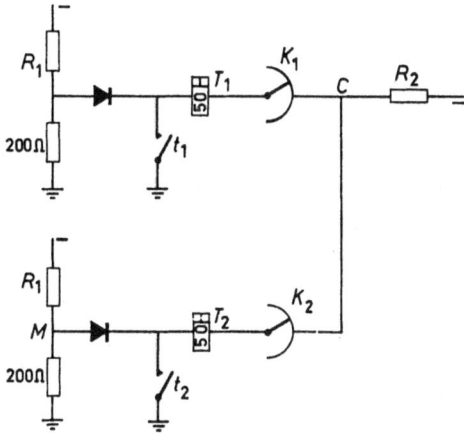

Chapter 10

CONNECTING CIRCUITS

10.1 Aim and design

The purpose of a connecting circuit is to connect one device of a certain sort with another device of a different sort. A selector, together with the relays which operate it, is in fact a connecting circuit. In this case a locking circuit is used to stop the selector at the right time, as described e.g. in Fig. 155 of Chapter 9.

Selectors can be used if the number of wires connecting the two devices in question is not too large; if a large number of connecting wires are required, relays have to be used.

The choice between selectors and relays is also determined by the time available for connecting the two devices. Suppose that in the circuit of Fig. 174 the connection between the device R and one of the group "A" of devices need only last 500 ms, while the mean period of rotation of the selector is also 500 ms, then the efficiency of device R will be reduced to 50% by this cause alone. If relays are used in this case, the switching time is only about 20 ms, so the operating time is only reduced by 4% in this case.

In the example considered above the "hold time" (i.e. the duration of the connection) is short. In some cases, a certain number of the connecting wires are in use for a short time, and the rest for a much longer time. In this case we can with advantage make use of relays for the short hold times and of a selector for the long hold times. We shall be returning to this in Section 10.2.

If selectors are used for connecting devices from two different groups (A and B) as shown in Fig. 175, then the contact wipers of the selectors are generally connected to the smaller group.

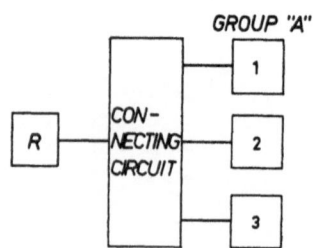

Fig. 174

In Fig. 175, thus, $m < n$. If one of the devices of group A requests to be connected, the selector (belonging to this device) is switched on and tests for a free device in group B. This test method has already been dealt with in Chapter 9 (Figures 169–170).

Device A is in this case "active". If however the request for a connection comes from a device in group B, then one of the devices of group A is switched on via a locking circuit.

At the same time in device B the "test voltage" is applied to the appropriate selector outputs, so that the selector of the device (A) which is switched on can test. Since in this case the initiative comes from device B, this device is "active" here.

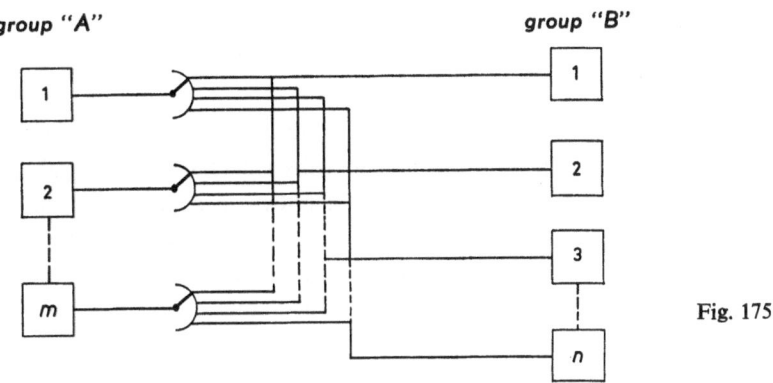

Fig. 175

There are various kinds of connecting circuits. As first we may mention the type which can connect *any* device of group A with *any* device of group B, and can continue to make connections as long as there are free devices in both groups. This type of connecting circuit can thus make as many connections as there are devices in the smaller group. The principle of such a circuit is sketched in Fig. 176. Devices 1 and 2 of group A can be connected to all devices in group B. The number of connections available is equal to the number of devices in the smaller group. In the following we shall call each contact unit which effects a connection between a horizontal and a vertical line a point of intersection. In Fig. 176 the number of points of intersection is $2 \times 3 = 6$. In Fig. 177 the contact units are omitted.

The minimum number of points of intersection is in general $m \times n$, where m and n are the number of devices in the two groups. In this kind of connecting circuits there is *no* internal blocking, i.e. as long as one device is free in the smaller group a connection can still be made with the other group, because there is still one connecting line free.

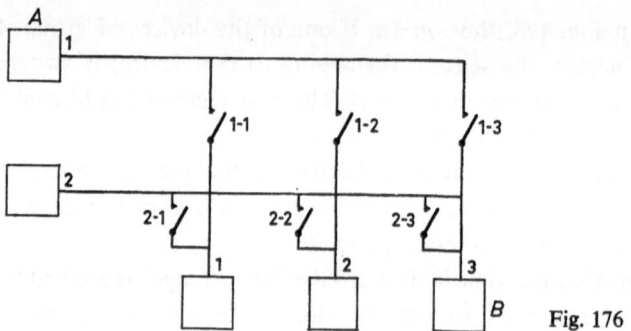

Fig. 176

This last condition is often not necessary. This means that the number of connecting lines is smaller than the number of devices in the smaller group. The possibility then exists that there are free devices in both groups, but that there are no more connecting lines available. One then speaks of internal blocking.

A practical example of the possibility of internal blocking is a telephone exchange. An exchange with e.g. 10 000 subscribers may have only 300 connecting lines. This means that 600 subscribers can be talking at any given moment. There is at that moment no possibility for any of the remaning 9400 subscribers to be connected, even if the party he wishes to call is free. It would be uneconomical to instal 5000 connection lines, as calculation and experience have shown that 300 connection lines are sufficient in certain cases. Internal blocking will then occur only rarely (e.g. once in 100 calls).

This situation is shown in Fig. 178 and 179. Here there are only 3 connecting lines, and each connecting line can make only one connection, so

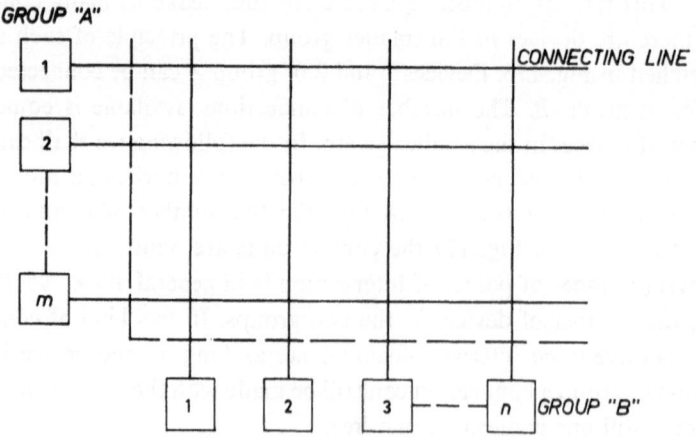

Fig. 177

that only 3 connections are possible at any given moment. The number of points of intersection is here $(m+n)v$, where v is the number of connecting lines. A connection between device 1 of group A and device N of group B can be made through any one of the connecting lines 1, 2 or 3.

The connecting circuits discussed so far all consist of a single stage. Multi-stage connecting circuits are also quite possible. All telephone exchanges have connecting circuits which consist of several states connected in series. This is illustrated in Fig. 180, where selectors (finders) are used as connecting units for the sake of simplicity.

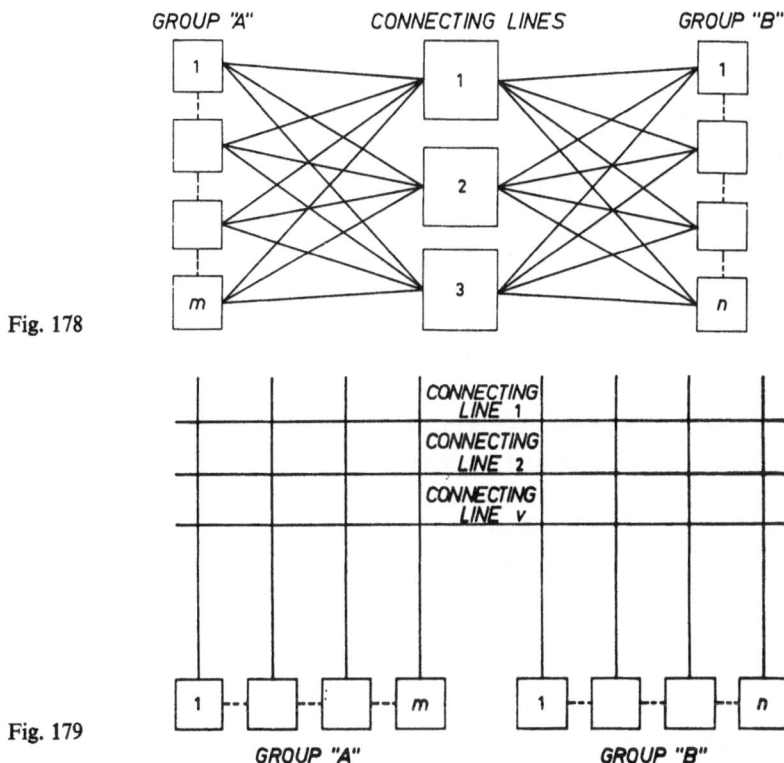

Fig. 178

Fig. 179

A thousand telephone connections are divided into 10 groups of 100. Each group of 100 subscribers has 10 finder switches Z_1 at its disposal. The 10 groups of finders Z_1 are connected to the 60 finders Z_2 (which are connected in parallel), and each finder Z_2 is connected to one connecting circuit V.

There is a successive reduction of the numbers involved at each stage (1000 telephone lines, 100 finders Z_1, 60 finders Z_2 and 60 possible con-

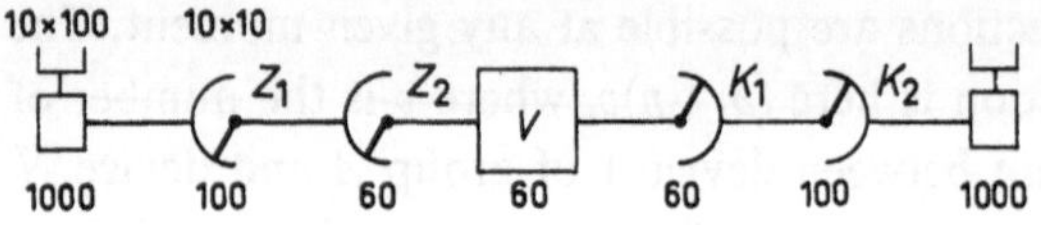

Fig. 180

nections) 10% of 100 subscribers and only 6% of 1000 subscribers can be connected as a caller at any given moment. The dialing of the first figure of a 3-figure number causes the corresponding selector K_1 to be connected to one of the selectors K_2 belonging to the desired group of 100 subscribers. The dialing of the last two figures causes the selector K_2 to be connected to the desired subscriber.

10.2 Combined connection circuits

In cases where a large amount of information has to pass between two devices in a short space of time, so that many connecting wires are necessary, but only a few wires need to be connected for a long time, it may be convenient to use a connecting circuit with relays for connecting the large number of wires for a short time, and a selector for connecting the small number of wires for a long time. The principle of this circuit is shown in Fig. 181.

If a connection is needed between a device A and a device B, the selector of the device A in question is connected to an output leading to a free device of group B. The long-term connection is now realized via selector arms a, b and c. Now a number of wires from the device A in question must be connected for a short time with the device B selected. For this purpose the contact unit b in A is closed, which causes a call relay K_1-K_m in a locking circuit to operate. As a result of this the "lock" relay SL operates. Relay V of the device A involved is switched on by means of a contact unit s_1. Relay Z in the desired device B is then switched on via selector wiper d. This thus produces a connection via contact units v and z. This connection is represented by only one wire in Fig. 181, but further connections can be realized by means of other contact units of relays V and Z. If a larger number of connections have to be made, relays V and Z have to be provided with "follow relays". As soon as the information has been transmitted from A to B via the relay connection, relay B releases, as a result of which the call relay (K_1-K_m) is also switched off. Relay V releases, and so therefore does relay Z. The relay connecting unit is now free for use by other devices. The connection via selector arms a, b and c remains intact.

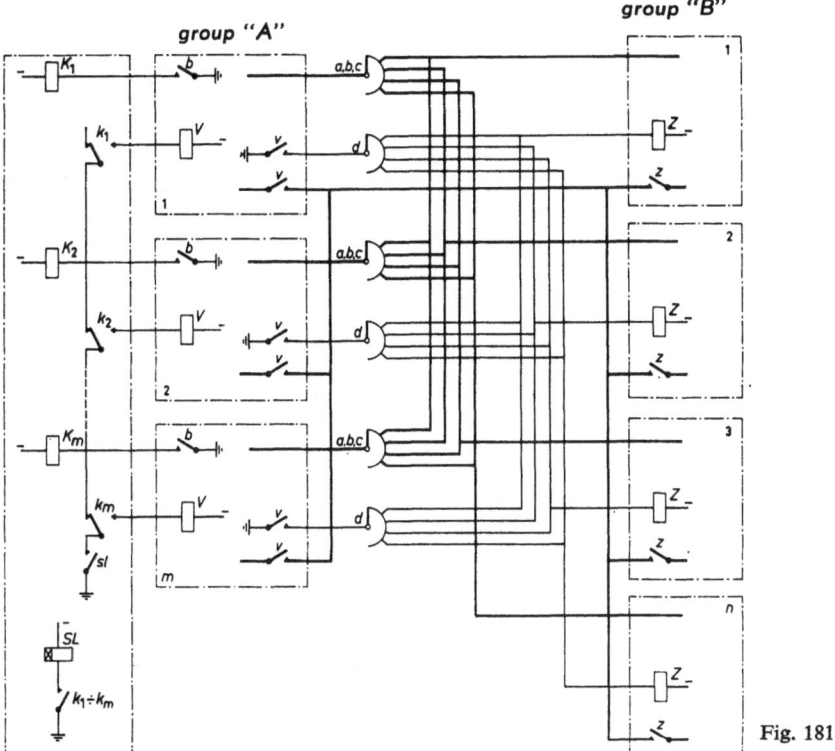

Fig. 181

10.3 Example of full connection possibilities

Fig. 182 shows in more or less complete detail a circuit for connecting any of 3 devices A with any one of 9 devices B, using a relay connecting circuit. In this example four connecting wires are drawn. The gate-lockout circuits A and B are used as auxiliary equipment; we could call the locking circuit belonging to the devices A the "allotter". Free devices A "present themselves" at the allotter, and as soon as a device B calls up it is connected to the selected device A; the allotter has already energized the corresponding relay V. The locking circuit A ensures that only one relay V is energized. If a device B wishes to be connected to the device which is already connected, B has to "enter" the "lock" B. The lock ensures that if several devices B call up at the same time, only one is connected. The other devices B are then connected in turn. As soon as the first one is connected, the response of the lock B causes a relay K to operate, connecting the selected devices A and B. Suppose that $B(1)$ and $A(2)$ are to be connected, then relay $2K1$

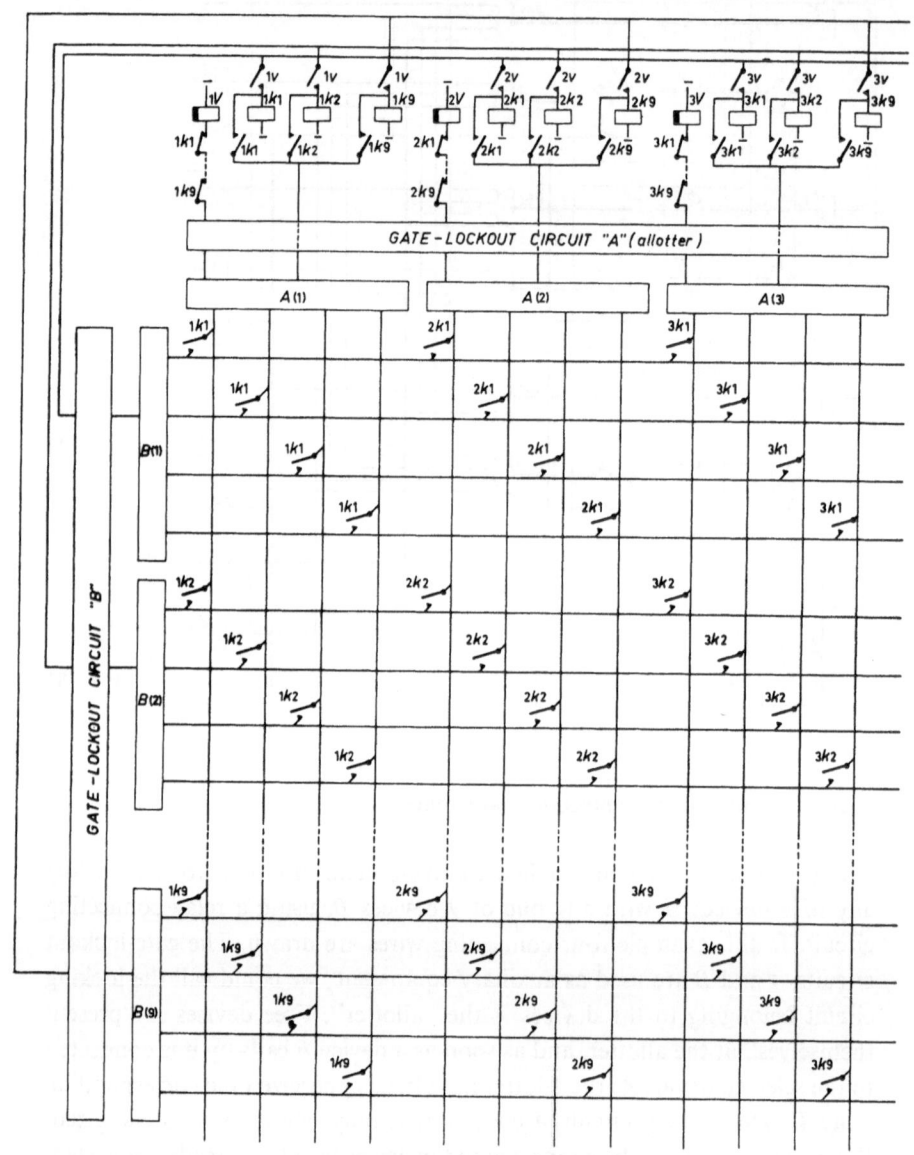

Fig. 182

operates (because $2V$ is energized). This relay will hold itself via a hold contact unit connected to device $A(2)$. When the connection between $B(1)$ and $A(2)$ can be broken, $A(2)$ will cut the hold circuit off from earth, so that $2K1$ can release again. Contacts of $2K1$ have effected the connection between $A(2)$ and $B(1)$ via 4 wires. The relay $2V$ releases as a result of the operation of $2K1$, and hence (not shown in Fig. 182) the allotter will be able to energize another relay V and thus select the device A to be connected next time. Device $B(1)$ also leaves the lock, so that other devices B can call up. In this way, it is possible to make three connections simultaneously with this connecting circuit, e.g. $A(2)$ to $B(1)$ via relay $2K1$, $A(1)$ to $B(9)$ via relay $1K9$ and $A(3)$ to $B(2)$ via relay $3K2$. If the number of connecting wires is insufficient, it is possible to make other relays follow the relay K in question, so that many more connections can be made. The number of devices in the two groups A and B can also be increased *ad lib*. However, if the number of devices becomes large, it is worth while considering whether a connecting circuit with e.g. selectors might not do the job better. Another possibility, as we have seen, is a combination of relays and selectors. No fixed rule can be given; the best solution must be determined for each individual case.

10.4 Problems

1. How many "points of intersection" are needed to connect 8 devices A to 9 devices B.

   a) with 3 connecting lines?

   b) with 8 connecting lines?

   c) What is internal blocking precisely? Give an example.

2. An installation involves connecting 4 devices A with 9 devices B. For reasons of probability, 2 connecting lines are just insufficient. Which of the remaining possibilities do you choose and why?

3. It must be possible to connect each one of a group A of 20 devices with each one of a group B of 50 devices.

   a) Give the *minimum* number of points of intersection for 1, 3, 4, 8, 12, 15 and 20 connecting lines.

   b) In which cases is the formula $m \times n$ used, and in which cases the formula $(m+n)v$?

4. A connecting circuit takes care of the connections between a group of 5 devices and a group of 10 devices.

   The following conditions must be satisfied:

   1. each device of the one group must be able to be connected to each device of the other group.

   2. Only 4 connections are needed at any one time.

      a) What type of connecting circuit do you choose?

      b) Would you choose the same method if only 3 connections were needed at any given time?

      c) What have you achieved by a correct choice of circuit according to question *a*?

Chapter 11

REGISTER CIRCUITS

The purpose of register circuits is to store one or more items of information (digits, letters, etc.) for a long or short period of time, and then to make this information available in the same or in a translated form.

Various means may be used to register information, e.g. relays or capacitors. Each item of information can be made up of one or more signal units. For example, a digit can be represented by a corresponding number of pulses, but can also be given directly in binary or some other coded form. There are also three different possible methods of reception, viz:

a) the signals for all items of information are received and recorded simultaneously (see Fig. 183*a*).

b) The different items of information are received in succession, but the signals of each item simultaneously (see Fig. 183*b*).

c) All the signals are received in succession (see Fig. 183*c*).

In case *a*) the information is received via 16 wires and registered directly. In case *b*) the different items of information are received one after the other via 4 wires, so that we need a distributor to connect the information source

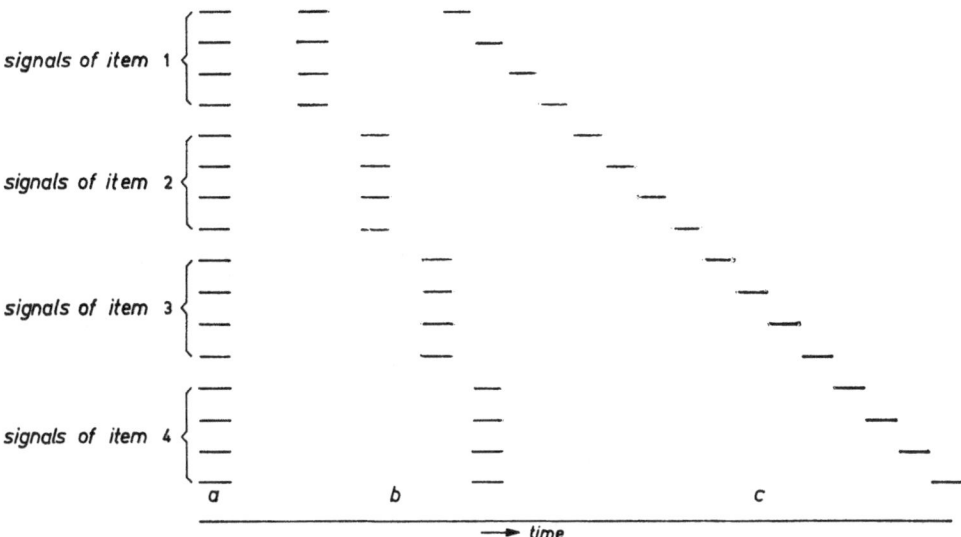

Fig. 183

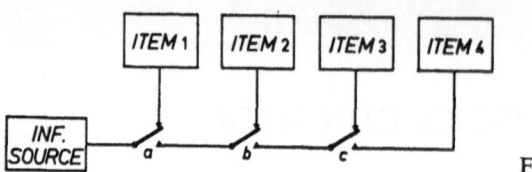

Fig. 184

to the different groups of register elements in turn. This is sketched in Fig. 184. The information source is first connected to the register elements for the first item of information. As soon as this has been received, the source is connected to the register elements for the second item of information by means of contact unit *a*. Contact elements *b* and *c* switch in a similar way to the recording elements for the third and fourth items of information. We shall be discussing the realization of this principle below.

In case *c*), all the information can be received via one wire. If the information is represented by pulses equal in number to the value of the digit to be transmitted and one wishes to store the information in coded form, one must make use of a counter circuit. Moreover, the pulse trains corresponding to the successive digits must be clearly distinguished from one another. This is possible by leaving a suitable interval of time between successive pulse trains. The basic principle of this method is shown in Fig. 185.

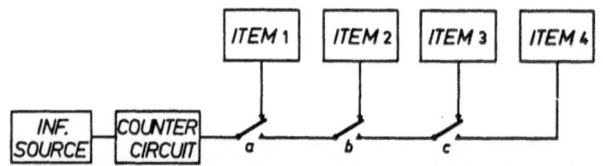

Fig. 185

11.1 Simultaneous transmission of signals for one item of information

Fig. 186 shows an arrangement for the simultaneous transmission of up to 4 signals, which thus allow the item of information to assume 15 different values.

The information source is formed by the contact units *a*, *b*, *c* and *d*, which can be closed in the above-mentioned 15 combinations.

The corresponding register relays *E*, *F*, *G* and *H* then operate, together with the relay *P* in series with them. Contact unit p^1 locks the switching-on circuit, while contact unit p^2 switches on relay *R*.

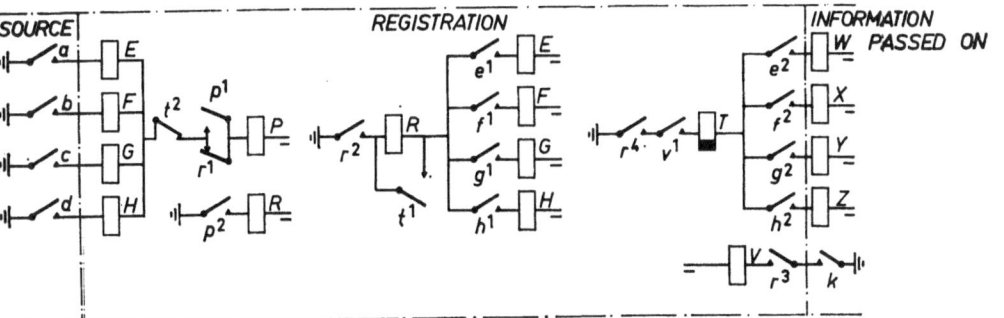

Fig. 186

After the information has been received and contact units a, b, c and d are open again, relay P releases. The register relays E, F, G and H, which are switched on, are held via contact units e^1, f^1, g^1 and h^1 respectively; relay R also remains switched on. After relay P has released, the information source is no longer connected to the register relays. The information stored in this way can be kept for an unlimited length of time. If the information has to be used, it can be requested by closing contact unit k, as a result of which relay V operates.

The relays W, X, Y and Z of a device to be controlled by the information receive the data via contact units e^2, f^2, g^2 and h^2, and thus assume the same positions as relays E, F, G and H. Relay T is also switched on, as a result of which relay R is short-circuited by contact unit t^1. After the opening of contact unit r^2, the recorded information is "erased" by the release of the combination E, F, G and H. Now the information can no longer be passed on, but the register unit is again able to take up new information, as a result of the closing of contact unit t^2.

If the information has to be passed on in another form from that in which it was received, use is made of translation. An example is shown in Fig. 187, where the information received is coded to a numerical value. The relays E, F, G and H then correspond to the successive binary digits, of value 1, 2, 4 and 8 respectively. The information received via four inputs is thus passed on through one of fifteen outputs.

Fig. 188 shows an example of a circuit which transmits a pulse train which, depending on the information, consists of at least 1 and at most 15 pulses.

In this figure the contact tree circuit of Fig. 187 is connected to the contact bank of a selector. As soon as information must be transmitted, contact

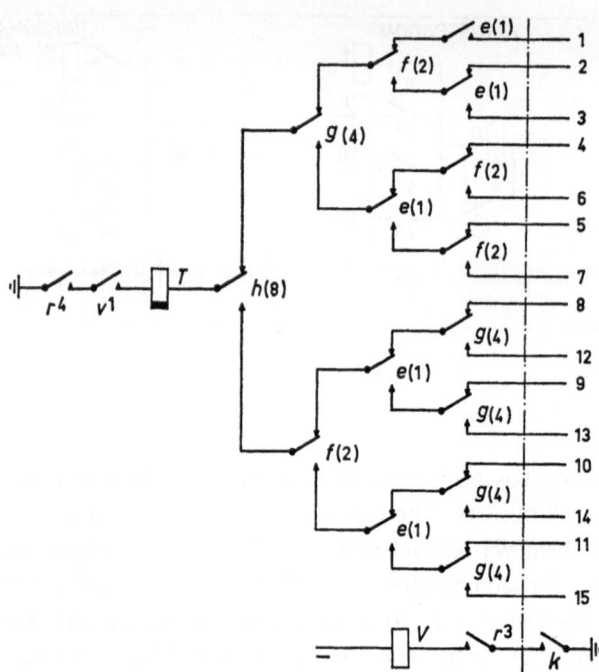

Fig. 187

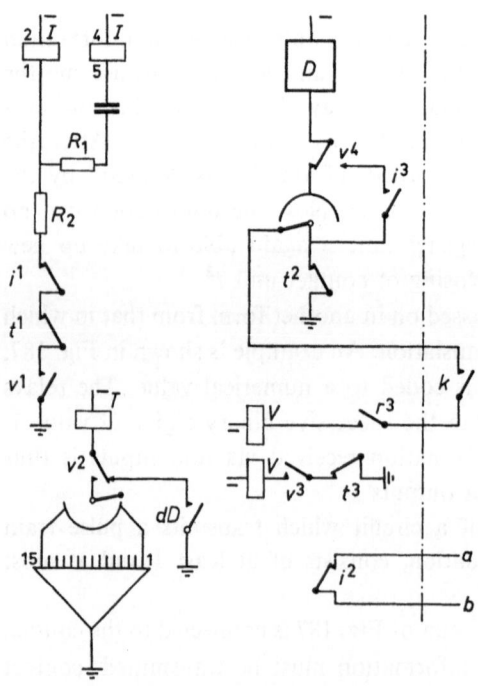

Fig. 188

unit k closes, as a result of which relay V operates (see Fig. 188). A circuit is then closed for relay I. The operation of this relay is retarded, because the two windings are opposed and the relay cannot operate until the capacitor in series with winding 4–5 is sufficiently charged. The circuit for relay I is broken by the opening of contact unit i^1.

The relay releases slowly, however, because the capacitor is discharged via both windings. In this situation the two windings do not oppose one another. The desired pulse frequency and pulse ratio can be obtained by a suitable choice of the capacitor and the resistances R_1 and R_2.

Meanwhile, a pulse is transmitted in the loop a–b via contact unit i^2. Contact unit i^3 switches on the rotation magnet of the selector, so that the selector wiper moves on one place. In this way a series of pulses are transmitted in the loop, and the selector moves on a corresponding number of places. As soon as this number corresponds to the number stored in the contact tree, relay T operates, as a result of which the pulse relay and the selector are switched off by contact units t^1 and t^2 respectively. Moreover, relay V is switched off by contact unit t^3. After relay V has released, the selector returns to its starting position with the aid of relay T.

Relay V can again be switched on when the selector has reached its starting position and more information is available or required.

11.2 Registering several items of information by means of relays

Fig. 189 shows a circuit for registering 3 digits. The signal for each digit is given in the binary code by a combination of the contact units 1, 2, 4 and 8. These contact units close simultaneously for a short time. Then the 2nd and finally the 3rd digit is given. The first digit is registered by relays HA–HD, which are held in series with the winding of relay X after contact units 1, 2, 4 and 8 open. This relay cannot operate until all contact units 1–8 are open, as it is short-circuited before this. The source is then connected by 4 change-over contact units of relay X with the register relays TA–TD. After the second digit has been registered and contact units 1–8 are open again, relay Y operates. The source is now connected by four change-over contact units of relay Y with the registering relays EA–ED. After the third digit has been registered, relay Z operates, as a result of which break contact units of relay Z cut off contact units 1–8, and no more information can be supplied.

When at some later moment the information has to be reproduced in the same form, the first digit is requested by closing contact unit k, which causes

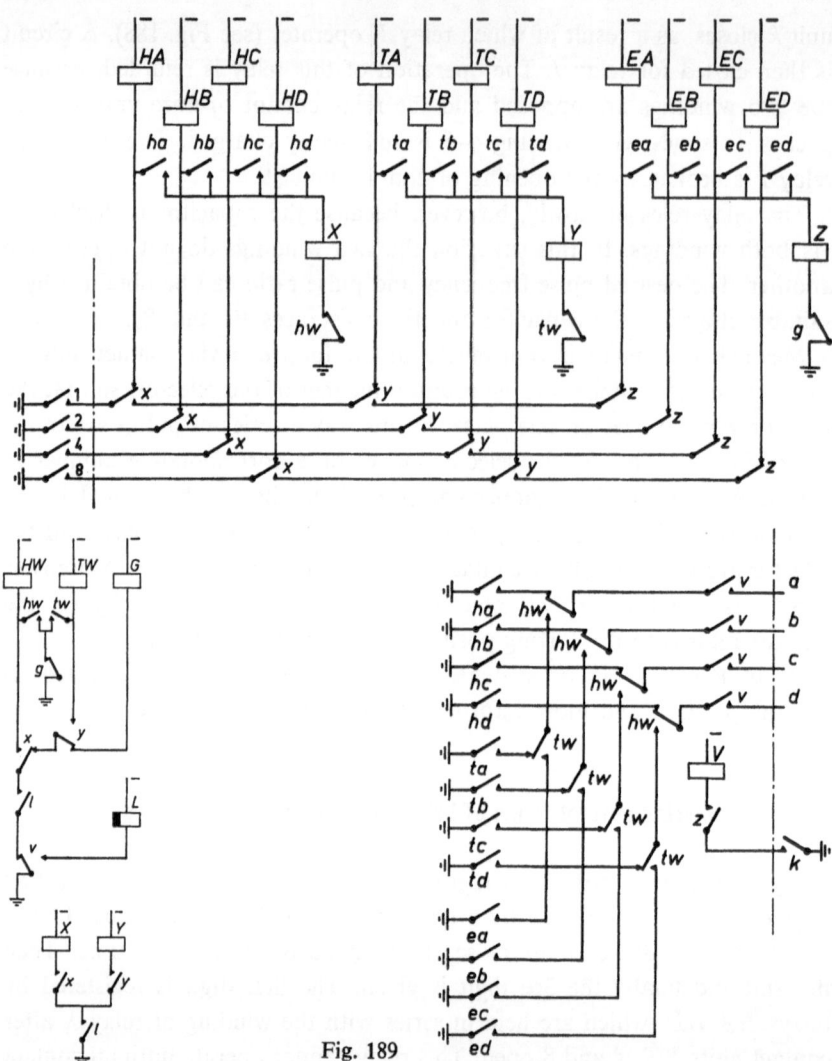

Fig. 189

relay V to operate, connecting the outputs a, b, c and d with contact units ha–hd.

Meanwhile, relay L has operated as a result of the switch-over of a change-over contact unit v. As soon as contact unit k is opened again, relay V releases, after which relay L releases (retarded). During the time that the change-over contact v has already been switched over, while the make contact l is still closed, relay HW is switched on. The hold circuit for relays HA–HD and X is broken by a break contact unit hw. Relay HW also closes a hold circuit via a contact unit *hw* and a break contact unit *g*. As soon as the second digit is requested, contact unit *k* closes again, so that relays *V* and *L* are again switched on.

The outputs *a–d* are now connected via the change-over contact units *hw* to contact units *ta–td*. After relay *V* has released as a result of the opening of contact unit *k*, while relay *L* (whose release is retarded) is still energized, relay *TW* is switched on. This relay is held via contact units *tw* and *g*. The hold circuit for relays *TA–TD* and *Y* is broken by a break contact unit of relay *TW*. The third digit is now requested, causing the outputs *a–d* to be connected to contact units *ea–ed* as described above. As soon as contact unit *k* allows relay *V* to release for the last time, a circuit is closed for relay *G*. As a result of this, the hold circuit for relays *EA–ED* is opened and relays *TW* and *HW* are switched off. After relay *L* has released, relay *G* is switched off again, and the whole circuit returns to its state of rest.

For the purposes of the circuit of Fig. 189, the value zero is represented by the binary number 1 0 1 0. This is necessary, because the normal representation 0 0 0 0 would not cause any of the recording relays to operate, so that the relay *X*, *Y* or *Z* would not operate either, and it would no longer be possible to switch over to the next recording unit.

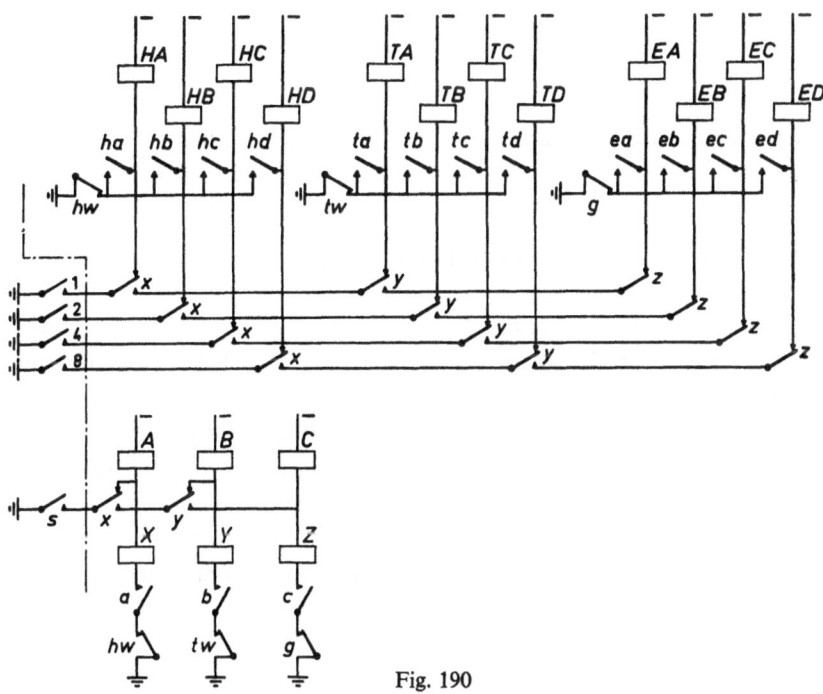

Fig. 190

If it is nevertheless desired to represent the value zero by 0 0 0 0, a counter circuit is added which is controlled by an extra contact of the information source.

This additional circuit is shown in Fig. 190. The relays X, Y and Z are here no longer in series with the recording relays, but form part of the counter circuit. Relays A, B and C act as auxiliary relays, and operate at the start of the transfer of the first, second and third digit respectivelty.

During the transmission of each combination by contact units 1, 2, 4 and 8, contact unit s also closes. When the value zero is transmitted, none of the contact units 1, 2, 4 or 8 closes, but contact units s does close, and takes care that relay A is switched on. After the opening of contact unit s, relay X operates and contact units 1, 2, 4 and 8 are connected with relays TA–TD, etc. Apart from the advantage that the value zero can be recorded in the normal binary notation 0 0 0 0, this circuit also gives the possibility of closing contact units 1, 2, 4 and 8 one by one. There are various possible ways of doing this in combination with contact unit s; some of these are shown in Fig. 191. The only condition which must be satisfied is that contact unit s should open *after* the last item of information has been presented.

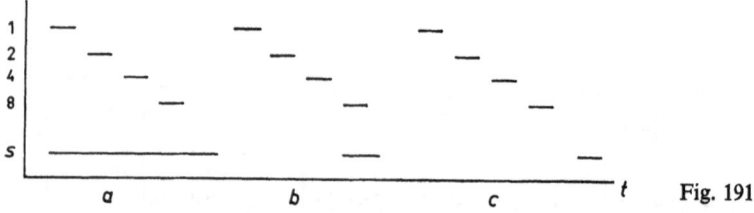

Fig. 191

11.3   Registering several items of information by means of capacitors

It is possible to use capacitors for registering purposes if the information does not have to be stored for a long time. Fig. 192 shows a circuit using capacitors for storing 3 digits. This circuit works as follows.

As soon as a combination of the contact units a, b, c and d is closed, the capacitors of the hundreds are charged. Relay A is switched on by contact unit s, so that the operation of relay X is prepared.

When the hundreds digit has been registered and contact unit s is open again, relay X operates, and contact units a, b, c and d are connected with the capacitors for registering the tens. As soon as this information is supplied, the capacitors are charged in the appropriate combination. Contact unit s now closes a circuit for relay B, which prepares a circuit for switching relay

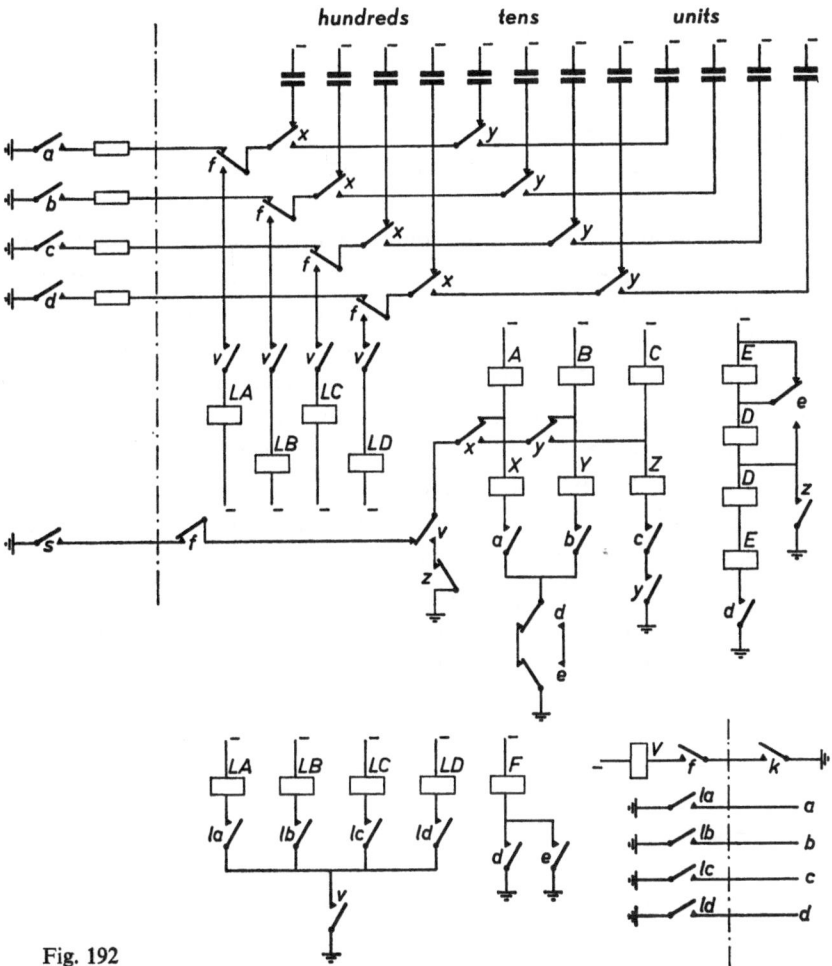

Fig. 192

Y on. This relay operates as soon as contact unit s is opened again. The units are then registered in the same way, while contact unit s now switches relay C on, which prepares the switching on of relay Z.

After contact unit s has been opened and relay Z has operated, a circuit is closed for relay D. As a result of this relay F operates, and contact units a, b, c and d are no longer connected to the register circuit. In the meantime, a contact unit of relay D breaks the circuits of relays A, B, X and Y. When relay Y has released, relays C and Z also release. The releasing of Z causes relay E to operate in series with relay D, preparing the hold circuits of relays A, B, X and Y again. In the meantime, a contact unit f in series with relay V has given permission to read out the registered information.

As soon as this information is needed, contact unit k is closed so that the first digit can be read out, causing relay V to operate so that relays LA, LB, LC and LD are connected with the capacitors for the hundreds. A hold circuit for these relays is closed via a contact unit v. As soon as contact unit k is opened again, relay V releases, so that relays $LA–LD$ are switched off too.

In the meantime, a change-over contact unit v has closed a circuit for the relay A. After relay V has released, relay X operates, selecting the capacitors of the tens for reading out. As soon as contact unit k is closed again, relays $LA–LD$ are connected with these capacitors and the information they contain is read out. While relay V is energized, relay B operates, and after contact unit k is opened and relay V has released relay Y operates. Now the units capacitors are selected for reading. As soon as contact unit k is closed again and relay V operates, these capacitors are read and give up their information. While relay V is energized this time, relay C operates, and after relay V releases, relay Z operates. As a result of this, relay D releases (short-circuited), while relay E is still held via contact unit z. The hold circuit for relays A, B, X and Y is broken and after relay Y releases, relays C and Z are switched off too. Finally, relay E releases, switching off relay F. The circuit is once again in its state of rest, and contact units a, b, c and d are again connected to the register circuit.

11.4 Problems

1. Selectors, relays and capacitors are available for registering a 2-figure number, which must be stored for several days. Each digit of the number is presented to the register circuit in the 2-out-of-5 code for 100 ms. The information must be produced by the register circuit as soon as contact unit k is closed. Both digits are then presented simultaneously, still in the 2-out-of-5 code.

   a) Which of the above-mentioned components do you intend to use for this register circuit?

   b) What is there against the use of the other components?

   c) Design a circuit for this purpose, which also meets the following requirements:

   — After two digits have been received, the register circuit must be closed for any more information until these digits have been read out again and all register and switching relays are at rest.

— The closing of contact unit k is only effective if two digits have been registered.

2. What are the functions of the following relays in the circuit of Fig. 189?

   a) *HA–HD, TA–TD* and *EA–ED*

   b) *X* and *Y*

   c) *Z*

   d) *V, L, HW, TW* and *G*.

   Why is the release of relay *L* retarded?

3. A register circuit must be able to register three digits in the binary-decimal code. The digits are fed in one after the other along 4 wires.

   a) Is a 5th wire needed for registering the digits?

   b) If so, why?

   c) Is the 5th wire needed for the 1245 code?

4. Name the functions of the relays or groups of relays in the register circuit of Fig. 192.

Chapter 12

TRANSLATION CIRCUITS

12.1 Purpose and design

Translation circuits are used to give data, originally in a form which is not suitable for further processing, a form which is suited to the processing involved. A good example of this is the conversion of letters and figures to the Morse code. The letters and figures are not suitable for transmission as such. This conversion to the Morse code is called a translation of the letters and figures into Morse characters. Conversely, in order to obtain a readable message we must ensure that the Morse characters are translated back into letters and figures at the other end.

The same is true of the characters used for teleprinter connections. The letters and figures presented are translated into a code with 5 elements, the "5-unit" code. This gives a total of $2^5 = 32$ different combinations. Two of these are used to prepare for the printing of letters on the one hand and figures (including punctuation marks) on the other. Three other combinations are used to indicate "carriage return", "line feed" (new line) and "space" (between words). The combination 00000 is not used, which leaves $2 \times 26 = 52$ possibilities for letters and figures. If it is desired to provide a check on the transmission, the 5-unit code (32 possibilities) must in its turn be translated into another code which allows this check to be made. The 3-out-of-7 code

$$\left(\frac{n!}{m!(n-m)!} = 35 \right)$$

is suitable for this purpose.

Another application of translation circuits is the positioning of selectors in a telephone exchange, where the selectors cannot be positioning with decimal digits (e.g. selector with 500 positions), while decimal digits are nevertheless fed in. In this case, the decimal code is transformed into a binary code, as much used in computers.

Many different kinds of switching devices can be used for making translation circuits. These may be electromechanical or electronic. In the translation circuits which we shall be discussing here, use is made of relays, diodes and selectors.

Some translation circuits have already been discussed during our treatment of counter circuits, such as the circuit for transforming a binary code into a linear code. This was done with the aid of contact trees. It is important that the translation circuit for translating from one given code into another should be designed as simply as possible.

In many cases, a translation can be made simply and clearly with the aid of notation in tabular form. An example is formed by the translation of the 1245 code, represented by relays A, B, C and D, into a 2-out-of-5 code, represented by the relays 0, 1, 2, 4 and 7. This code is given in Table IL below. The functions for 0, 1, 2, 4 and 7 are given separately in the following smaller tables.

TABLE IL

	A	B	C	D	0	1	2	4	7
1	1	0	0	0	1	1	0	0	0
2	0	1	0	0	1	0	1	0	0
3	1	1	0	0	0	1	1	0	0
4	0	0	1	0	1	0	0	1	0
5	0	0	0	1	0	1	0	1	0
6	1	0	0	1	0	0	1	1	0
7	0	1	0	1	1	0	0	0	1
8	1	1	0	1	0	1	0	0	1
9	0	0	1	1	0	0	1	0	1
0	1	0	1	0	0	0	0	1	1

$f(0) =$

	A	B	C	D
1	1	0	0	0
2	0	1	0	0
4	0	0	1	0
7	0	1	0	1

$f(1) =$

	A	B	C	D
1	1	0	0	0
3	1	1	0	0
5	0	0	0	1
8	1	1	0	1

$f(2) =$

	A	B	C	D
2	0	1	0	0
3	1	1	0	0
6	1	0	0	1
9	0	0	1	1

$f(4) =$

	A	B	C	D
4	0	0	1	0
5	0	0	0	1
6	1	0	0	1
0	1	0	1	0

$f(7) =$

	A	B	C	D
7	0	1	0	1
8	1	1	0	1
9	0	0	1	1
0	1	0	1	0

In this translation circuit, there is no point in realizing such combinations that e.g. $f(0)$ and $f(1)$ have one group of contact units completely in common. This merely leads to a permanent coupling. It is better to use a group of contact units which, apart from one contact unit, is common to the two functions. This latter contact unit can then be used to distinguish between the two outputs.

If we start by simplifying the functions (0) to (7) as far as possible, we obtain:

	A	B	C	D
1	1	0	0	0
$f(0)2+7$	0	1	0	–
4	0	0	1	0

	A	B	C	D
1	1	0	0	0
$f(1)3+8$	1	1	0	–
5	0	0	0	1

	A	B	C	D
2+3	–	1	0	0
$f(2)$ 6	1	0	0	1
9	0	0	1	1

	A	B	C	D
4+0	–	0	1	0
$f(4)5+6$	–	0	0	1

	A	B	C	D
7+8	–	1	0	1
$f(7)$ 9	0	0	1	1
0	1	0	1	0

The next step is to look for correspondances between functions: if we want to save contacts, we have to find a condition for e.g. $f(0)$ which has something in common with a condition for $f(1)$. We find the following combinations:

$$\begin{array}{cccc} & A & B & C & D \\ \end{array}$$

$f(0)$ has 0 1 0 – ⎱ here A has a change-over contact unit
$f(1)$ has 1 1 0 – ⎰

$$\begin{array}{cccc} & A & B & C & D \\ \end{array}$$

$f(0)$ has 1 0 0 0 ⎱ D=change-over contact unit
$f(2)$ has 1 0 0 1 ⎰

The third term of $f(0)$ must now be combined. This is done as follows:

$$\begin{array}{cccc} & A & B & C & D \\ \end{array}$$

$f(0)$ also has 0 0 1 0
$f(2)$ has   0 0 1 1

The following combinations can also be found:

$$\begin{array}{cccc} A & B & C & D \\ \end{array} \qquad \begin{array}{cccc} A & B & C & D \\ \end{array}$$

$f(2)$ has – 1 0 0 $f(1)$ has 1 0 0 0
$f(7)$ has – 1 0 1 $f(7)$ has 1 0 1 0

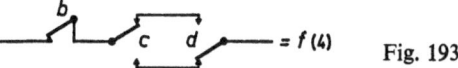

	A	B	C	D		A	B	C	D
$f(1)$ has	0	0	0	1	$f(4)$ has	–	0	1	0
$f(7)$ has	0	0	1	1	$f(4)$ has	–	0	0	1

With the two conditions for $f(4)$ we can make a circuit in which the change-over contact units of C and D form the circuit of Fig. 193.

$$b \quad c \quad d \quad = f(4)$$

Fig. 193

The conditions for $f(0)$ and $f(2)$ reveal a similar possibility:

	A	B	C	D					
$f(0)$	1	0	0	0	and	0	0	1	0
$f(2)$	1	0	0	1		0	0	1	1

It may be seen that these two groups of contact units have a change-over contact unit of relay D in common, viz the break side for $f(0)$ and the make side for $f(2)$. The two functions also have the break contact unit of relay B in common, which thus only leaves the conditions $A=1$, $C=0$ and $A=0$, $C=1$. This is the above-mentioned circuit again, so that the entire circuit for functions (0) and (2) is as shown in Fig. 194. The entire translation circuit is shown in Fig. 195.

It is also possible to design a translation circuit by means of switching algebra. We will be giving an example of this in Section 12.4.

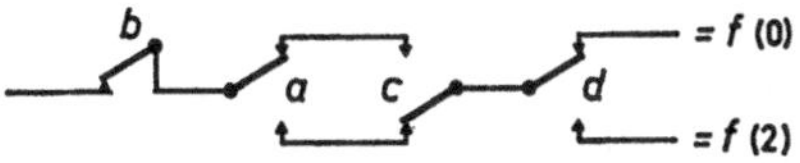

$$b \quad a \quad c \quad d \quad = f(0)$$
$$= f(2)$$

Fig. 194

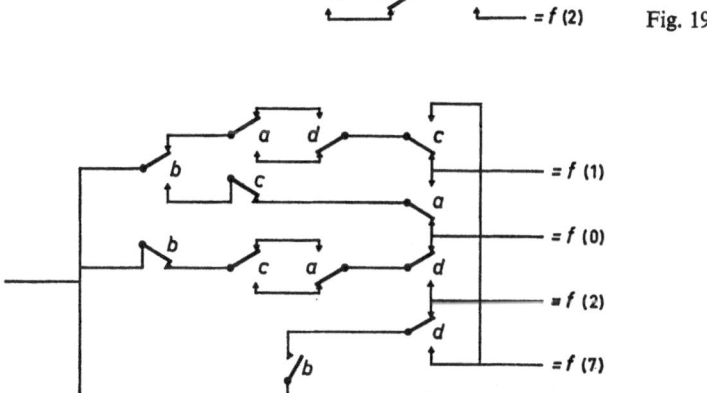

= f (1)
= f (0)
= f (2)
= f (7)
= f (4)

Fig. 195

12.2 Translation of binary-decimal numbers into 1-out-of-n

In a telephone system where the number of the subscriber with whom one wishes to speak is dialed, the last 2 figures of the number indicate which subscriber of the particular group of 100 is wanted. In general, these two digits are translated into the appropriate one of the 100 outputs by mechanical switching units (100-position selectors).

If the selector is specially designed for translating from $2 \times (1$ out of 10) to 1 out of 100, no further translation is needed. This is done by controlling the selector in the vertical direction by the penultimate digit, the horizontal movement then being brought about by the pulse train of the last digit. One then speaks of a two-motion selector.

The translation is in this case thus carried out mechanically, which however demands a special construction of the selector.

		1	2	3	4	5	6	7	8	9	0
	0	01	02	–	–	–	–	–	–	–	00
	9	–	–								–
	8	–	–								–
penulti-	7	–	–								–
mate	6	–	–								–
digit	5	–	–								–
	4	–	–								–
	3	–	–								–
	2	21	–								–
	1	11	–	–	–	–	–	–	–	–	–

$\longrightarrow$ last digit

It must however also be possible to move 100-position selectors which can only move in one direction (uniselectors) to the desired one out of 100 outputs by means of the last 2 digits of the dialed number. One possible way of doing this is to translate the 2 digits received into linear form. However, as we have seen, a lot of relays are needed to register numbers in linear form. It is therefore usual to store these digits separately in binary form (4 relays per digit), so that it is then necessary to carry out the translation from binary-decimal to 1-out-of-100 (1-out-of-n). The principle of this translation is shown in Fig. 196, where the contact tree TT takes care of the translation from the binary of the tens to 1-out-of-10. The contact tree EH in series with each of the outputs 1–0 then translates the units from binary to 1-out-of-10. This thus gives the desired total of 100 outputs.

The distribution of the contact units over the relays used in the binary counter circuit is however very unsuitable in this circuit. The *TT* contact tree occurs only once, while the *EH* trees occur ten times. A better distribution is shown in Fig. 197, where *TT* occurs eleven times and *EH* also eleven times. Both *TT* and *EH* have one contact tree with 10 outputs and all the rest with only 5 outputs.

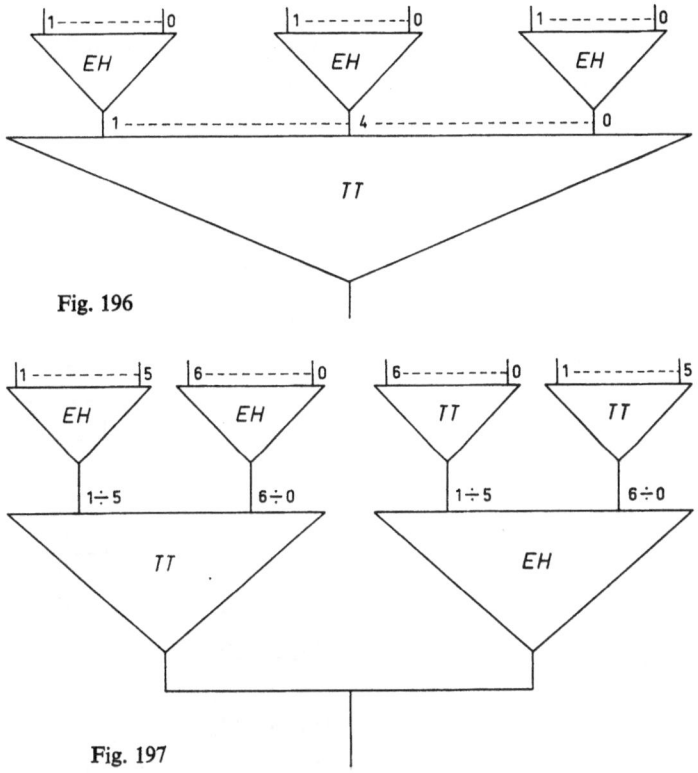

Fig. 196

Fig. 197

Despite the better distribution, the number of contacts needed for the trees is enormous, as a result of which one relay per binary element is not enough. A better approach is first to translate from binary to linear, and then to make the translation to 1-out-of-100 with 2 linear groups of relays. Now this seems illogical, because the 2 digits in question were presented in linear from in the first place, then translated to binary – and now we want to translate them back to linear.

However, since the reception of two digits takes much more time than the translation from binary to linear, one translation circuit can in many cases translate the binary information from a number of callers in succession (with the aid of a lock-out circuit). This is in fact what is done in telephone exchanges where the last 2 digits are recorded in the final selector circuit.

We now get the following circuit (Fig. 198). The 2 digits are registered by binary counters of 4 relays each. When the two digits have been received and permission has been given to position the selector, $2 \times (1$ out of 10) relays are energized via the TT and EH contact trees, which are formed of the contact units of relays T_1–T_8 and E_1–E_8 respectively; 2×10 relays are sufficient for a complete translation to 1-out-of-100, since as may be seen from Fig. 198, a springset of 6 make contact units suffices for each of the relays 1–0 and 10–00, so that extra relays are unnecessary.

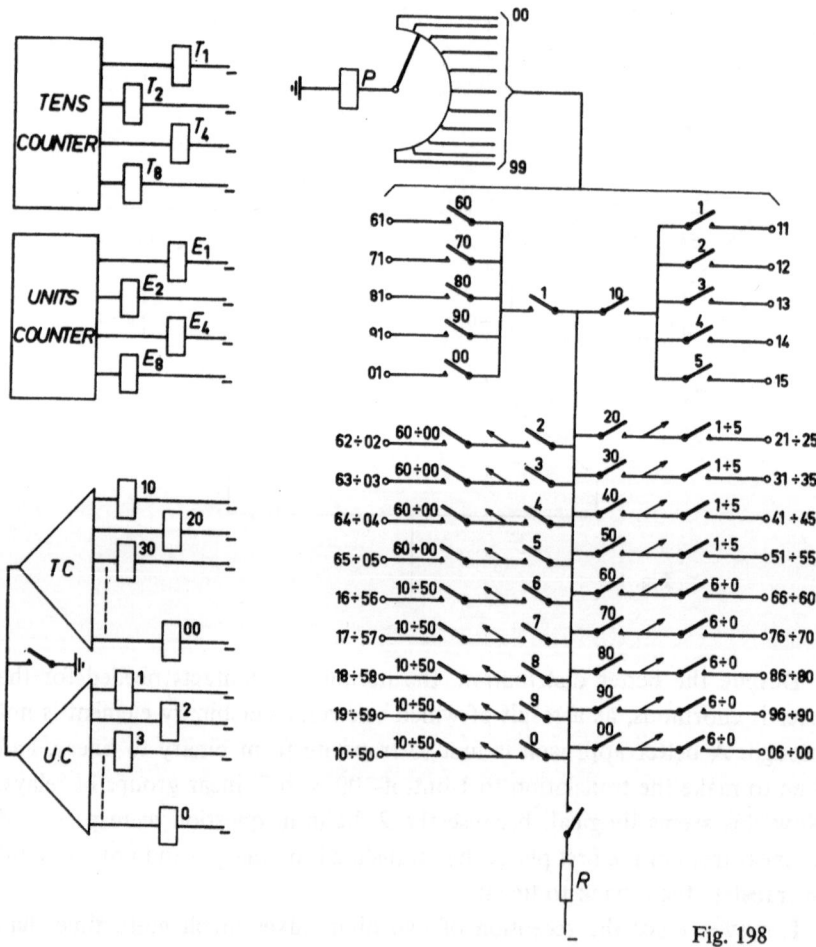

Fig. 198

12.3 Translation from decimal numbers to 1-out-of-n

A telephone number consists of a number of digits which are presented to the exchange by mean of the caller's dial. By making use of 100-position selectors, we can simply bring the selectors involved to the desired output, since we are working with the decimal system, for which the 100-position selector is especially well suited. As has already been mentioned in the previous section, the last two digits are used in such a case to position the last selector (final selector, FS) to one of the 100 outputs.

A telephone exchange with 10000 subscribers uses 4-figure numbers, from 0000 to 9999. 500-position two-motion selectors are also in use; these selectors have a 25×20 contact bank instead of 10×10. The number of subscribers connected to a FS which makes use of a selector of this type is thus 500 instead of 100. It is then necessary to translate the decimal numbers presented into the 1-out-of-500 code. The 10000 subscribers are assigned to 20 FS groups. The first FS group may e.g. contain the numbers 0000–0499, the second FS group 0500–0999, the third FS group 1000–1499 and so on. The 4-figure number presented must thus be translated into:

1. one of the 20 groups of 500 subscribers
2. one of the 25 groups of 20 subscribers and
3. one of the 20 subscribers in the selected group of 20.

The subscriber is precisely defined by this means. We must thus translate the 4 decimal digits into 3 other units, and thus bring the selector to the desired position; and this translation must of course be so designed as to give the desired result for each one of the 10000 numbers which can be presented. In the following example of the realization of this translation, we shall make use of 10-position uniselectors, one per digit, which are operated by the dialing pulses so that they take up a position corresponding to the digit in question; we shall also use a number of relays, to store the various items of information and pass them on to the following piece of equipment.

Let us suppose that we receive the number 3425. We must now determine the position where this subscriber is attached to the 500-position selector. The separate digits of this number are received by a thousands selector (D), a hundreds selector (H), a tens selector (T) and a units selector (E). Each of these selectors is provided with several wipers. Now in order to determine the group of 500 subscribers involved, we can make use of the D selector and the H selector as shown in Fig. 199.

Here the relays $0D$ to $9D$ represent the thousands, and the 2 relays

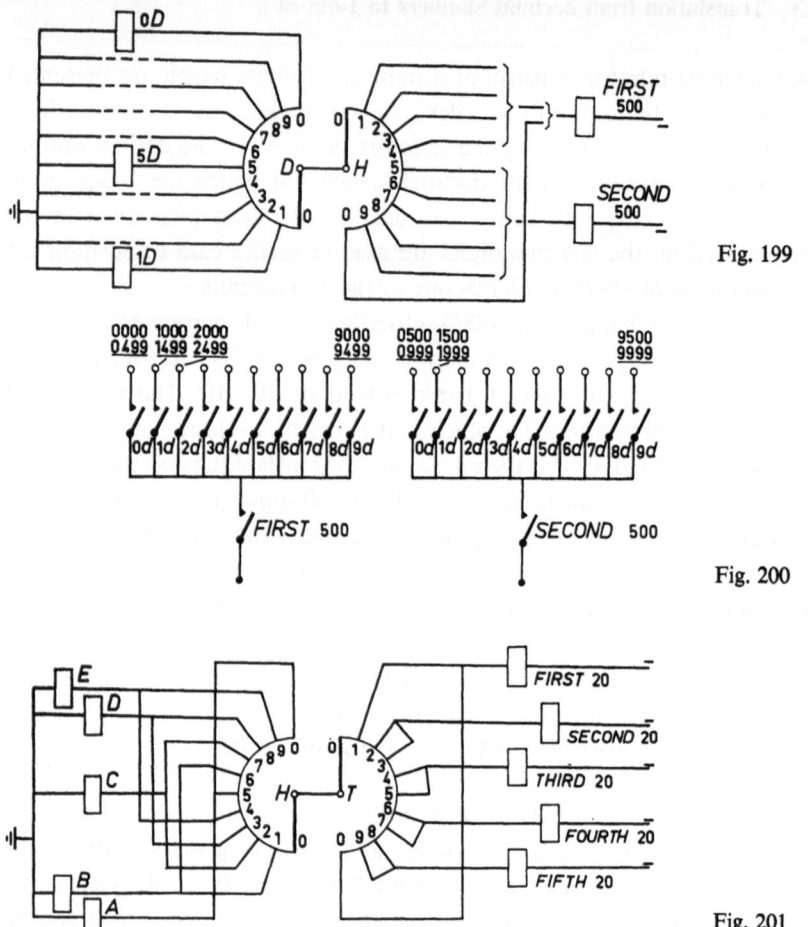

Fig. 199

Fig. 200

Fig. 201

connected to the *H* selectors represent the first and second 500 of each thousand. We can now indicate with the aid of these 12 relays the *FS* where the subscriber is to be found, by means of the translation shown in Fig. 200. Now that we have determined the *FS*, we must determine in which of the 25 groups of 20 the subscriber is to be found. This is done with the aid of the *H* and *T* selectors, as shown in Fig. 201.

The relays *A–E* select the hundred, and with the selector *T* we select the appropriate group of 20 out of that hundred. The *FS*, which we have already determined, contains the hundreds 0–4 or 5–9. The hundreds 0 and 5 have thus acquired the same value, which may also be seen from the fact that the points 0 and 5 of the *H* selector are connected in parallel (to relay *A*).

With the aid of these ten relays, we can now determine in which of the 25 groups of 20 we must look for the subscriber. The circuit for this is indicated in Fig. 202. Now all that remains to be done is to select the right one of the group of 20 subscribers. For this purpose we make use of selectors T and E, as shown in Fig. 203, where the right output is found with the aid of twelve relays. The tens are indicated by relays $1TT$ and $2TT$, and the units by relays $E0$–$E9$. The 20 subscribers can now be selected with the aid of the translation circuit of Fig. 204.

The subscriber 3425 is now determined by the following data. Relay $3D$ and the first 500 indicate that the subscriber lies in the range 3000–3499. Further, relay E and the second 20 indicate that the subscriber is in the 22nd group of 20, i.e. 3420–3439. Finally, relays $1TT$ and 5 indicate that the desired number is 3425. It is possible to assign any desired number between 0000 and 9999 to a given output by means of this circuit.

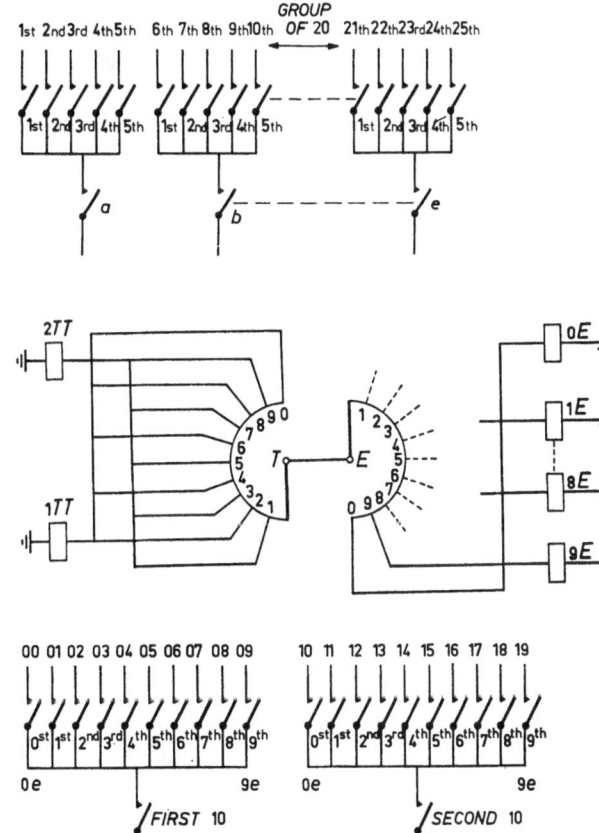

Fig. 202

Fig. 203

Fig. 204

12.4 Translation from the binary code to the 2-out-of-5 code

Here again, writing the codes in tabular form gives a good picture of the desired translation (Table L). The translation is only needed for the 10 cases given in Table L; the other combinations of the relays A, B, C and D cannot thus occur.

TABLE L

	A	B	C	D	0	1	2	4	7
1	1	0	0	0	1	1	0	0	0
2	0	1	0	0	1	0	1	0	0
3	1	1	0	0	0	1	1	0	0
4	0	0	1	0	1	0	0	1	0
5	1	0	1	0	0	1	0	1	0
6	0	1	1	0	0	0	1	1	0
7	1	1	1	0	1	0	0	0	1
8	0	0	0	1	0	1	0	0	1
9	1	0	0	1	0	0	1	0	1
10	0	1	0	1	0	0	0	1	1

The translation to the value 0 of the 2-out-of-5 code may be written as follows:

$$f(0)=ab'c'd' \quad +a'bc'd'+a'b'cd'+abcd'=$$
$$d'(ab'c' \quad +a'bc' \quad +a'b'c \quad +abc) \;=$$
$$d'\{a(b'c'+bc) \quad +a'(bc' \quad +b'c)\}$$

This expression gives the circuit of Fig. 205.

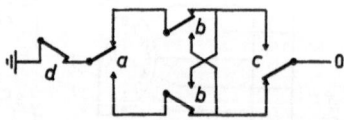

Fig. 205

$$f(1)=ab'c'd'+abc'd'+ab'cd'+a'b'c'd=$$
$$ac'd' + \qquad ab'cd'+a'b'c'd=$$
$$ad'(c' +b') \quad +a'b'c'd$$

In the last term, c' may be omitted, because according to the code relay D is never switched on together with relay C, so that we find:

$$f(1)=ad'(c'+b')+a'b'd$$

This expression leads to Fig. 206.

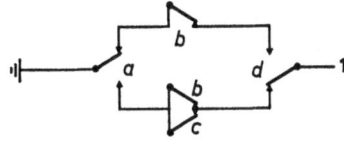

Fig. 206

$$f(2) = a'bc'd' + abc'd' + a'bcd' + ab'c'd =$$
$$a'd'(bc' + bc) \quad + abc'd' + ab'c'd =$$
$$a'bd' \quad + abc'd' + ab'c'd =$$
$$bd'(a' + ac') \quad + ab'c'd =$$
$$bd'(a' + c') \quad + ab'c'd$$

In the last term, c' is superfluous, because relays C and D are never switched on at the same time. We thus obtain:

$$f(2) = bd'(a' + c') + ab'd.$$

We can eliminate b' from the last term, because the combination abd does not occur.

$f(2) = bd'(a' + c') + ad$. The circuit is shown in Fig. 207.

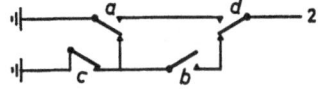

Fig. 207

$$f(4) = a'b'cd' + ab'cd' + a'bcd' + a'bc'd$$

In the first, 2nd and 3rd terms, d' can be left out, because relays C and D never operate at the same time. In the 4th term, c' can be omitted for the same reason, so that we get:

$$f(4) = a'b'c + ab'c + a'bc + a'bd =$$
$$b'c \quad + a'bc + a'bd =$$
$$b'c \quad + a'c \quad + a'bd.$$

In the last term we can omit a', because according to the code the relays A, B and D are never switched on together, so that:

$$f(4) = b'c + a'c + bd = c(a' + b') + bd.$$

The circuit is shown in Fig. 208.

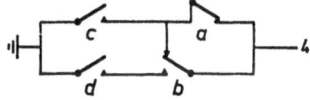

Fig. 208

$$f(7) = abcd' + a'b'c'd + ab'c'd + a'bc'd$$

According to the code, d' may be omitted from the first term. In the 2nd, 3rd and 4th terms, c' is superfluous, so that we obtain:

$$f(7) = abc + a'b'd + ab'd + a'bd$$

In the 2nd, 3rd and 4th terms, d occurs with all possible combinations of a and b except ab. Since however the combination abd is not used in the code, a and b may be omitted from these terms. We are thus left with:

$$f(7) = abc + d.$$

Fig. 209 gives the circuit. The full translation circuit is shown in Fig. 210.

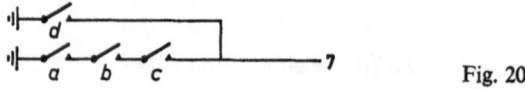

Fig. 209

A certain further saving of contacts may be achieved by combining make and break contact units to change-over contact units. The result of this is shown in Fig. 211.

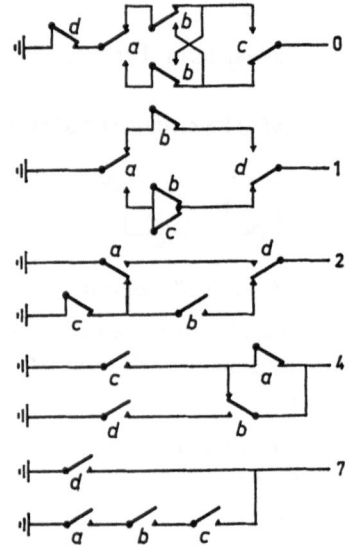

Fig. 210

12.5 Translation of the 5-unit code to the 3-out-of-7 code

The "5-unit" code is used e.g. in telegraphy for the transmission of letters and figures. The character is presented by pushing a key of the transmitting machine, which transmits the coded character along the telegraph line to the receiving machine. Each character consists of 5 bits. The receiving machine has a decoding device which transforms the signal received into a legible character, which is printed on paper.

The 5-unit code is in fact a binary code. 5 bits in the binary code allow 32 different possible combinations to be made. These 32 different combinations can be used to transmit all the letters of the alphabet and certain

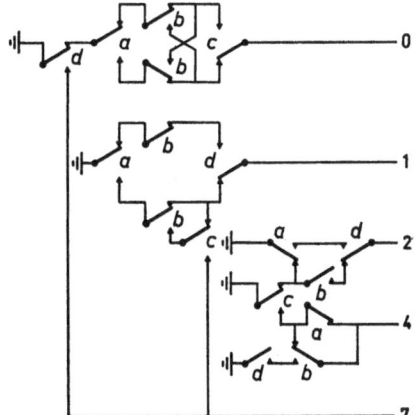

Fig. 211

more general signs such as space, line feed, carriage return, etc. The combinations used for the digits and punctuation marks are the same as those used for the letters. If a digit has to be printed instead of the corresponding letter, the combination representing it is preceded by one of the 32 combinations (figure shift), while if the letter has to be printed it is preceded by another combination (letter shift). This brings the type wheel into the digit position and the letter position respectively. This thus allows 26 combinations of the 5-unit code to do double duty. Of the remaining 6 combinations, five are used in both positions of the type wheel (space, line feed, letter shift, figure shift, and carriage return).

The binary code has the disadvantage that it is not really possible to check whether a character has been received correctly. If telegraphy characters are transmitted via radio links, it is quite likely that some of the characters will be mutilated by atmospheric interference. This is one of the reasons why the characters are now transmitted in a code which can be checked. One checkable code with at least as many combinations as the 5-unit code is the 3-out-of-7 code, for which

$$\binom{n}{m} = \frac{n!}{m!\,(n-m)!} = \frac{7!}{3!\,(7-3)!} = 35$$

This thus leaves 3 combinations spare in the 3-out-of-7 code. These combinations can be used for special purposes if necessary.

The translation from the binary code to the 3-out-of-7 code will be worked out in the rest of this section, by way of example. The combinations for both codes are numbered 0–31, corresponding to the binary code (see Table LI).

TABLE LI

	A	B	C	D	E	1	2	3	4	5	6	7		A	B	C	D	E	1	2	3	4	5	6	7
0	0	0	0	0	0	0	0	0	0	1	1	1	16	1	0	0	0	0	0	1	1	1	0	0	0
1	0	0	0	0	1	1	0	0	0	1	0	1	17	1	0	0	0	1	0	1	1	0	0	0	1
2	0	0	0	1	0	1	0	0	0	0	1	1	18	1	0	0	1	0	0	0	1	1	1	0	0
3	0	0	0	1	1	1	0	0	0	1	1	0	19	1	0	0	1	1	0	0	1	1	0	0	1
4	0	0	1	0	0	1	1	0	1	0	0	0	20	1	0	1	0	0	0	1	0	1	0	1	0
5	0	0	1	0	1	1	0	1	0	0	1	0	21	1	0	1	0	1	0	0	1	0	1	0	1
6	0	0	1	1	0	1	0	1	0	1	0	0	22	1	0	1	1	0	0	0	1	0	0	1	1
7	0	0	1	1	1	1	0	1	0	0	0	1	23	1	0	1	1	1	0	0	1	0	1	1	0
8	0	1	0	0	0	1	0	1	1	0	0	0	24	1	1	0	0	0	0	0	1	1	0	1	0
9	0	1	0	0	1	1	1	0	0	0	1	0	25	1	1	0	0	1	0	1	0	0	1	0	1
10	0	1	0	1	0	1	1	0	0	1	0	0	26	1	1	0	1	0	0	1	0	0	0	1	1
11	0	1	0	1	1	1	1	0	0	0	0	1	27	1	1	0	1	1	0	1	0	0	1	1	0
12	0	1	1	0	0	1	1	1	0	0	0	0	28	1	1	1	0	0	0	1	1	0	0	1	0
13	0	1	1	0	1	1	0	0	1	0	1	0	29	1	1	1	0	1	0	0	0	1	1	0	1
14	0	1	1	1	0	1	0	0	1	1	0	0	30	1	1	1	1	0	0	0	0	1	0	1	1
15	0	1	1	1	1	1	0	0	1	0	0	1	31	1	1	1	1	1	0	0	0	1	1	1	0

If we inspect all the outputs from 1 to 7 separately, we see that $f(1)$ is formed by combinations 1–15. In all these combinations, $A=0$. In combination 0, however, A is also 0. In 15 of the 16 combinations were $A=0$, a circuit must thus be closed for $f(1)$ of the 3-out-of-7 code. This can be realized by a series circuit of a break contact unit a with a parallel circuit of make contact units b, c, d and e (see Fig. 212).

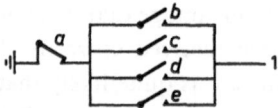

Fig. 212

For $f(2)$ we note the following combinations (Table LII). The circuit for $f(2)$ can be simplified as far as possible by the elimination method. In order to do this, we arrange the combinations as far as possible with like near like, which gives a better picture of the situation (see Table LIII).

TABLE LII

	A	B	C	D	E		A	B	C	D	E		A	B	C	D	E
4	0	0	1	0	0	12	0	1	1	0	0	25	1	1	0	0	1
9	0	1	0	0	1	16	1	0	0	0	0	26	1	1	0	1	0
10	0	1	0	1	0	17	1	0	0	0	1	27	1	1	0	1	1
11	0	1	0	1	1	20	1	0	1	0	0	28	1	1	1	0	0

TABLE LIII

		A	B	C	D	E			A	B	C	D	E			A	B	C	D	E
	4	0	0	1	0	0		10	0	1	0	1	0	III	9	0	1	0	0	1
I	12	0	1	1	0	0	II	11	0	1	0	1	1		25	1	1	0	0	1
	20	1	0	1	0	0		26	1	1	0	1	0	IV	16	1	0	0	0	0
	28	1	1	1	0	0		27	1	1	0	1	1		17	1	0	0	0	1

In combinations 4, 12, 20 and 28, relays C, D and E are represented by the code 100. In these 4 cases, relays A and B have all possible conditions, viz 00, 01, 10 and 11. As far as relays A and B are concerned, therefore, the circuit found for these 4 combinations will always be closed. The conditions of A and B may therefore be omitted for these combinations (see Chapter 3). These four combinations may thus be replaced by $- - 1\ 0\ 0$.

We can carry out the same process for the codes denoted by II–IV in Table LIII, thus simplifying the original 12 combinations to the 4 combinations I–IV of Table LIV.

If we consider codes II and III, we see that we may write for them:

$$bc'd + bc'd'e = bc'(d + d'e) = bc'(d + e) = bc'd + bc'e$$

This gives a further simplification of combinations I–IV, leading to the final result shown in Table LV. The circuit of Fig. 213 follows from this.

TABLE LIV

	A	B	C	D	E
I	–	–	1	0	0
II	–	1	0	1	–
III	–	1	0	0	1
IV	1	0	0	0	–

TABLE LV

	A	B	C	D	E
I	–	–	1	0	0
II	–	1	0	1	–
III	–	1	0	–	1
IV	1	0	0	0	–

It will be apparent that in this figure we have three times a break contact unit c, twice in series with a make contact unit b and once in series with a break contact unit b. This part of the circuit can therefore be simplified as shown in Fig. 214. All that remains to be done is to add the upper branch of Fig. 213 to this circuit. This gives rise to a change-over contact unit c, while we can also make use of the break side of the change-over contact unit d, already shown in Fig. 214. The complete circuit is thus as shown in Fig. 215.

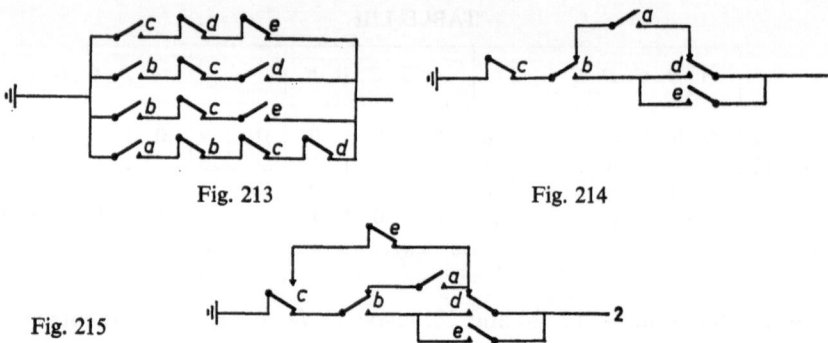

Fig. 213   Fig. 214

Fig. 215

We can try to simplify $f(3)$ in the same way, so as to give the simplest possible circuit. The combinations for $f(3)$ are shown in Table LVI.

TABLE LVI

	A B C D E		A B C D E		A B C D E		A B C D E
5	0 0 1 0 1	12	0 1 1 0 0	19	1 0 0 1 1	24	1 1 0 0 0
6	0 0 1 1 0	16	1 0 0 0 0	21	1 0 1 0 1	28	1 1 1 0 0
7	0 0 1 1 1	17	1 0 0 0 1	22	1 0 1 1 0		
8	0 1 0 0 0	18	1 0 0 1 0	23	1 0 1 1 1		

The pattern can more suitably be arranged as shown in Table LVII. Contraction of combinations 6, 7, 22 and 23 (I) gives – 0 1 1 –, since all possible conditions of A and E occur together with the same values of B, C and D. Groups II, III and IV can be simplified in a similar way. The

TABLE LVII

		A B C D E			A B C D E			A B C D E			A B C D E
	6	0 0 1 1 0		8	0 1 0 0 0		16	1 0 0 0 0	IV	5	0 0 1 0 1
I	7	0 0 1 1 1	II	12	0 1 1 0 0	III	17	1 0 0 0 1		21	1 0 1 0 1
	22	1 0 1 1 0		24	1 1 0 0 0		18	1 0 0 1 0			
	23	1 0 1 1 1		28	1 1 1 0 0		19	1 0 0 1 1			

	TABLE LVIII			TABLE LIX

	A B C D E			A B C D E
I	– 0 1 1 –		I	– 0 1 1 –
II	– 1 – 0 0		II	– 1 – 0 0
III	1 0 0 – –		III	1 0 0 – –
IV	– 0 1 0 1		IV	– 0 1 – 1

result is given in Table LVIII. This can be simplified even further, since I and IV of Table LVIII can be written:

$$b'cd + b'cd'e = b'c(d + d'e) = b'c(d + e) = b'cd + b'ce$$

giving the result shown in Table LIX. The circuit may be constructed from the data of Table LIX in the same way as was done for $f(2)$, and is shown in Fig. 216.

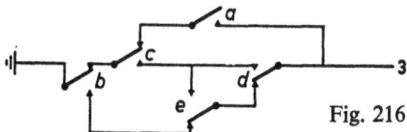

Fig. 216

The combinations for $f(4)$ are shown in Table LX:

TABLE LX

	A	B	C	D	E		A	B	C	D	E		A	B	C	D	E		A	B	C	D	E
4	0	0	1	0	0	15	0	1	1	1	1	20	1	0	1	0	0	31	1	1	1	1	1
8	0	1	0	0	0	16	1	0	0	0	0	24	1	1	0	0	0						
13	0	1	1	0	1	18	1	0	0	1	0	29	1	1	1	0	1						
14	0	1	1	1	0	19	1	0	0	1	1	30	1	1	1	1	0						

TABLE LXI

		A	B	C	D	E			A	B	C	D	E			A	B	C	D	E			A	B	C	D	E
I	13	0	1	1	0	1	II	14	0	1	1	1	0	IV	24	1	1	0	0	0	VI	4	0	0	1	0	0
	15	0	1	1	1	1		30	1	1	1	1	0		8	0	1	0	0	0							
	29	1	1	1	0	1	III	16	1	0	0	0	0	V	18	1	0	0	1	0							
	31	1	1	1	1	1		20	1	0	1	0	0		19	1	0	0	1	1							

TABLE LXII

	A	B	C	D	E
I	–	1	1	–	1
II	–	1	1	1	0
III	1	0	–	0	0
IV	–	1	0	0	0
V	1	0	0	1	–
VI	0	0	1	0	0

TABLE LXIII

	A	B	C	D	E
I	–	1	1	–	1
II	–	1	1	1	–
III	1	0	–	0	0
IV	–	1	0	0	0
V	1	0	0	1	–
VI	–	0	1	0	0

These are shown more suitably arranged in Table LXI. Groups I–VI are each reduced to one line by means of the elimination method. The result of the elimination is shown in Table LXII. Further simplification leads to

Table LXIII. This may be seen as follows. Combinations I and II of Table LXII give the expression:

$$bce + bcde' = bc(e + de') = bc(e + d)$$

The value 0 of relay E may thus be omitted from combination II.

Combinations III and VI give:

$$ab'd'e' + a'b'cd'e' = b'd'e'(a + a'c) = b'd'e'(a + c)$$

whence it follows that the value 0 of A in combination VI may be omitted. The circuit for $f(4)$ can be derived from Table LXIII, and is shown in Fig. 217.

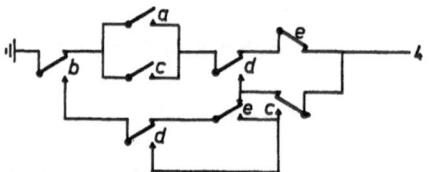

Fig. 217

The combinations for $f(5)$ are as shown in Table LXIV. Here again, a better arrangement gives possibilities of elimination. The combinations arranged in groups are shown in Table LXV.

TABLE LXIV

	A B C D E		A B C D E		A B C D E		A B C D E
1	0 0 0 0 1	14	0 1 1 1 0	25	1 1 0 0 1	0	0 0 0 0 0
3	0 0 0 1 1	18	1 0 0 1 0	27	1 1 0 1 1		
6	0 0 1 1 0	21	1 0 1 0 1	29	1 1 1 0 1		
10	0 1 0 1 0	23	1 0 1 1 1	31	1 1 1 1 1		

TABLE LXV

		A B C D E			A B C D E			A B C D E			A B C D E
	25	1 1 0 0 1	II	21	1 0 1 0 1	IV	1	0 0 0 0 1	VII	18	1 0 0 1 0
I	27	1 1 0 1 1		23	1 0 1 1 1		0	0 0 0 0 0			
	29	1 1 1 0 1	III	10	0 1 0 1 0	V	3	0 0 0 1 1			
	31	1 1 1 1 1		14	0 1 1 1 0	VI	6	0 0 1 1 0			

The result of the simplification of the various groups is shown in Table LXVI, which can be further simplified to Table LXVII: it follows from combinations I and II of Table LXVI that the value 0 of B may be omitted. Similarly, the value 1 of D can be omitted from combinations IV and V.

TABLE LXVI	A	B	C	D	E
I	1	1	–	–	1
II	1	0	1	–	1
III	0	1	–	1	0
IV	0	0	0	0	–
V	0	0	0	1	1
VI	0	0	1	1	0
VII	1	0	0	1	0

TABLE LXVII	A	B	C	D	E
I	1	1	–	–	1
II	1	–	1	–	1
III	0	1	–	1	0
IV	0	0	0	0	–
V	0	0	0	–	1
VI	0	–	1	1	0
VII	1	0	0	1	0

Finally, it follows from combinations III and VI that the value 0 of B is superfluous. The circuit of Fig. 218 realizes the conditions of Table LXVII.

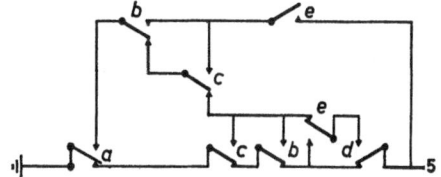

Fig. 218

$f(6)$ is given by the combinations of Table LXVIII.

When rearranging the combinations in this case, it is worth while using a number of combinations more than once. There is naturally nothing against this, since two or more identical branches in parallel will not alter the final result, while a greater degree of simplification can be achieved in this way.

TABLE LXVIII

	A	B	C	D	E		A	B	C	D	E		A	B	C	D	E		A	B	C	D	E
2	0	0	0	1	0	13	0	1	1	0	1	24	1	1	0	0	0	30	1	1	1	1	0
3	0	0	0	1	1	20	1	0	1	0	0	26	1	1	0	1	0	31	1	1	1	1	1
5	0	0	1	0	1	22	1	0	1	1	0	27	1	1	0	1	1	0	0	0	0	0	0
9	0	1	0	0	1	23	1	0	1	1	1	28	1	1	1	0	0						

TABLE LXIX

		A	B	C	D	E			A	B	C	D	E			A	B	C	D	E			A	B	C	D	E
I	22	1	0	1	1	0	III	24	1	1	0	0	0	V	20	1	0	1	0	0	VII	26	1	1	0	1	0
	23	1	0	1	1	1		26	1	1	0	1	0		22	1	0	1	1	0		27	1	1	0	1	1
	30	1	1	1	1	0		28	1	1	1	0	0		28	1	1	1	0	0		30	1	1	1	1	0
	31	1	1	1	1	1		30	1	1	1	1	0		30	1	1	1	1	0		31	1	1	1	1	1
II	2	0	0	0	1	0	IV	2	0	0	0	1	0	VI	13	0	1	1	0	1	VIII	13	0	1	1	0	1
	0	0	0	0	0	0		3	0	0	0	1	1		5	0	0	1	0	1		9	0	1	0	0	1

TABLE LXX

	A	B	C	D	E
I	1	–	1	1	–
II	0	0	0	–	0
III	1	1	–	–	1
IV	0	0	0	1	–
V	1	–	1	–	0
VI	0	–	1	0	1
VII	1	1	–	1	–
VIII	0	1	–	0	1

The combinations are rearranged in Table LXIX, where combinations 2, 13, 22, 26, 28 and 31 occur twice, and combination 30 four times.

Groups I–VIII can be simplified by elimination; the result is shown in Table LXX. The data of this table lead to the circuit of Fig. 219.

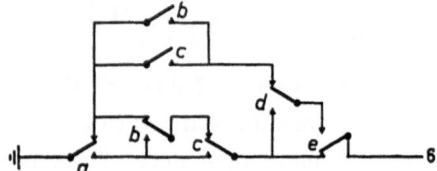

Fig. 219

Finally, $f(7)$ is represented by the combinations of Table LXXI.

TABLE LXXI

	A	B	C	D	E		A	B	C	D	E		A	B	C	D	E		A	B	C	D	E
1	0	0	0	0	1	15	0	1	1	1	1	22	1	0	1	1	0	30	1	1	1	1	0
2	0	0	0	1	0	17	1	0	0	0	1	25	1	1	0	0	1	0	0	0	0	0	0
7	0	0	1	1	1	19	1	0	0	1	1	26	1	1	0	1	0						
11	0	1	0	1	1	21	1	0	1	0	1	29	1	1	1	0	1						

TABLE LXXII

		A	B	C	D	E			A	B	C	D	E			A	B	C	D	E
I	1	0	0	0	0	1	IV	25	1	1	0	0	1	VII	7	0	0	1	1	1
	0	0	0	0	0	0		29	1	1	1	0	1		15	0	1	1	1	1
II	11	0	1	0	1	1	V	26	1	1	0	1	0	VIII	21	1	0	1	0	1
	15	0	1	1	1	1		30	1	1	1	1	0		29	1	1	1	0	1
III	17	1	0	0	0	1	VI	0	0	0	0	0	0	IX	22	1	0	1	1	0
	19	1	0	0	1	1		2	0	0	0	1	0		30	1	1	1	1	0

These combinations are again arranged so that they can be simplified by elimination of bits. The arrangement in groups is shown in Table LXXII, combinations 0, 15, 29 and 30 being used twice.

TABLE LXXIII

	A	B	C	D	E
I	0	0	0	0	–
II	0	1	–	1	1
III	1	0	0	–	1
IV	1	1	–	0	1
V	1	1	–	1	0
VI	0	0	0	–	0
VII	0	–	1	1	1
VIII	1	–	1	0	1
IX	1	–	1	1	0

Fig. 220

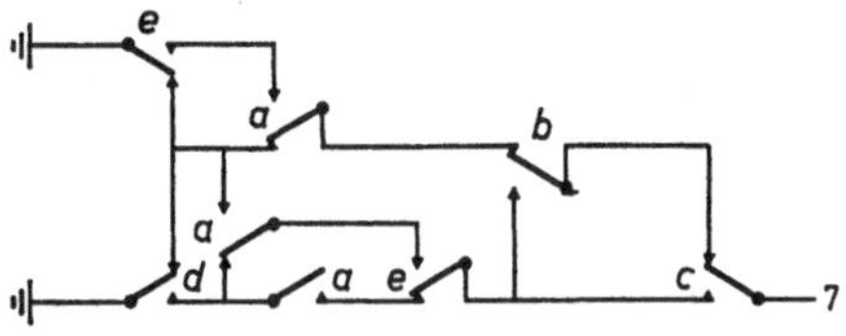

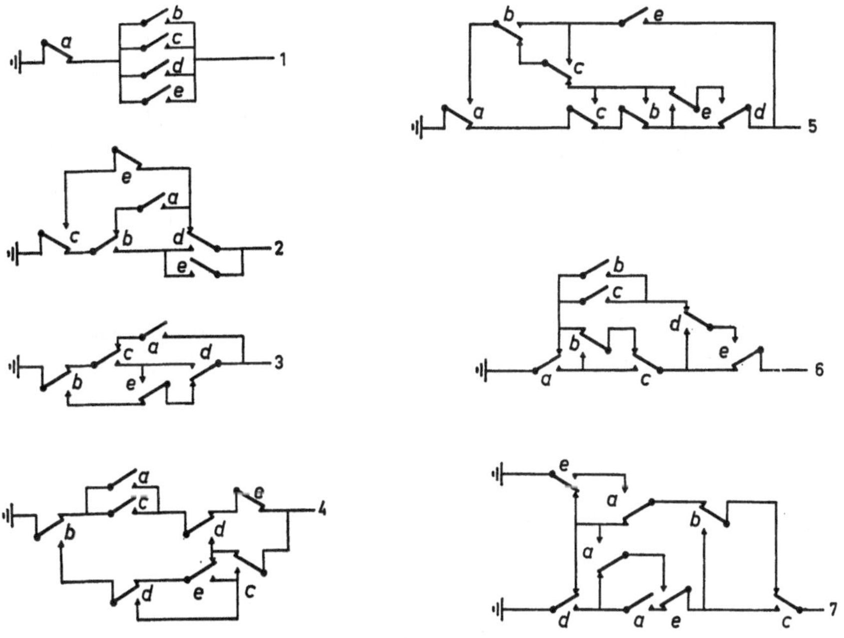

Fig. 221

Simplification of the groups leads to Table LXXIII, on the basis of which we can design the circuit of Fig. 220. The complete translation circuit is finally given in Fig. 221.

12.6 Translation of the 3-out-of-7 code to the 5-unit code

This translation is simpler than that described in Section 12.5, because the 3-out-of-7 code has a more regular pattern: 3 relays are always energized, and 4 released.

The code list is given below (Table LXXIV), with the combinations of the 3-out-of-7 code placed next to the desired combinations of the 5-units code.

TABLE LXXIV

	1	2	3	4	5	6	7	A	B	C	D	E		1	2	3	4	5	6	7	A	B	C	D	E
0	0	0	0	0	1	1	1	0	0	0	0	0	16	0	1	1	1	0	0	0	1	0	0	0	0
1	1	0	0	0	1	0	1	0	0	0	0	1	17	0	1	1	0	0	0	1	1	0	0	0	1
2	1	0	0	0	0	1	1	0	0	0	1	0	18	0	0	1	1	1	0	0	1	0	0	1	0
3	1	0	0	0	1	1	0	0	0	0	1	1	19	0	0	1	1	0	0	1	1	0	0	1	1
4	1	1	0	1	0	0	0	0	0	1	0	0	20	0	1	0	1	0	1	0	1	0	1	0	0
5	1	0	1	0	0	1	0	0	0	1	0	1	21	0	0	1	0	1	0	1	1	0	1	0	1
6	1	0	1	0	1	0	0	0	0	1	1	0	22	0	0	1	0	0	1	1	1	0	1	1	0
7	1	0	1	0	0	0	1	0	0	1	1	1	23	0	0	1	0	1	1	0	1	0	1	1	1
8	1	0	1	1	0	0	0	0	1	0	0	0	24	0	0	1	1	0	1	0	1	1	0	0	0
9	1	1	0	0	0	1	0	0	1	0	0	1	25	0	1	0	0	1	0	1	1	1	0	0	1
10	1	1	0	0	1	0	0	0	1	0	1	0	26	0	1	0	0	0	1	1	1	1	0	1	0
11	1	1	0	0	0	0	1	0	1	0	1	1	27	0	1	0	0	1	1	0	1	1	0	1	1
12	1	1	1	0	0	0	0	0	1	1	0	0	28	0	1	1	0	0	1	0	1	1	1	0	0
13	1	0	0	1	0	1	0	0	1	1	0	1	29	0	0	0	1	1	0	1	1	1	1	0	1
14	1	0	0	1	1	0	0	0	1	1	1	0	30	0	0	0	1	0	1	1	1	1	1	1	0
15	1	0	0	1	0	0	1	0	1	1	1	1	31	0	0	0	1	1	1	0	1	1	1	1	1

It may be seen from the code table that relay A is always switched on when relay 1 is switched off (combinations 16–31), except in combination 0, where relay 1 is also released, but relays 5, 6 and 7 are energized. We may therefore write:

$$f(A) = 1'(5' + 6' + 7')$$

This expression is realized by the circuit of Fig. 222.

Fig. 222

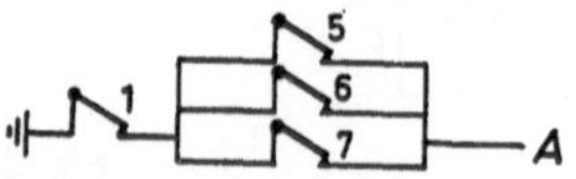

Relay B is switched on when one of the relays 2 and 4 is energized, *except* in combinations:

$$17 \quad 0\ 1\ 1\ 0\ 0\ 0\ 1$$
$$18 \quad 0\ 0\ 1\ 1\ 1\ 0\ 0$$
$$19 \quad 0\ 0\ 1\ 1\ 0\ 0\ 1$$

Here relay 3 is always energized, and relays 1 and 6 are always released. The circuit of B must thus be opened by these combinations.

$$f(B) = 2'4(1+3'+6) + 24'(1+3'+6) = (24'+2'4)(1+3'+6)$$

The circuit is given in Fig. 223.

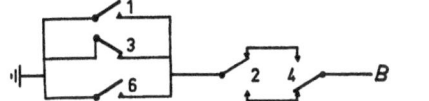

Fig. 223

According to the code list, relay C only operates if one of the relays 3 and 4 operates, apart from combination 17. Combination 17 is 0 1 1 0 0 0 1. The switching on of relays 2 and 7 must thus open the circuit.

$$f(C) = (3'4+34')(2'+7')$$

Fig. 224 shows the circuit.

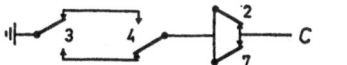

Fig. 224

According to the code list, relay D only operates when one of the relays 5 and 7 operates, apart from combination 17 (0 1 1 0 0 0 1), so that relays 2 *and* 3 must both be energized to prevent a closed circuit for D.

$$f(D) = (5'7+57')(2'+3')$$

The circuit is shown in Fig. 225.

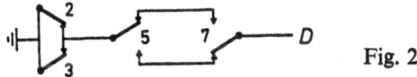

Fig. 225

Finally, relay E operates according to the code list if relay 6 is off and relay 7 is on and if relay 7 is off and relay 6 is on, apart from combinations 20, 24 and 28. These are:

$$0\ 1\ 0\ 1\ 0\ 1\ 0$$
$$0\ 0\ 1\ 1\ 0\ 1\ 0$$
$$0\ 1\ 1\ 0\ 0\ 1\ 0$$

$$f(E) = 6'7+67'(2'4+3'4+2'3') = 6'7+67'\{4(2'+3')+2'3'\}$$

The circuit is given in Fig. 226.

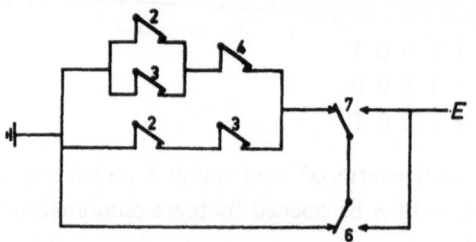

Fig. 226

When the circuits for relays *A* to *E* as derived above are combined, some further simplification is possible. The complete translation circuit is given in Fig. 227.

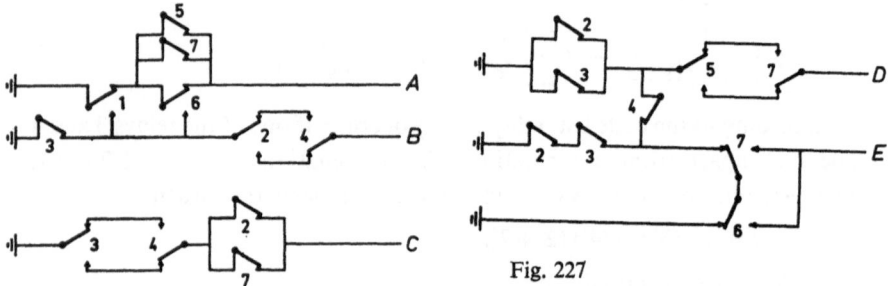

Fig. 227

12.7 Translation from the 1245 code to the 01247 code

In the previous sections we have dealt with translations where the bits of the desired code are expressed as functions of the values of the bits in the original code. These functions are simplified as far as possible by means of switching algebra and, where possible, further contact springs are eliminated when combining the partial circuits to the final translation circuit.

TABLE LXXV

	1	2	4	5		0	1	2	4	7
0	1	0	1	0		0	0	0	1	1
1	1	0	0	0		1	1	0	0	0
2	0	1	0	0		1	0	1	0	0
3	1	1	0	0		0	1	1	0	0
4	0	0	1	0		1	0	0	1	0
5	0	0	0	1		0	1	0	1	0
6	1	0	0	1		0	0	1	1	0
7	0	1	0	1		1	0	0	0	1
8	1	1	0	1		0	1	0	0	1
9	0	0	1	1		0	0	1	0	1

However, switching algebra also allows us to combine terms from the various functions before the total circuit is assembled. As an example of this we shall consider a translation from the 1 2 4 5 code to the 0 1 2 4 7 code. The two codes are given alongside one another in Table LXXV.

With the aid of the data from the table, the operation of the relays 0 1 2 4 7 can be expressed as a function of the relay combinations 1 2 4 5:

$$f(0) = 1\ 2'\ 4'\ 5' + 1'\ 2\ 4'\ 5' + 1'\ 2'\ 4\ 5' + 1'\ 2\ 4'\ 5$$

It follows from the 2nd and 4th terms that:

$$1'\ 2\ 4'\ 5' + 1'\ 2\ 4'\ 5 = 1'\ 2\ 4'(5' + 5) = 1'\ 2\ 4'$$

and from the 1st and 3rd terms:

$$1\ 2'\ 4'\ 5' + 1'\ 2'\ 4\ 5' = 2'\ 5'(1\ 4' + 1'\ 4)$$

so that:

$$f(0) = 1'\ 2\ 4' + 2'\ 5'(1\ 4' + 1'\ 4).$$
$$f(1) = 1\ 2'\ 4'\ 5' + 1\ 2\ 4'\ 5' + 1'\ 2'\ 4'\ 5 + 1\ 2\ 4'\ 5 =$$
$$2'\ 4'(1\ 5' + 1'\ 5) + 1\ 2\ 4'$$
$$f(2) = 1'\ 2\ 4'\ 5' + 1\ 2\ 4'\ 5' + 1\ 2'\ 4'\ 5 + 1'\ 2'\ 4\ 5 =$$
$$2\ 4'\ 5' + 2'\ 5(1\ 4' + 1'\ 4)$$
$$f(4) = 1\ 2'\ 4\ 5' + 1'\ 2'\ 4\ 5' + 1'\ 2'\ 4'\ 5 + 1\ 2'\ 4'\ 5 =$$
$$2'(1\ 4\ 5' + 1'\ 4\ 5' + 1'\ 4'\ 5 + 1\ 4'\ 5)$$

It follows from the 1st and 2nd terms that:

$$1\ 4\ 5' + 1'\ 4\ 5' = 4\ 5'(1 + 1') = 4\ 5'$$

and from the 3rd and 4th terms:

$$1'\ 4'\ 5 + 1\ 4'\ 5 = 4'\ 5(1' + 1) = 4'\ 5$$

so that:

$$f(4) = 2'(4\ 5' + 4'\ 5)$$
$$f(7) = 1\ 2'\ 4\ 5' + 1'\ 2\ 4'\ 5 + 1\ 2\ 4'\ 5 + 1'\ 2'\ 4\ 5 =$$
$$2'\ 4\ (1\ 5' + 1'\ 5) + 2\ 4'\ 5$$

All the functions are repeated below:

$$f(0) = 1'\ 2\ 4' \qquad + 2'\ 5'\ (1\ 4' + 1'\ 4)$$
$$f(1) = 2'\ 4'\ (1'\ 5 + 1\ 5') \qquad + 1\ 2\ 4'$$
$$f(2) = 2\ 4'\ 5' \qquad + 2'\ 5\ (1\ 4' + 1'\ 4)$$
$$f(4) = 2'\ (4\ 5' \qquad + 4'\ 5)$$
$$f(7) = 2'\ 4\ (1\ 5' \qquad + 1'5) + 2\ 4'\ 5$$

It is now our aim to combine as many terms as possible from these 5 functions, to arrive at a simple circuit.

The part 2' 4'(1' 5+1 5') from function 1 and the part 2' 4(1 5'+1' 5) from function 7 can simply be combined as shown in Fig. 228.

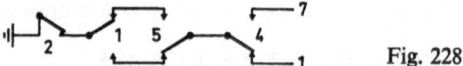

Fig. 228

The rest of $f(1)$ is 1 2 4', and
the rest of $f(7)$ is 2 4' 5.

The 1 2 4' from $f(1)$ can be combined with the part 1' 2 4' of $f(0)$. This is shown in Fig. 229.

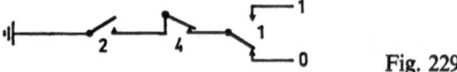

Fig. 229

Combining the above two circuits gives Fig. 230, which satisfies $f(1)$.

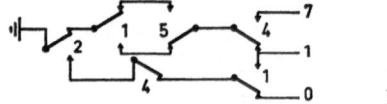

Fig. 230

The rest of $f(0)$ is 2' 5'(1 4'+1' 4) and
the rest of $f(7)$ is 2 4' 5.

We can now quite well combine the part 2' 5'(1 4'+1' 4) of $f(0)$ with 2' 5(1 4'+1' 4) from $f(2)$. This gives the circuit shown in Fig. 231. If this result is combined with that obtained above, we obtain the circuit of Fig. 232. $f(0)$ is now also completely satisfied.

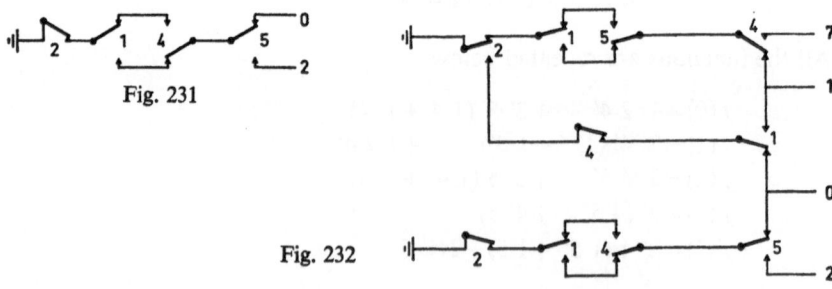

Fig. 231

Fig. 232

The rest of $f(2)$ is 2 4′ 5′, and
The rest of $f(7)$ is 2 4′ 5.

These can easily be combined, as shown in Fig. 233. Combining this with the previous results gives the circuit of Fig. 234. $f(2)$ and $f(7)$ are thus both satisfied too. Only $f(4)$ remains to be included in the circuit. $f(4)$ is shown in its entirety in Fig. 235; the complete circuit can now be assembled, and is shown in Fig. 236.

Fig. 233

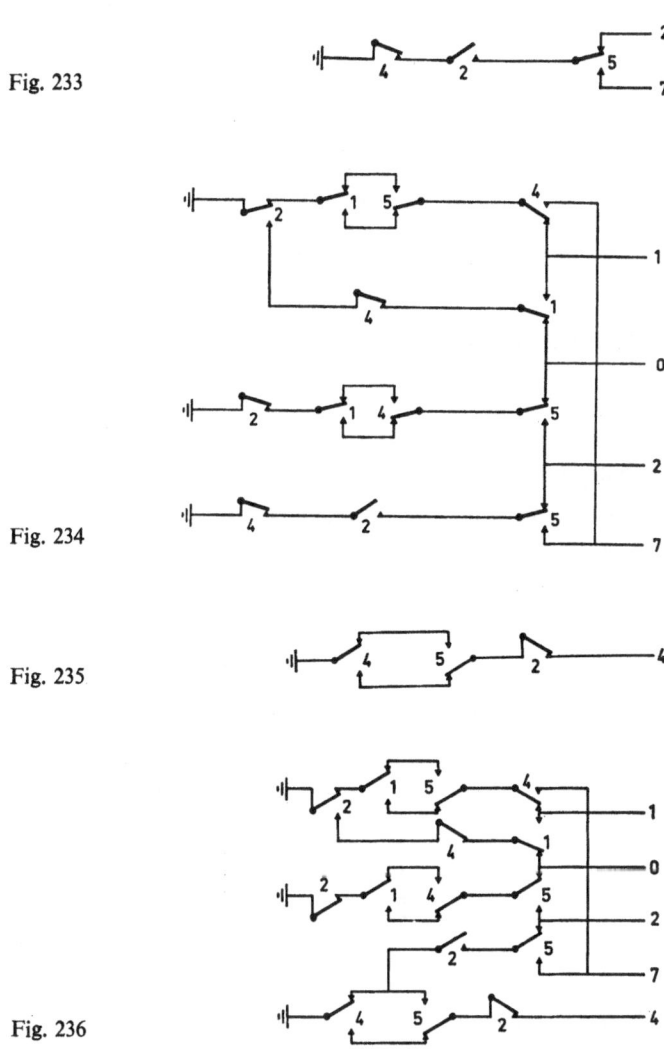

Fig. 234

Fig. 235

Fig. 236

12.8 Problems

1. Given the following 2-out-of-6 code.

	A	B	C	D	E	F
1	0	0	1	0	0	1
2	0	0	0	1	0	1
3	0	0	0	0	1	1
4	0	1	0	0	1	0
5	0	0	1	0	1	0
6	0	0	0	1	1	0
7	0	1	0	1	0	0
8	0	0	1	1	0	0
9	0	1	1	0	0	0
10	1	1	0	0	0	0
11	1	0	1	0	0	0
12	1	0	0	1	0	0
13	1	0	0	0	1	0
14	1	0	0	0	0	1
15	0	1	0	0	0	1

Translate this code into a binary code with 4 relays, where the 0-position is not used. Make use of switching algebra, and use no more than 47 contact springs.

2. The relays A, B and C must be switched on in accordance with the reflecting code by the contact units of relays 1, 2 and 4 (which are operated according to the binary code).

Design a circuit for this purpose, with 14 contact springs.

	1	2	4		A	B	C
0	0	0	0		0	0	0
1	1	0	0		1	0	0
2	0	1	0		1	1	0
3	1	1	0		0	1	0
4	0	0	1		0	1	1
5	1	0	1		1	1	1
6	0	1	1		1	0	1
7	1	1	1		0	0	1

3. Make a translation from 1-out-of-10 to 2-out-of-5, and vice versa.

4. The 2-out-of-5 code with the relays 0, 1, 2, 4 and 7 must be obtained by translation from a 4-unit code. For the 10 states of the 2-out-of-5 code the combinations of the 4-unit code with 1 relay out of 4 energized, and 2 relays out of 4 energized are used as shown, below.

$$f(0) = abc'd' + ab'cd' + ab'c'd' + ab'c'd$$

$$f(1) = abc'd' + a'bcd' + a'bc'd' + a'bc'd$$

$$f(2) = ab'cd' + a'bcd' + a'b'cd' + a'b'cd$$

$$f(4) = ab'c'd' + a'bc'd' + a'b'cd' + a'b'c'd$$

$$f(7) = ab'c'd + a'bc'd + a'b'cd + a'b'c'd$$

a) Write the code in tabular form.

b) Draw the circuit, using not more than 22 contact springs.

c) Design a circuit for the translation from the 2-out-of-5 code to the 4-unit code, using not more than 8 contact springs.

5. A linear counter circuit with 10 positions must, 200 ms after the reception of a pulse train, display the number of these pulses by means of 4 lamps (binary). Design the translation circuit for these 4 lamps.

6. Translate the 2-out-of-6 code given below to the reflecting code. The 0-position of the reflecting code is not used.

1	2	3	4	5	6	A	B	C	D
1	0	0	0	1	0	1	0	0	0
1	1	0	0	0	0	1	1	0	0
0	1	0	0	1	0	0	1	0	0
0	1	1	0	0	0	0	1	1	0
0	1	0	0	0	1	1	1	1	0
1	0	1	0	0	0	1	0	1	0
0	0	1	0	1	0	0	0	1	0
0	0	1	1	0	0	0	0	1	1

1	2	3	4	5	6	A	B	C	D
1	0	0	0	0	1	1	0	1	1
0	0	0	0	1	1	1	1	1	1
0	0	1	0	0	1	0	1	1	1
0	1	0	1	0	0	0	1	0	1
0	0	0	1	0	1	1	1	0	1
1	0	0	1	0	0	1	0	0	1
0	0	0	1	1	0	0	0	0	1

Not more than 8 contact springs may be used per relay, and the total number of contact springs should not exceed 20.

7. A 2-out-of-5 code must be translated to the 1 2 4 5 code. Design the circuit for the code given below. Use not more than 16 contact springs.

	A	B	C	D	E	1	2	4	5
0	1	0	1	0	0	1	0	1	0
1	1	0	0	1	0	1	0	0	0
2	0	1	1	0	0	0	1	0	0
3	1	1	0	0	0	1	1	0	0
4	0	0	1	1	0	0	0	1	0
5	0	0	0	1	1	0	0	0	1
6	1	0	0	0	1	1	0	0	1
7	0	1	0	1	0	0	1	0	1
8	0	1	0	0	1	1	1	0	1
9	0	0	1	0	1	0	0	1	1

8. $f(X)$ is given by the following table. Give the circuit for $f(X)$, using not more than 11 contact springs. Make use of the elimination method when working out the results.

	A	B	C	D	E
1	1	1	0	1	0
2	0	0	1	0	1
3	1	1	0	0	1
4	1	0	1	0	1
5	1	1	0	1	1
6	0	1	1	0	1
7	1	1	0	0	0
8	1	1	1	0	1

Chapter 13

IDENTIFICATION AND ANALYSIS CIRCUITS

13.1 Definitions

Identification circuits are used to determine the name or number of a line or a device.

By way of example, we may consider the following problem (see Fig. 237).

A receiver R is connected via a locking circuit V with various lines which carry meteorological information. The character of the information differs for all the lines: line 1 gives information about the air pressure, line 2 about the wind direction and line 3 about the amount of precipitation. The receiver, which has to decode the information which it receives, can only do this if this information is interpreted in the correct manner. The receiver must thus "know" with which line it is connected, identifying it as line 1, 2 or 3.

Another example is given in Fig. 238. Positions A, B and C are connected by lines, while position D is only connected to position C. Information can be transmitted in both directions along the lines. If information must be transmitted from A to C and D, this information must be preceeded by the designation of the positions involved. The information received at C is then also transmitted to D. Position C is responsible for information received from A which has to be passed on to D.

If, however, information is transmitted from A to B and C, then the designations of the positions must again occur in the information. In this

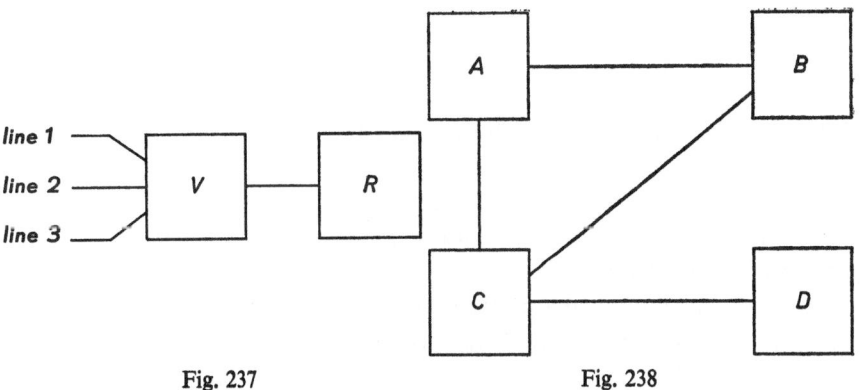

Fig. 237   Fig. 238

case, we must prevent information being transmitted from *C* to *B* or from *B* to *C*, since this would lead to the same information being received twice. Position *C* is thus not responsible for transmission to *B* if information is received from *A*, while position *B* is not responsible for passing information received from *A* on to *C*.

Position *C* is however responsible for the transmission of information from *D* meant for *A* and/or *B*. The question of responsibility can be decided by identification of the line via which the information is received.

Analysis circuits are used to draw certain conclusions from information received. Examples are e.g. the making of connections between telephone exchanges and the determination of the proper rate from the digit combinations received, or the solution of the similar problem for telegraph exchanges from the letter combinations received.

13.2   Identification of a given point by means of relays

The simplest form of identification is obtained by connecting each point which has to be identified with a relay, as shown in Fig. 239. After contact unit 1 is closed, point 1 is identified by the operation of relay *A*. The other points are identified in a similar way.

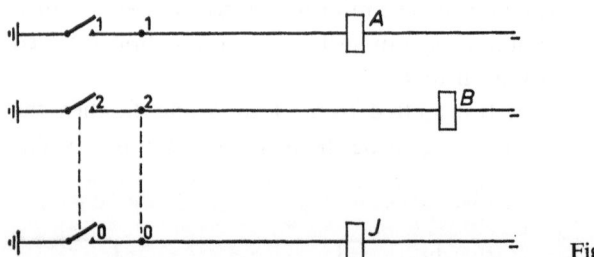

Fig. 239

13.3   Identification by means of diodes and relays

In order to save relays, it is usually desirable to store the identification in coded form. The points to be identified are then connected to a smaller number of relays. Diodes may be used to decouple these points. Fig. 240 shows an example where the position of a selector must be identified, and the result stored in the 2-out-of-5 code. If the selector is in position 1, relays 0 and 1 are switched on via the diodes. All the other positions also result in

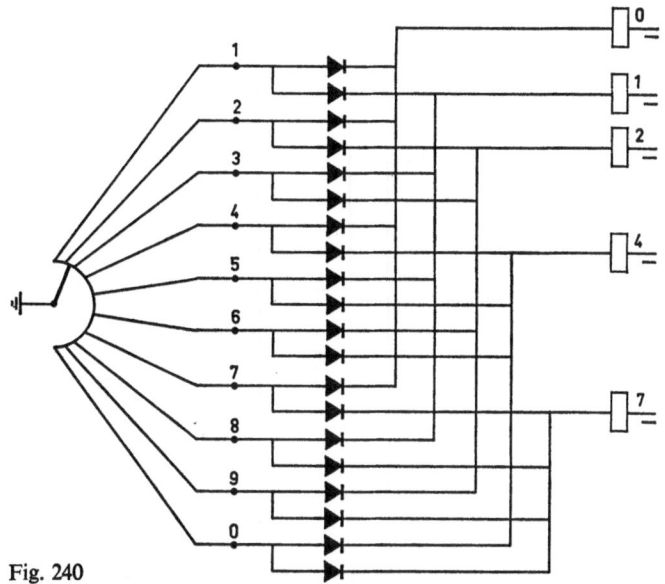

Fig. 240

the switching on of two out of the five relays according to the 0 1 2 4 7 code. In this way the identification given at one point is stored in a coded numerical form.

If the identification must be stored in the form of one or more letters, the circuit of Fig. 241 may be used. This figure also shows how the name of a line can simply be altered by means of plugs with diodes mounted on them.

TABLE LXXVI

5 unit code

	1	2	3	4	5		1	2	3	4	5
A	1	1	0	0	0	N	0	0	1	1	0
B	1	0	0	1	1	O	0	0	0	1	1
C	0	1	1	1	0	P	0	1	1	0	1
D	1	0	0	1	0	Q	1	1	1	0	1
E	1	0	0	0	0	R	0	1	0	1	0
F	1	0	1	1	0	S	1	0	1	0	0
G	0	1	0	1	1	T	0	0	0	0	1
H	0	0	1	0	1	U	1	1	1	0	0
I	0	1	1	0	0	V	0	1	1	1	1
J	1	1	0	1	0	W	1	1	0	0	1
K	1	1	1	1	0	X	1	0	1	1	1
L	0	1	0	0	1	Y	1	0	1	0	1
M	0	0	1	1	1	Z	1	0	0	0	1

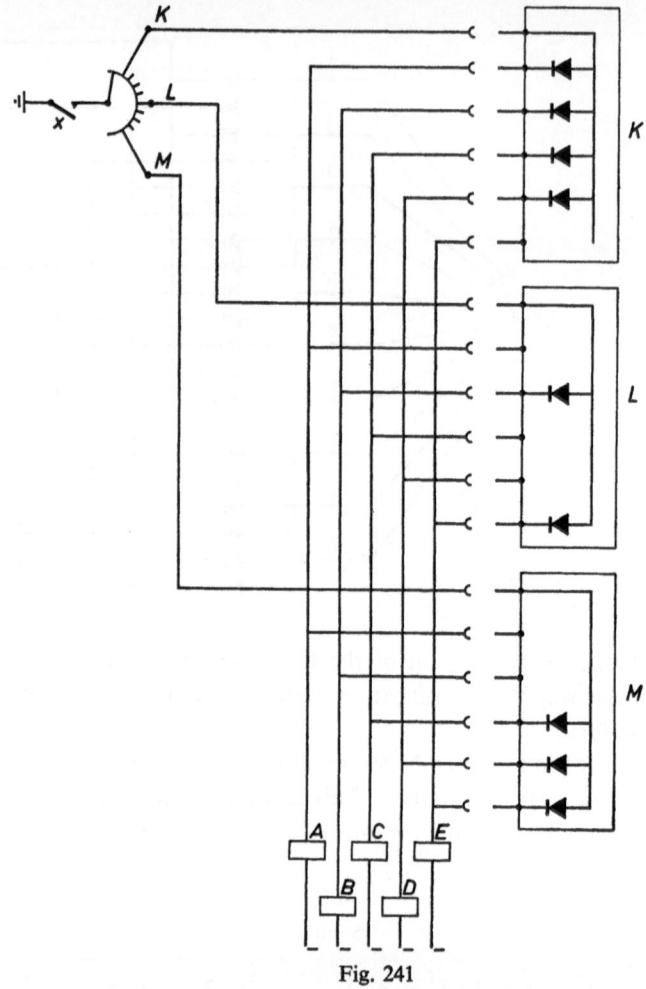

Fig. 241

In this example use is made of the 5-unit code (see Table LXXVI) as used in telegraphy. The letters K, L and M are repeated below.

$$K = 1\ 1\ 1\ 1\ 0$$
$$L = 0\ 1\ 0\ 0\ 1$$
$$M = 0\ 0\ 1\ 1\ 1$$

If the selector is connected to point K and contact unit x is closed, then a circuit is closed for relays A, B, C and D via the diodes of plug K. Points L and M can similarly be connected via plugs L and M respectively with a combination of the relays $A-E$, as soon as the selector has reached the position in question and contact unit x has been closed.

13.4 Analysis of information received by means of relays

Fig. 242 shows a circuit by means of which information is analyzed according to the 5-unit code and a circuit closed to one of a number of outputs. Use is made of a contact tree for this purpose; Fig. 242 only shows the contact units leading to the output L.

The analysis is not so simple if it has to be carried out on information consisting of two or more letters. If we were to use the same system as in Fig. 242 for this problem, we would have to follow each output of the contact tree for the first letter by a contact tree for the second letter. Fig. 243 however shows a solution to this problem using only two contact trees for the determination of both letters.

The information for the first letter is stored in a combination of the relays AE–EE. Only two outputs of the contact tree are drawn, those connected

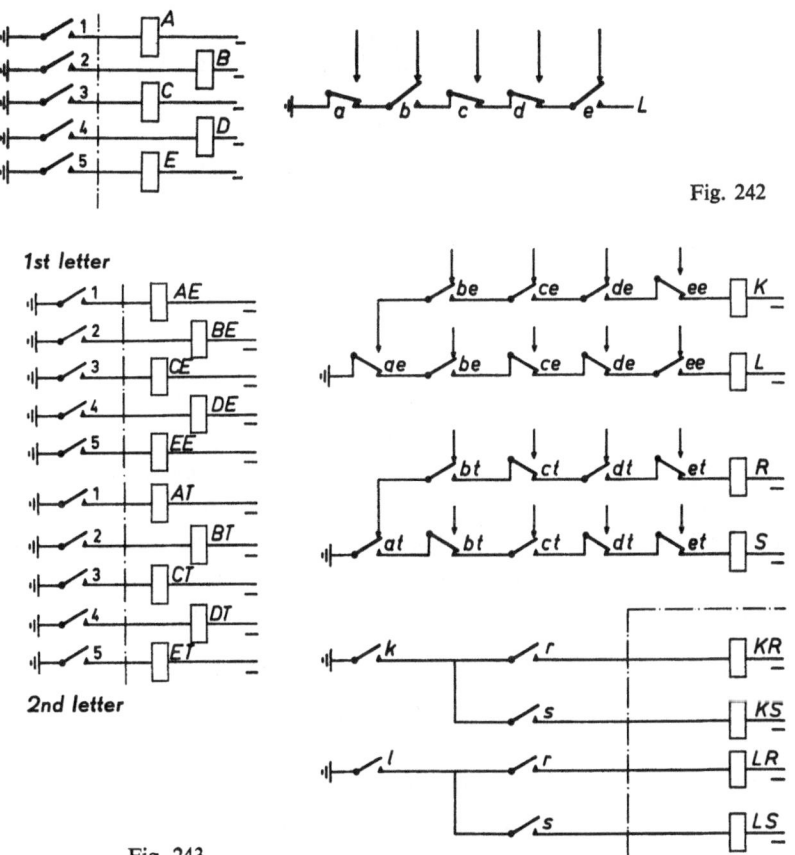

Fig. 242

Fig. 243

to relays K and L. The information for the second letter is stored in a combination of relays AT–ET. Two outputs of this contact tree are drawn too, those connected to relays R and S.

If e.g. the information KR must be analyzed, relays K and R operate, closing a circuit for relay KR. The three other letter combinations shown in the figure, KS, LR and LS, are analyzed in a similar way.

If we use for relays KR, KS, LR and LS a type which operates at a current i, but not at a current of about $\frac{1}{2}i$, the relays K, L, R and S of Fig. 243 can be dispensed with. This is illustrated in Fig. 244. The resistance of relays KR–LS is low compared to that of the resistors connected before them. They can only operate if switched on via both resistors. The diodes serve to decouple the relays.

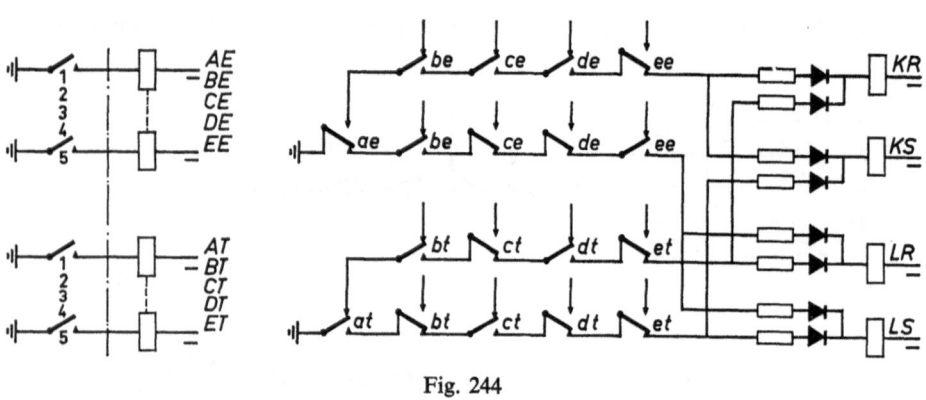

Fig. 244

13.5   Analysis by means of diodes, transistors and relays

In Section 13.4 we have already touched on the difficulty of analyzing information consisting of a combination of several letters.

These difficulties are basically due to the fact that the results for each decoded letter must be combined in series, which entails the use of an enormous amount of relays if a group of several letters has to be analyzed.

A more efficient method can be found by combining the decoded information for each letter in parallel.

Fig. 245 shows in some detail an analyzing circuit using diodes, transistors and relays. Diodes are mounted on the plug S, which is provided with a printed circuit. The diodes are mounted corresponding to the letter involved: diodes mounted one way round correspond to the "ones" of the code, and those mounted the other way round to the "zeroes".

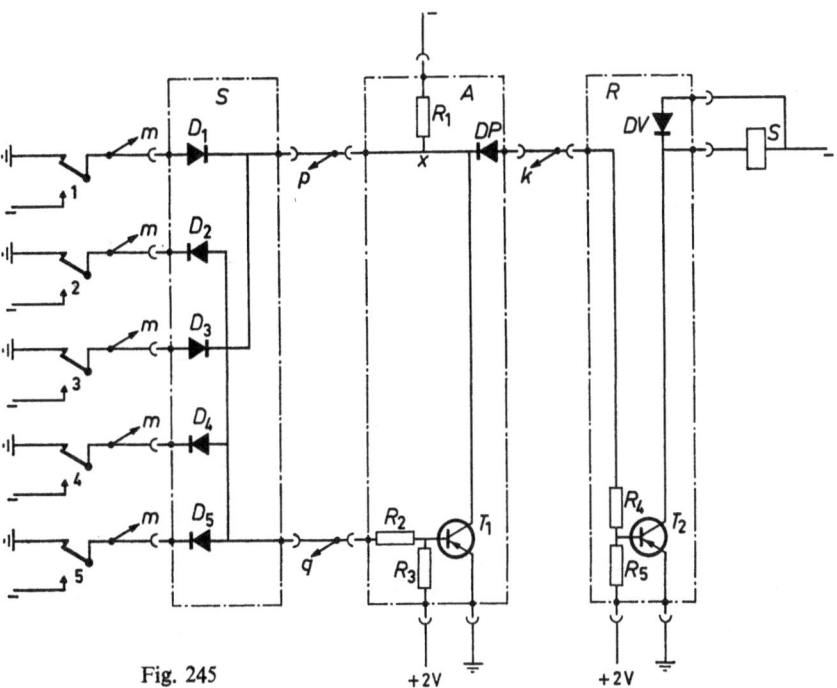

Fig. 245

+2V   +2V

In the example shown in Fig. 245, diodes D_1 and D_3 represent "ones" and diodes D_2, D_4 and D_5 represent "zeroes".

The code group presented by contact units 1 to 5 must agree with the code group represented by the diodes of the plug S. In the present case, this is the code group 1 0 1 0 0, corresponding to the letter S. If contact units 1 and 3 are indeed the only ones to be switched over, all the diodes are cut off, as a result of which the point X of plug A has a negative potential.

As a result of this, a circuit is formed from +2V via the resistances R_5 and R_4 of plug R, diode DP and the resistance R_1 to minus. The base of transistor T_2 thus becomes negative with respect to the emitter, making this transistor conducting so that the relay S is switched on via the collector.

If one or both of contact units 1 and 3 are in the rest position, the diode D_1 and/or D_3 is conducting, as a result of which the point X is at earth potential and the above-mentioned circuit via the resistances R_5, R_4 and R_1 cannot be formed. The transistor T_2 then remains cut off, and relay S does not operate.

If apart from contact units 1 and 3, one or more of the contact units 2, 4 and 5 are switched over, one or more of the diodes D_2, D_4 and D_5 becomes conducting and a circuit is formed from +2V via the resistances R_3 and R_2, the above-mentioned diodes D_2, D_4 and D_5 and the contact units in question to minus. The base of transistor T_1 thus becomes negative with respect to the emitter, so that T_1 becomes conducting and point X comes to be at earth potential. As a result of this, transistor T_2 cannot become conducting, and relay S does not operate.

If the information consists of more than one letter, the programmed plugs are connected in parallel at the points p and q, and are all connected with the same plug A.

Plug A may thus be regarded as an "address", denoted by the letter group in question. It is also possible that several addresses should result in the operation of the same relay S. This can be arranged by connecting several plugs in parallel to the multiple point K. Relay S will then operate each time one of the addresses in question is given. If however n addresses are intended for energizing relay S and one of these n addresses is given, then the other $(n-1)$ addresses will form an instruction *not* to energize relay S, while this relay should however operate in response to the address given. This is the reason for the inclusion of diode DP, which allows transistor T_2 to become conducting even though the point X in $(n-1)$ of the plugs A is at earth potential, because point X in one of the plugs A is negative.

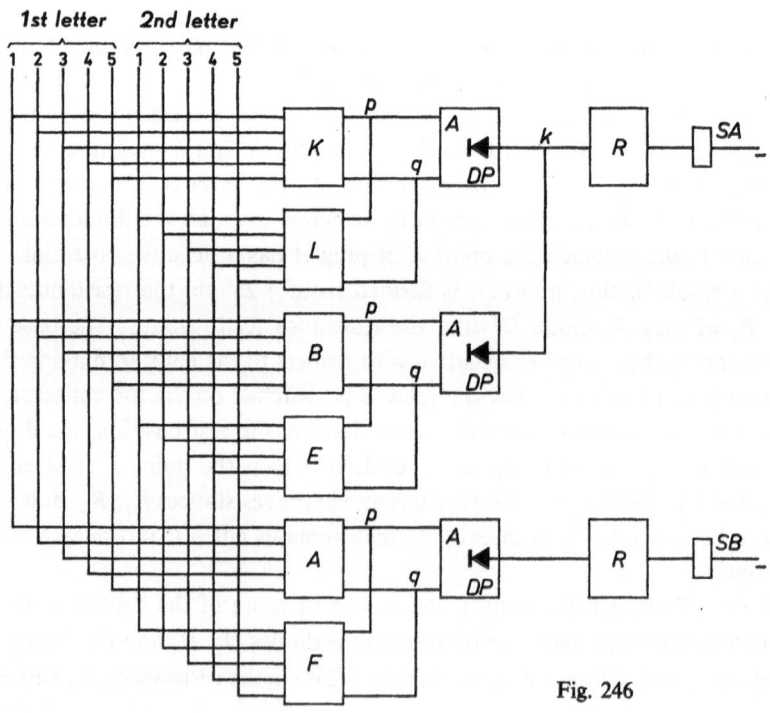

Fig. 246

This is worked out by way of example in Fig. 246 for three two-letter addresses, viz *KL*, *BE* and *AF*. The addresses *KL* and *BE* must cause relay *SA* to operate, while the address *AF* must only switch on relay *SB*. The multiple points shown in Fig. 245, viz *m*, *p*, *q* and *k*, are drawn in full in Fig. 246. The advantage of this method is that it is much faster than that illustrated in Fig. 243 and 244, because only one relay operate time is needed to arrive at the result. A second advantage is the increased flexibility of programming: if for example we want to replace address *KL* in Fig. 246 by *RS*, then all we have to do is to replace the plugs *K* and *L* by plugs *R* and *S*. Summing up, we may state that this solution of the analysis is based on the use of AND gates $(D_1$ to $D_5)$ and OR gates (DP).

13.6 Problems

1. Two points *A* and *B* are connected to the break and make sides respectively of a change-over contact unit *w*. The moving spring of the contact unit *w* is connected to earth. If contact unit *w* is in the rest position, point *A* must be identified by the display of the number 96 in the binary-decimal code by means of twice four lamps. If the change-over contact is switched over, point *B* must be identified by the display of the number 79 by the same lamps in the same code. Design the circuit for this, using not more than 2 diodes.

2. How many diodes are needed if the points *A* and *B* of question 1 must be denoted by the numbers 87 and 97?

3. A rotary selector has 10 positions. Each position must be identified by the switching on of two out of five lamps according to the 0 1 2 4 7 code. Design a circuit for this, using only 20 diodes, 5 lamps and one selector wiper.

4. Solve the above problem with two selector wipers but without diodes.

5. Two digits can be registered in the binary-decimal code by means of two groups of 4 relays. These relays are denoted by the letters *TA–TD* and *EA–ED* $(A=1, B=2, C=4$ and $D=8)$. Design an analysis circuit which switches on a red, white, yellow or green lamp on receipt of the following numbers:
25 switches on the red lamp

52 switches on the white lamp
26 switches on the yellow lamp
59 switches on the green lamp.
Use not more than 28 contact springs.

6. The lamps mentioned in the previous question must now be switched on according to the following scheme:
red lamp for odd tens and odd units
white lamp for odd tens and even units
yellow lamp for even tens and odd units
green lamp for even tens and even units.
Use not more than 9 contact springs for this circuit.

Chapter 14

SYMBOLS, CIRCUIT DIAGRAMS AND
SEQUENCE DIAGRAMS

14.1 Symbols

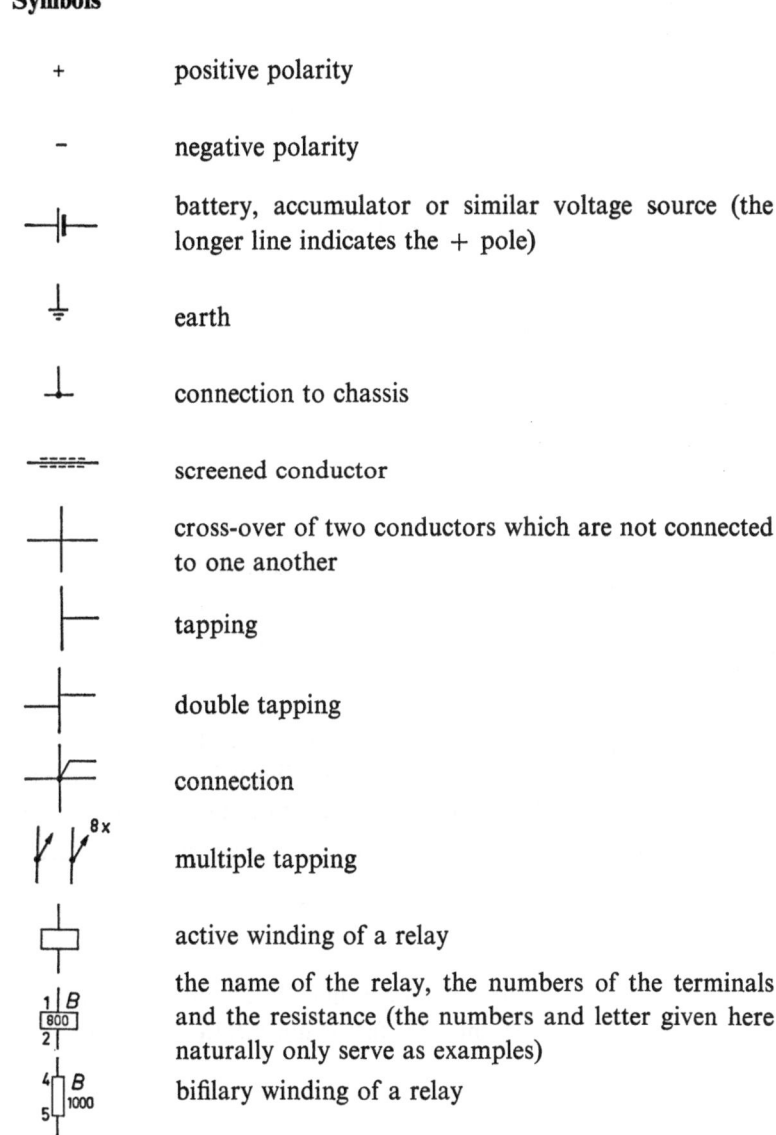

+   positive polarity

−   negative polarity

battery, accumulator or similar voltage source (the longer line indicates the + pole)

earth

connection to chassis

screened conductor

cross-over of two conductors which are not connected to one another

tapping

double tapping

connection

multiple tapping

active winding of a relay

the name of the relay, the numbers of the terminals and the resistance (the numbers and letter given here naturally only serve as examples)

bifilary winding of a relay

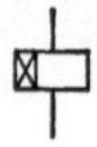

active winding of a relay with retarded operation

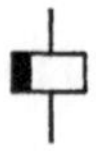

active winding of a relay with retarded release

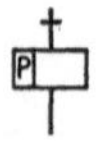

active winding of a polarized relay

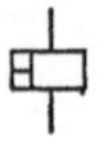

active winding of a high-speed relay

energizing coil of a selector, etc.

make contact unit

break contact unit

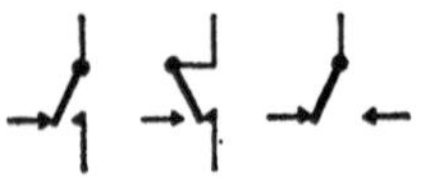

change-over contact unit

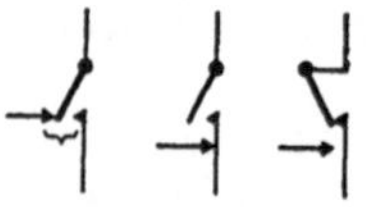

make-before-break contact unit

Relay windings are indicated by a capital letter, the contact units of the relay by the corresponding small letter.

change-over before change-over contact unit of relay A. Contact unit a^5 is not switched over until contact unit a^2 has switched over. This is indicated by the numbers 1 and 2 in brackets after the names of the contact units.

side stable change-over contact unit of a polarized relay

centre stable change-over contact unit of a polarized relay

tungsten contact unit. When silver is used for the contact unit, this fact is not mentioned.

contact bank

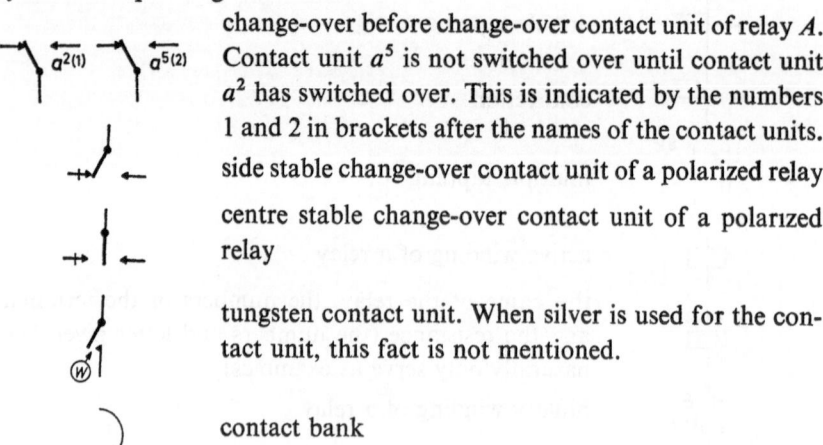

selector with a given initial position

selector with arbitrary initial position

two-motion selector

normal selector wiper
bridging wiper

resistor

resistor with fixed tapping

variable resistor

variable voltage divider (potentiometer)

resistor in tube

capacitor

polarized electrolytic capacitor

non-polarized electrolytic capacitor

non-locking push button A (make contact unit)

non-locking push button A (break contact unit)

locking push button B

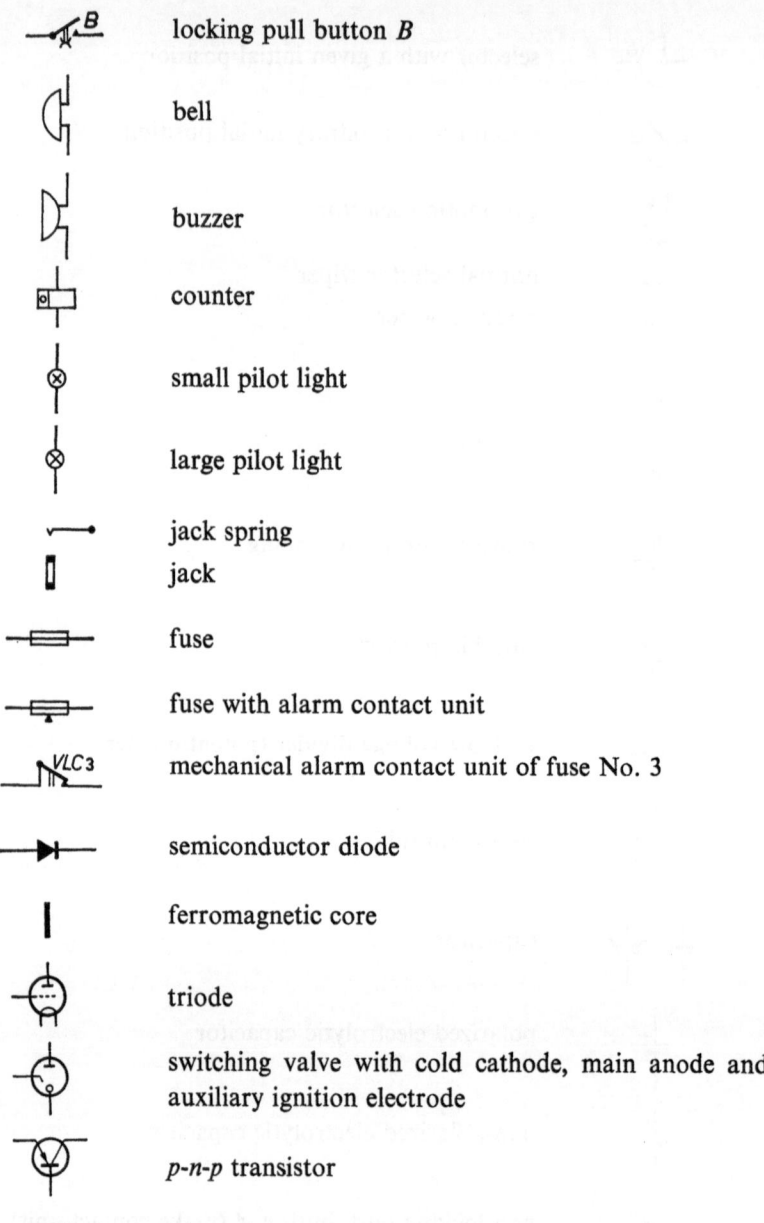

locking pull button *B*

bell

buzzer

counter

small pilot light

large pilot light

jack spring
jack

fuse

fuse with alarm contact unit

mechanical alarm contact unit of fuse No. 3

semiconductor diode

ferromagnetic core

triode

switching valve with cold cathode, main anode and auxiliary ignition electrode

p-n-p transistor

repeater

14.2   Circuit diagrams

The circuit diagram is intended to show the operation of a device or installation as clearly as possible. The principle used in this book for drawing circuit diagrams is the "separate contacts method". In this method, the mechanical connection between the various elements (e.g. relays and the corresponding contact units) is disregarded. The position of the circuit elements and their components in the diagram is determined by the order in which they are included in the circuit and by the desire to make the diagram as clear and as simple as possible. As far as possible, one avoids intersecting lines in the circuit diagram. The various parts of one and the same element, spread out throughout the circuit diagram in this way, must be denoted by suitable letters or numbers (e.g. relay A, with contact units a).

In order to obtain a clear circuit diagram, one must sometimes include a part, e.g. a contact unit of a relay, 2 or more times. This is then indicated by a number in a box beside the part in question.

$/a^3$ ▣   contact unit a^3 occurs 3 times in the circuit diagram.

If the make and break sides of a change-over contact unit must be drawn separately in the circuit diagram for the sake of simplicity, this is indicated as follows:

$|a^3$--   means that this also occurs in the diagram: --$a^3|$

The broken line indicates the presence of a contact shown in some other part of the circuit diagram.

The above two notations can be combined:

▣ $|a^1$---   The break contact unit is drawn twice, and the corresponding make contact unit is drawn elsewhere in the circuit diagram.

Wires which despite all precautions must cross one another are drawn as follows:

This indicates that there is no electrical connection between the wires. If there is an electrical connection, one draws:

14.3 Sequence diagrams

The object of a sequence diagram is to show simply and clearly the order in which a number of elements come into operation. It forms a valuable aid for the reading of complicated circuit diagrams. The making of a sequence diagram is also a first requirement for the designing of circuits.

In a time sequence diagram, the time axis is graduated so that the duration of the various operations can also be read off from the diagram.

The basic symbol for the (time) sequence diagram is a rectangle. A letter or number placed in this rectangle indicates the element in question. The top of the rectangle indicates the moment at which the element begins to operate (e.g. operation of relay), while the bottom of the rectangle indicates the moment at which the operation takes effect (e.g. in a relay the moment when the contact units have switched over).

A full line means that a certain state prevails (e.g. that a relay is energized). We shall illustrate this with reference to a couple of examples.

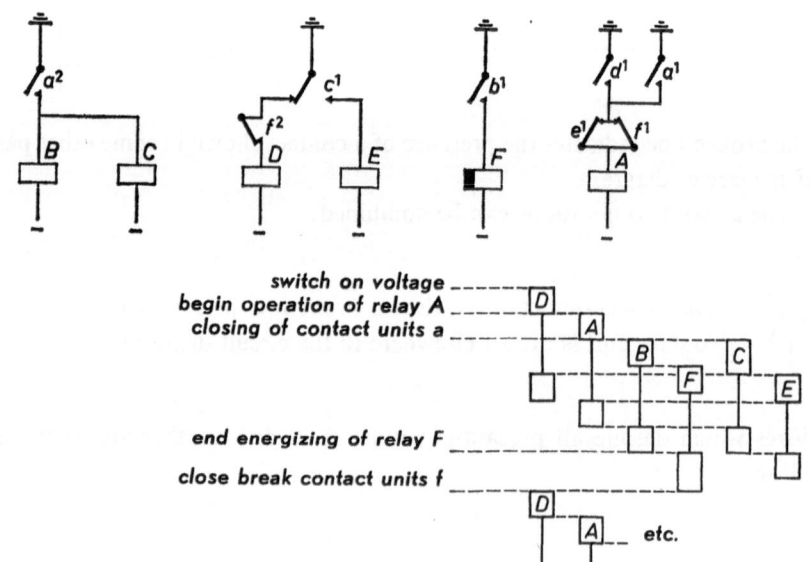

The situation shown in the above figure (p. 284) can be described in words as follows (certain self-evident aspects of the situation will not be mentioned): when the voltage is switched on, relay D will operate via contact units c^1 and f^2. The make contact unit d^1 closes a circuit for the relay A, which thus operates, and contact unit a^2 energizes relays B and C. Relay F can now operate via b^1, and relay E via c^1. The contact unit c^1 also breaks the circuit for relay D, which thus releases. Contact unit d^1 now tries to break the circuit of relay A, but contact unit a^1 has shorted d^1, so that relay A will not release until both e^1 and f^1 are open. In the sequence diagram we see that e^1 causes A to release. When A releases, the circuits for relays B and C are broken, so that these release too. The opening of make contact unit c^1 causes relay E to release, but the closing of break contact unit c^1 does not yet cause relay D to operate, since contact unit f^2 is still open. Relay F is cut off by b^1, but its release is retarded (indicated in the sequence diagram by a longer rectangle). When the release of F is complete, D can operate again, and so on.

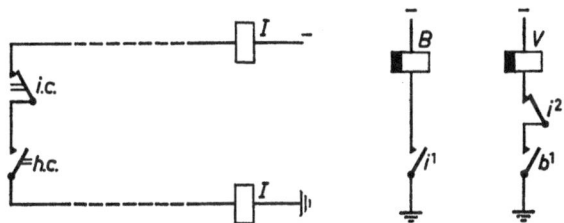

The closing of $h.c.$ (hook contact unit) causes relay I to operate. Contact unit i^1 makes relay B operate, and i^2 ensures that although contact unit b is closed relay V cannot operate yet. The release of relay B is very strongly retarded. If now the circuit for relay I is broken by the pulse contact unit ($i.c.$), the frequency of this interruption can be chosen so that relay I can follow it, while relay B cannot. Relay I will thus alternately release and operate, while relay B will remain energized because of its high inertia. The release of relay I will cause relay V to operate via contact units b^1 (still closed) and i^2. Relay V will not release between two successive interruptions of the circuit of relay I, but only at the end of the series of interruptions. Two successive pulse trains can be distinguished from one another in this way, since the interval between the two pulse trains is greater than the interval between successive pulses of the train. The sequence diagram now becomes:

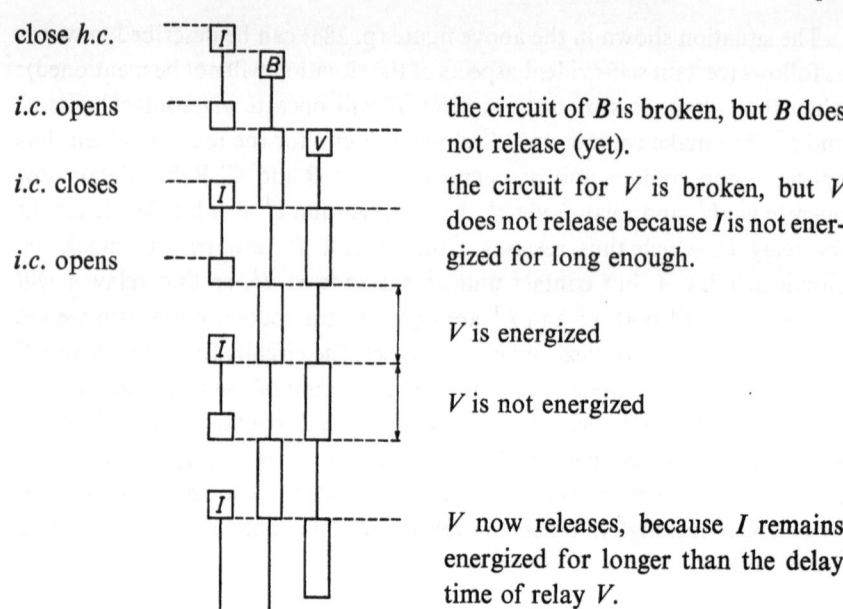

close *h.c.*

i.c. opens — the circuit of *B* is broken, but *B* does not release (yet).

i.c. closes — the circuit for *V* is broken, but *V* does not release because *I* is not energized for long enough.

i.c. opens —

V is energized

V is not energized

V now releases, because *I* remains energized for longer than the delay time of relay *V*.

If a relay has more than one winding, this can be indicated in the (time) sequence diagram by drawing more than one vertical line under the rectangle concerned. One can then denote the winding corresponding to each vertical line by two digits.

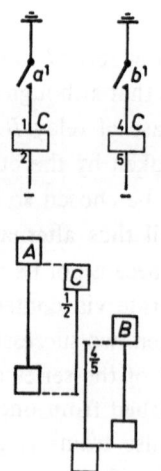

Chapter 15

ANSWERS TO PROBLEMS

CHAPTER 1

1. $a = 1$ A; $b = 114$ mA; $c = 65.5$ mA.
2. $a = 7460$ turns $b = 7650$ turns; 0.07 mm; 14.2% ohm;
   21.32% turns $c = 23.2\%$ packing.
3. The minimum energization is 107 AT.
   A relay with 3 make contact units and a 0.2 or 0.1 mm residual pin.
4. $a = 145$ AT; 3720 turns. Number of hold $AT = 77$.

$$R = \frac{\dfrac{43.3720}{77} - 1100}{1.2} = 816\,\Omega.$$

5. See theory
6. The first circuit needs a relay A with 3 break and 6 make contact units.
   Necessary number of operate $AT = 1.2 \times 179 = 215$.
   The second circuit needs a relay A with 3 change-over and 3 make contact
   units. Necessary number of operate $AT = 1.2 \times 157 = 187$.
   The number of operate AT which can be obtained, taking the given
   tolerances into account, is:

$$\frac{43}{1200 + 120} \cdot 6000 = 195.$$

   The given relay is thus suitable for the second circuit.
7. See theory.
8. See theory.

CHAPTER 2

1.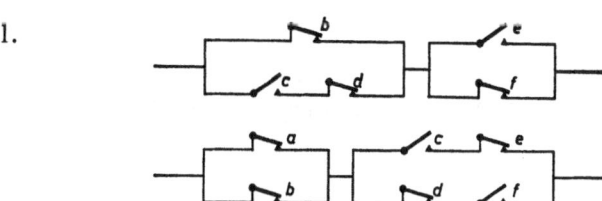

2.

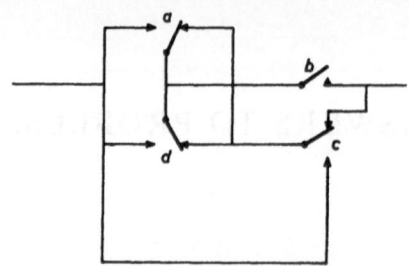

3.

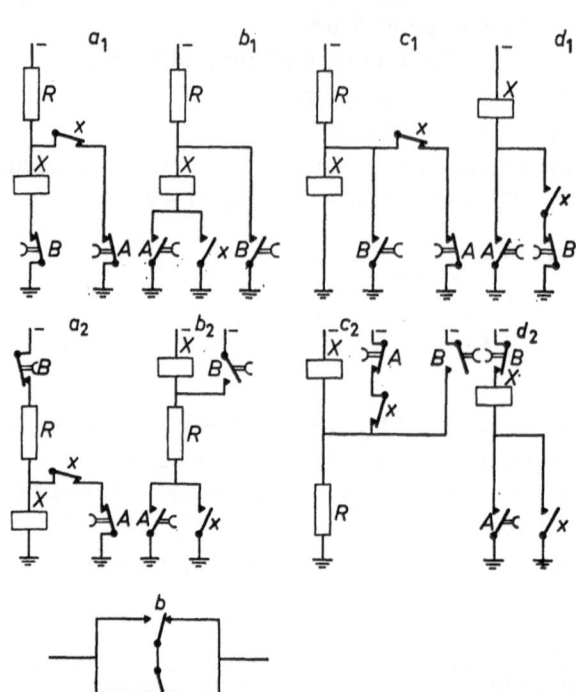

4.

5. When one of the circuits is broken, the number of operate AT reached under unfavourable circumstances is

$$\frac{44}{1200} \cdot 4200 = 154AT$$

A relay with one change-over contact unit and a 0.2 mm residual pin needs $1.2 \times 90 = 108\ AT$.

If both circuits are switched on, then winding 1–2 can have 154 operate AT, while winding 5–4 can reach the following number of operate AT:

$$\frac{56}{1200} \cdot 4200 = 196$$

if the resistances of the resistor and of winding 5–4 have both the minimum value. The difference between the above two values is $196 - 154 = 42$ AT. Relay X with one change-over contact unit and a 0.2 mm residual stop may have only $0.8 \times 48 = 38.4$ AT if one wishes to be sure that it will not operate. Relay X does not thus work properly under all possible circumstances in this circuit.

1st solution

Give relay X 3 change-over contacts and a 0.1 mm residual pin. The number of AT now needed for operation (Table I) is $1.2 \times 127 = 152.2$. The number of AT at which one can be certain that the relay will not operate is $0.8 \times 71 = 56.8$. Relay X will now work satisfactorily in the given circuit.

2nd solution

Leave relay X as it is, and increase the resistance in the circuit to 1400 Ω. The number of operate AT now becomes

$$\frac{44}{1600} \cdot 4200 = 115.5,$$

which is enough to energize relay X. The number of non-operate AT is here

$$\frac{56}{1600} \cdot 4200 - \frac{44}{1600} \cdot 4200 = 32.5.$$

This alteration thus also causes relay X to work satisfactorily.

 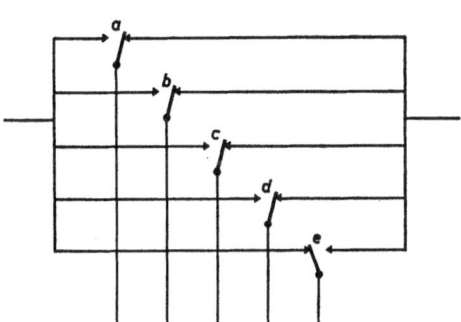

8.

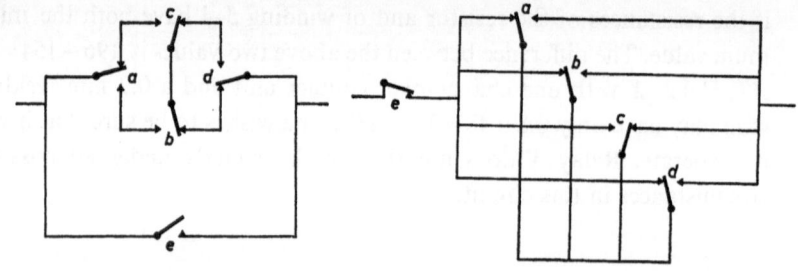

CHAPTER 3

1. $x(y+z)+yz$ or $y(x+z)+xz$ or $z(x+y)+xy$
2. $z(x+y')$
3. $xyz+x'y'z'$
4. $x'y'+yz+w'z'$ (see example 4).
5. $a'(b+c+d')+(a+d)\{f'+(b'+c)e'\}$
6. $a'd(b'+c)$
7. $a'b+b'c+ac'=a'b+b'c+ac'+a'c+bc'+ab'$ (see example 4)
   $a'b+b'c+ac'+a'c+bc'+ab'=a'c+bc'+ab'$ (theorem 22).
8. Multiplying out gives:
   $ab'c+ab'c'+a'bc'+abc'+a'bc+a'b'c=$
   $ab' \qquad +a'b \ +abc' \qquad +a'b'c=$
   $ab' \qquad +a'b \ +ac' \qquad +a'c \ =a'(b+c)+a(b'+c')$

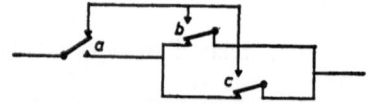

9. a) $(a+b+eg+fg+f')(a+b'+e+f)(g'+e+f)(g+f+h')(g+h)$
   b) $g(e+f)+hf(a+b)$
   c)

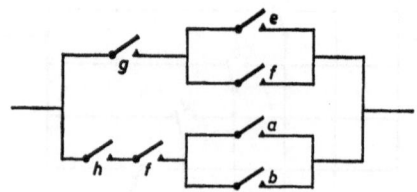

10. $a+b+cd'+c'd$

11. Multiplying out, we find:
$$a'b'+a'c'+a'd+ab+bc'+bd+ac+b'c+cd+ad'+b'd'+c'd'.$$
Application of theorem 22 to the above can yield both:
$$ab+b'c+c'd'+a'd$$
and:
$$a'b'+bc'+cd+ad'$$
whence the desired equation follows.

12. a) We must prove that:
$$ac+bd+b'c+bc'=cd+ab+b'c+bc'.$$
This can be done with the aid of theorem 22.

b) $b'c'+a'bcd'$

c)

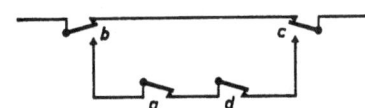

13. $bd'f(a+g)$

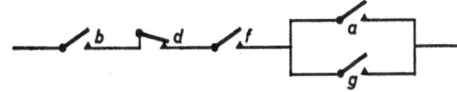

14. a) $ab(e+g)(g+f)f'+ab'ef+g'ef+gh'f+gh$
    b) $g(ab+f+h)+ef$
    c)

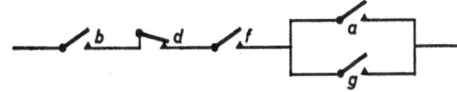

d) $(a'+b'+e'g'+g'f'+f)(a'+b+e'+f')(g+e'+f')(g'+h+f')$
   $(g'+h')$

15. We can write the following expression for the left-hand figure:
$$a'bdf+ac'df'+ac'd'e'f'+ab'c'd'e$$
while for the right-hand figure we have the algebraic form:
$$a'bdf+ab'c'f'+ac'e'f'+ac'def'+ab'c'd'e$$

If we can now prove that these two expressions are equivalent, then the simplification is justified.

a) $a'bdf + ac'df' + ac'd'e'f' + ab'c'd'e =$
   $a'bdf + ac'(df' + d'e'f' + b'd'e) =$
   $a'bdf + ac'(df' + e'f' + b'd'e + b'f'e + b'f') =$
   $a'bdf + ac'(df' + e'f' + b'd'e + b'f')$

b) $a'bdf + ab'c'f' + ac'e'f' + ac'def' + ab'c'd'e =$
   $a'bdf + ac'(b'f' + e'f' + def' + b'd'e) =$
   $a'bdf + ac'(b'f' + e'f' + df' + b'd'e)$

16. a)

b)

c)

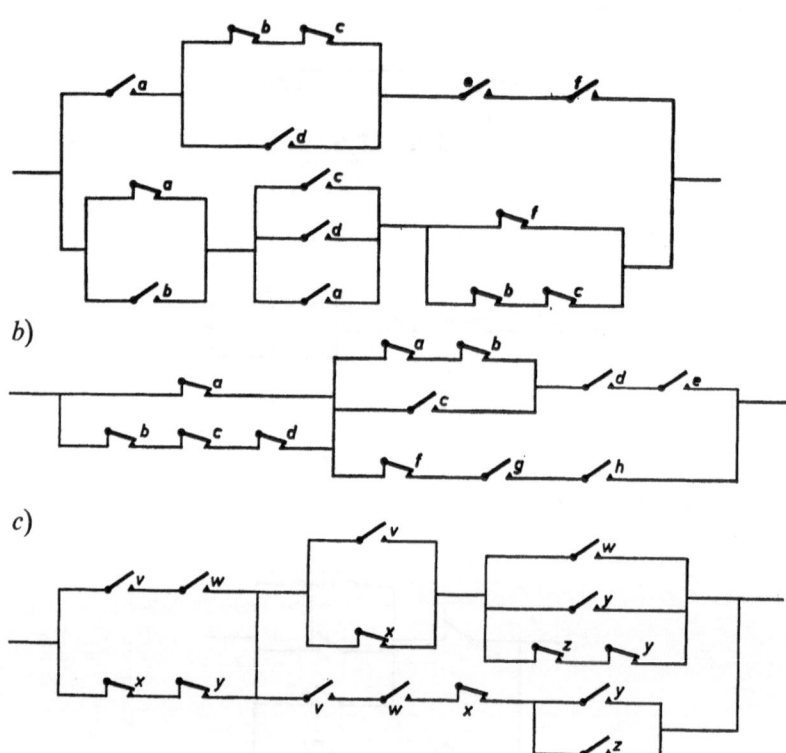

17. a) $a'b(c+d+e') + c'\{b'(c+d+e') + ade\}$
    $\{(v+x')(w+y+z'y') + vwx'(y+z)\}(vw+x'y')$

18. a) $ab' + a'd + bc' + cd'$

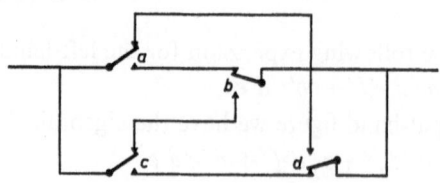

b) $abc+a'b'c'$

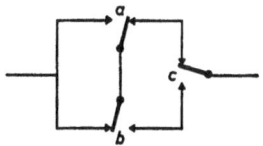

c) $(a+b)(a'+c')$ or $ac'+a'b$

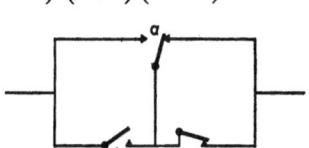

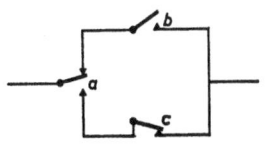

19. a) $(a'+b'+c'+d+e'+f')(a+b+c+d'+e+f)$

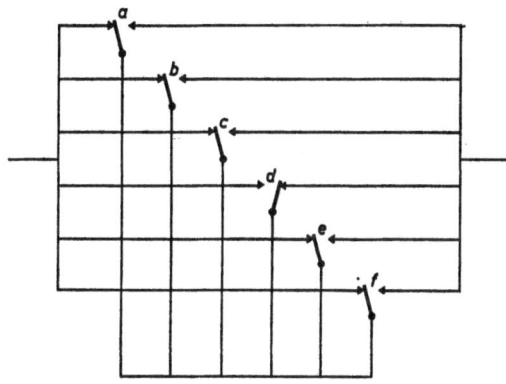

b) $(a'+b)(a+c'+e)(a+c+d)$

20. $x(w+y')+w'y'$ or $y'(x+w')+wx$

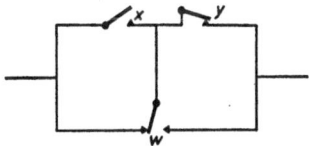

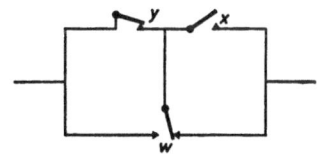

21. $(a+b+c'+d')(a+b'+c+d')(a+b+c'+d)(a+b+c+d')$
$(a'+b+c'+d)(a+b+c+d)$
Multiplication of the 2nd and 4th terms gives: $(a+c+d')$
Multiplication of the 3rd and 6th terms gives: $(a+b+d)$
Multiplication of the 1st and 5th terms gives: $(b+c'+ad+a'd')$

$$(a+c+d')(a+b+d)(b+c'+ad+a'd')=$$
$$(a+c+d')(b+ac'+ad+c'd)=$$
$$ab+ac'+ad+bc+bd'=ac'+ad+bc+bd'=a(c'+d)+b(c+d')=$$

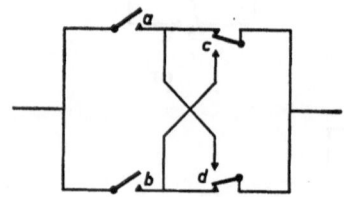

22. $(a'+b+c'+d'+e'+f)(a+b'+c+d+e+f')$

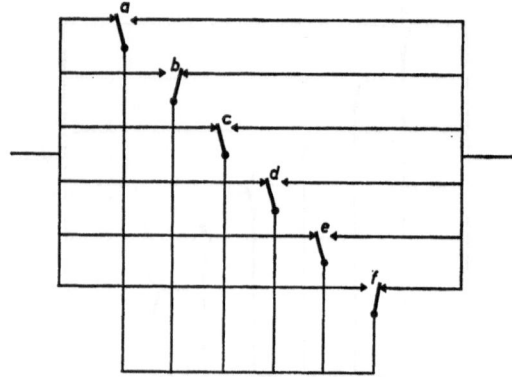

23. This can be proved with the aid of theorem 22.

24. $d'(ac'e'+b)+b'd$

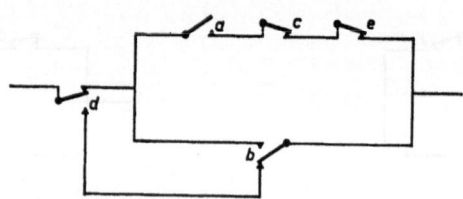

25. $ac'+bd'+d(c+e)$

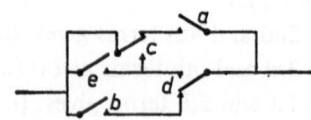

26. For the green lamp: $ab' + a'b + bc$
    For the red lamp   : $c'(ab' + a'b)$

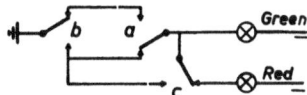

27. $abcd + a'b'c'd'$

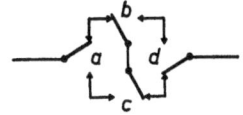

28. This can be proved by inversion and application of theorem 22.

CHAPTER 4

1. 1 1 0 1 1 1 1 1 1

2. $\dfrac{12!}{5!7!} = 792$

3. | A B C D E | A B C D E | A B C D E | A B C D E |
|---|---|---|---|
| 0 0 0 0 0 | 0 0 1 1 0 | 0 0 0 1 1 | 0 0 1 0 1 |
| 1 0 0 0 0 | 1 0 1 1 0 | 1 0 0 1 1 | 1 0 1 0 1 |
| 1 1 0 0 0 | 1 1 1 1 0 | 1 1 0 1 1 | 1 1 1 0 1 |
| 0 1 0 0 0 | 0 1 1 1 0 | 0 1 0 1 1 | 0 1 1 0 1 |
| 0 1 1 0 0 | 0 1 0 1 0 | 0 1 1 1 1 | 0 1 0 0 1 |
| 1 1 1 0 0 | 1 1 0 1 0 | 1 1 1 1 1 | 1 1 0 0 1 |
| 1 0 1 0 0 | 1 0 0 1 0 | 1 0 1 1 1 | 1 0 0 0 1 |
| 0 0 1 0 0 | 0 0 0 1 0 | 0 0 1 1 1 | 0 0 0 0 1 |

4. | 1 1 0 1 1 0 | 54 |
|---|---|
| 1 0 1 1 1 1 | 47 |
| 1 1 0 0 1 0 1 | 101 |

5. $2^7 = 128$

6. $\dfrac{8!}{8!} + \dfrac{8!}{7!1!} + \dfrac{8!}{6!2!} + \dfrac{8!}{5!3!} + \dfrac{8!}{4!4!} + \dfrac{8!}{3!5!} =$

   $1 + 8 + 28 + 56 + 70 + 56 = 219$

   Another solution:

   $2^8 - \left(\dfrac{8!}{8!} + \dfrac{8!}{1!7!} + \dfrac{8!}{2!6!} \right) = 256 - (1 + 8 + 28) = 219$

7. *A B*
   B C
      C D
         D E
            E F
 A   *F*
 A   *E*
 A   *D*
 A   *C*
     C   *F*
    C   *E*
  B   *E*
  B *D*
     D   *F*
  B   *F*

8. *a)* 0. 1 1 0 1 0 0 1 1 1 0 1
   b) 0. 1 1 1 1 1 0̸ 0̸ 1̸ 1̸
   c) 0. 1 0 1 1 0 1
   d) 0. 0 0 1 1 0 1

9. *a)* 0.7265625
   b) 1.8515625
   c) 0.6953125
   d) 0.171875

10.

	0	1	2	4	7
4	1	0	0	1	0
9	0	0	1	0	1
4	1	0	0	1	0
8	0	1	0	0	1
9	0	0	1	0	1
P.C.	1	0	1	1	0

CHAPTER 5

1. The red, yellow, green and white lamps burn in all positions, except positions 4, 3, 2 and 1 respectively. The circuit is as follows:

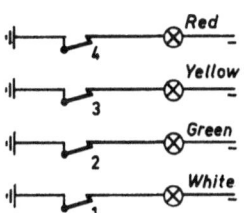

2. *a*) 28 relays
   b) 15 relays
   c) 8 relays or 4 relays (+ 4 capacitors).

3.

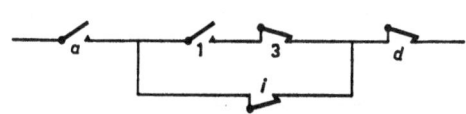

4.

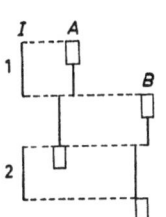

Relay *B*
releases after
the 2nd pulse,
when it should
still remain
energized

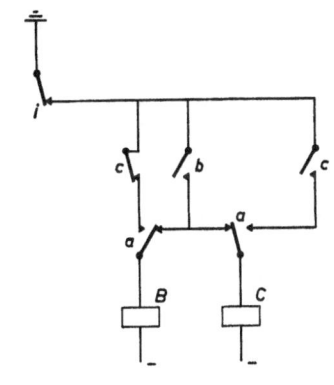

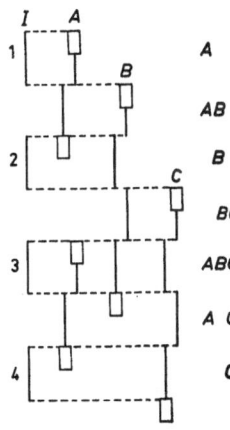

5.

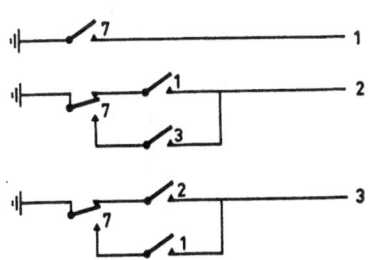

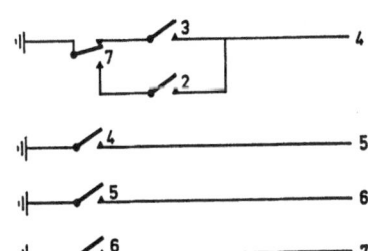

6.

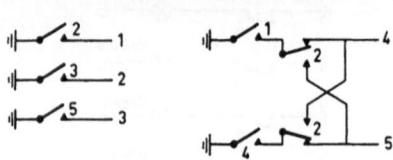

7.

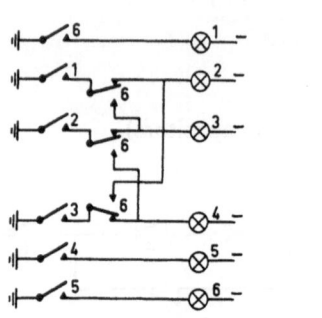

8. *a)*

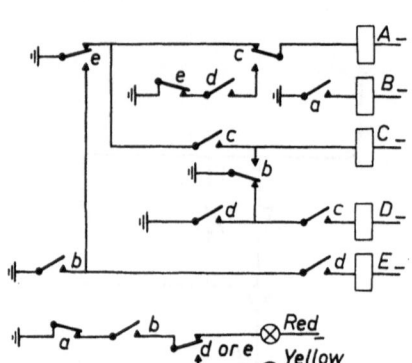

b)

CHAPTER 6

1.

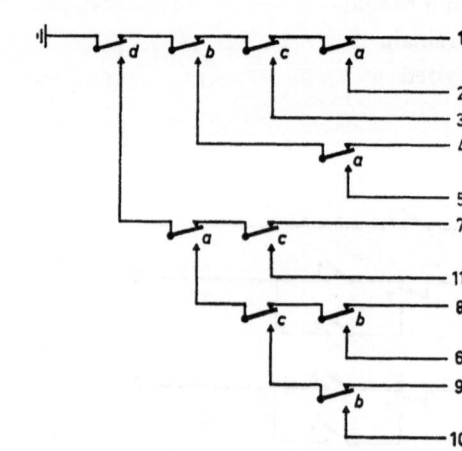

2. 1 auxiliary relay

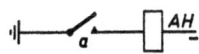

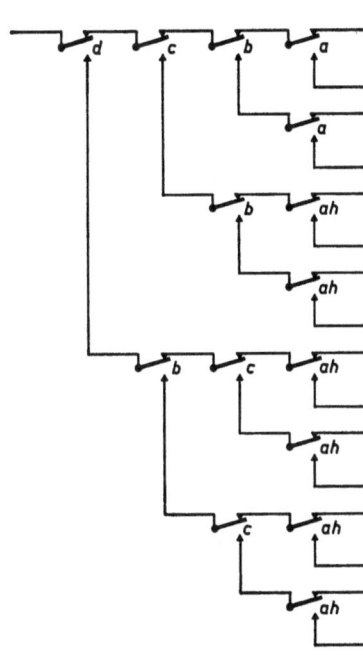

3. a = reflecting code

$b =$

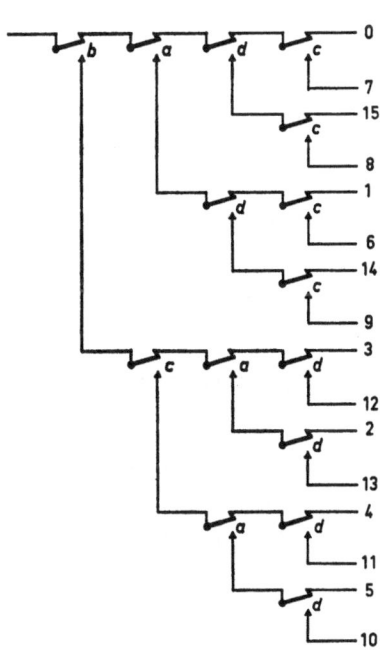

4.

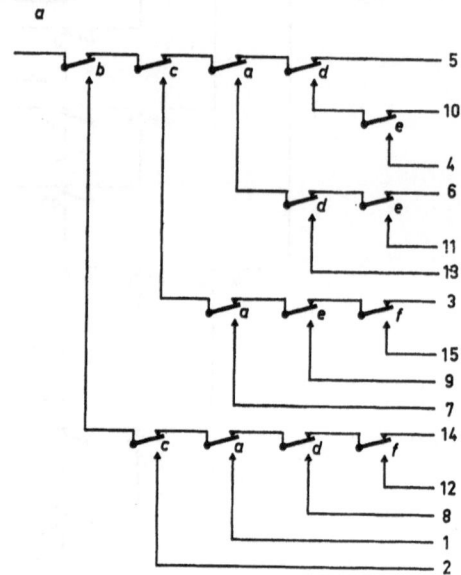

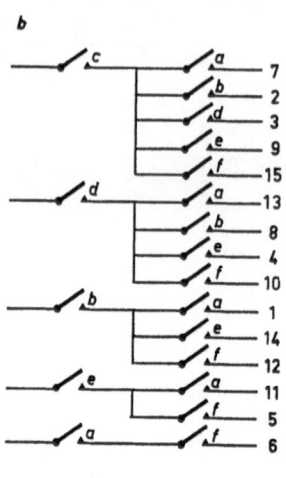

5. See page 301.

6. *a)*   *b)*

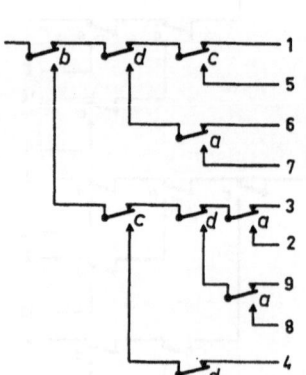

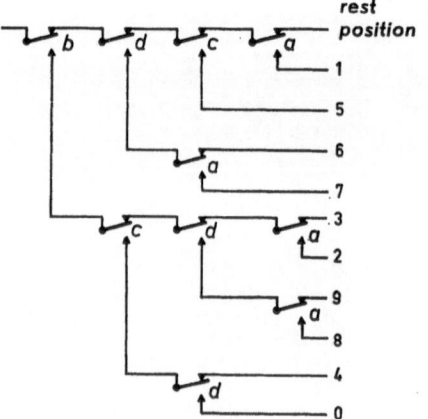

5.

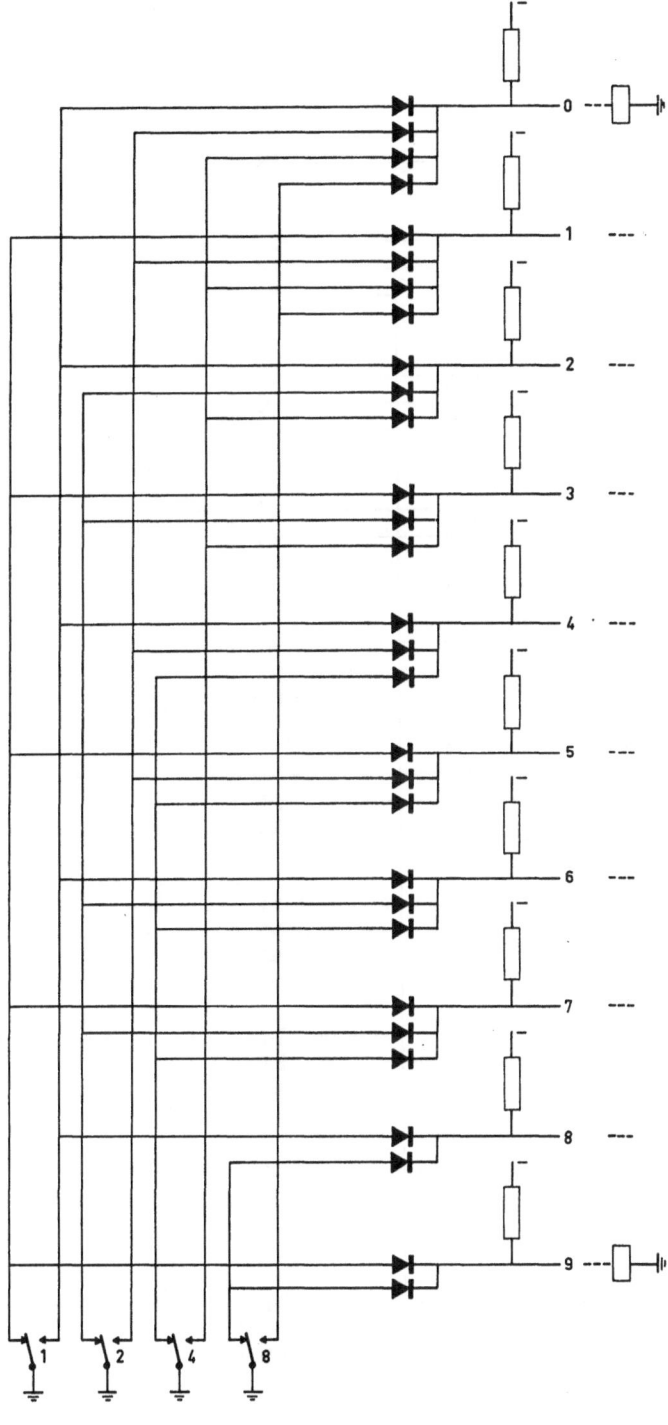

7.

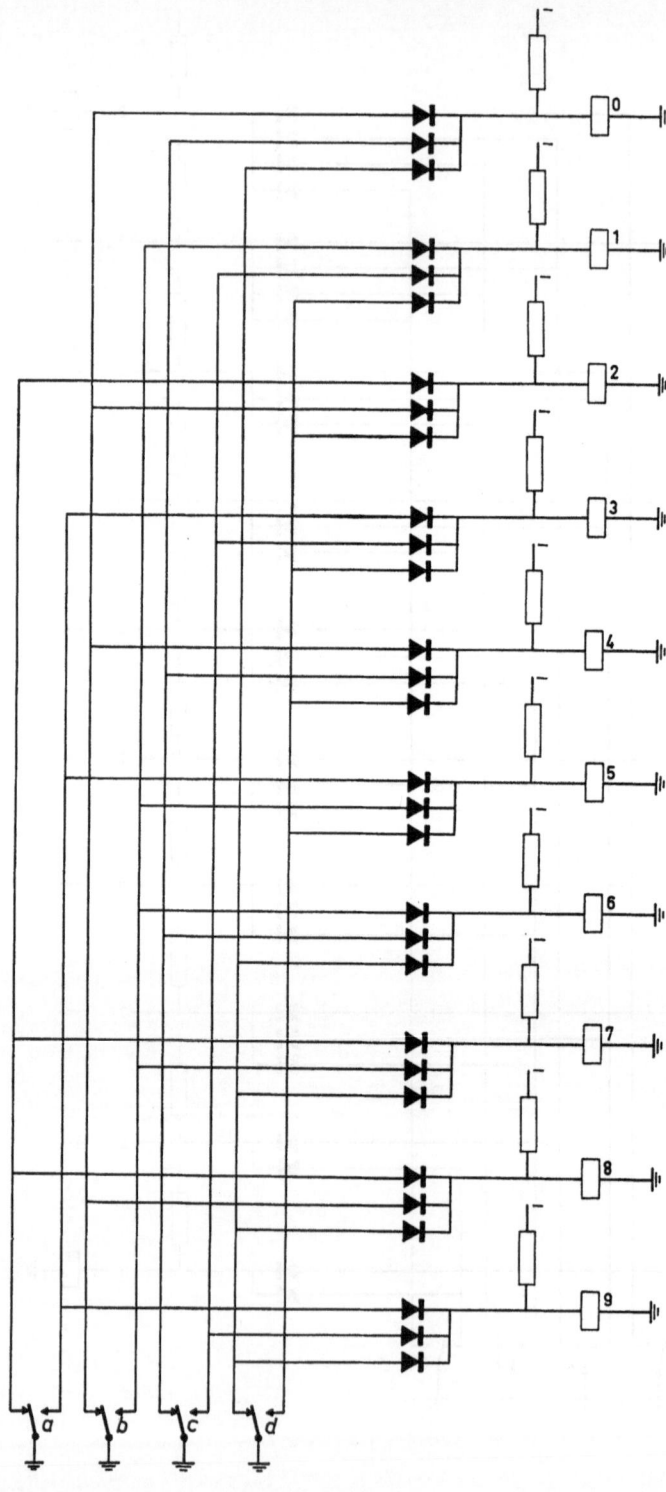

8.

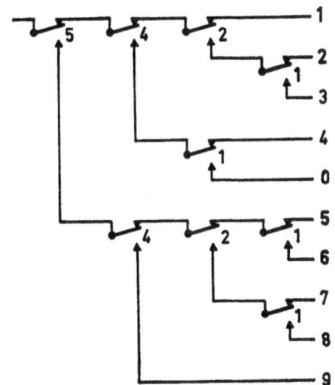

CHAPTER 7

1.

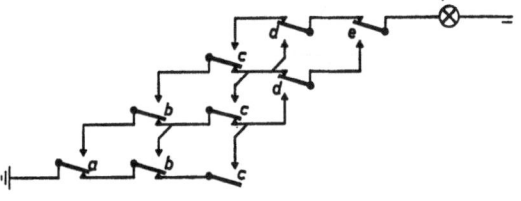

2.

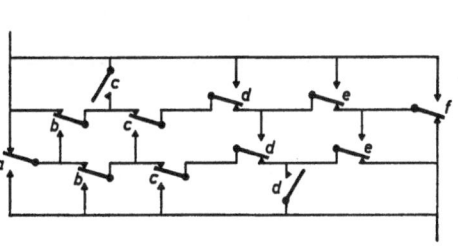

3.   4.

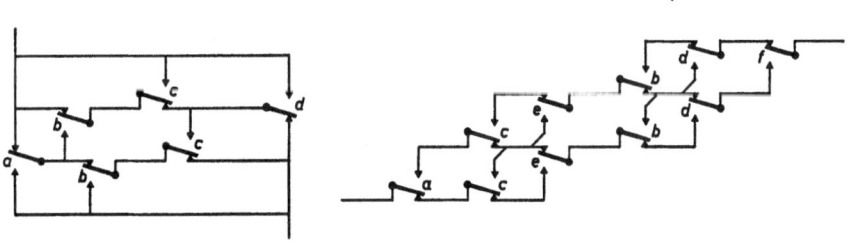

5. *a*) =6 *b*) Yes, with a 2-out-of-6-code.

c.

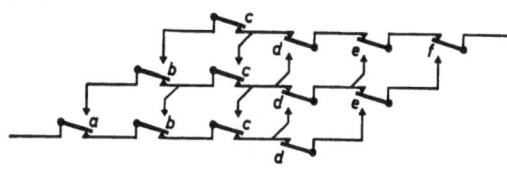

6.

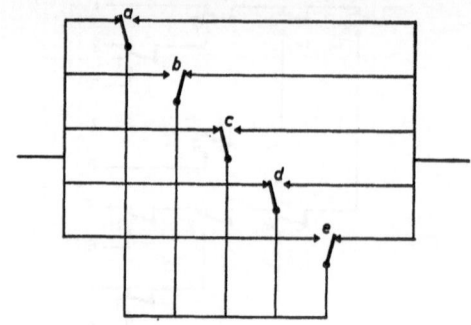

7.

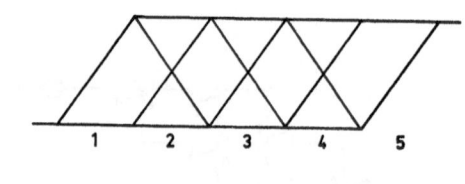

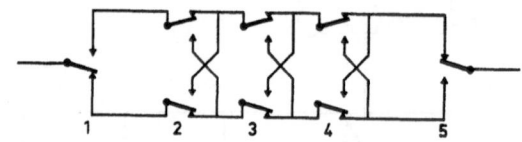

8.

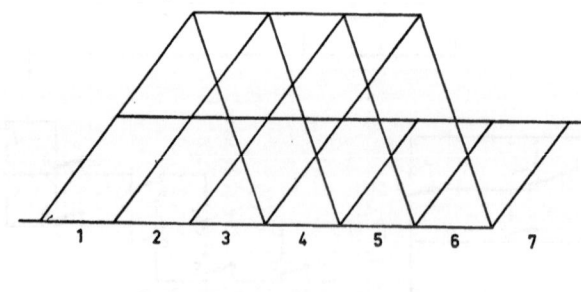

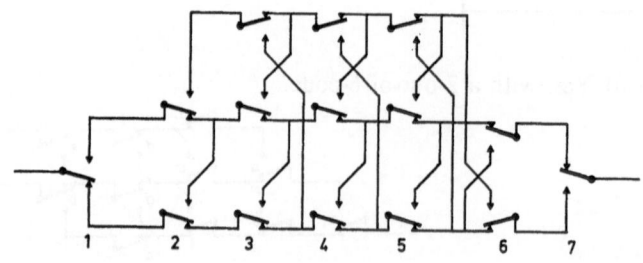

9.

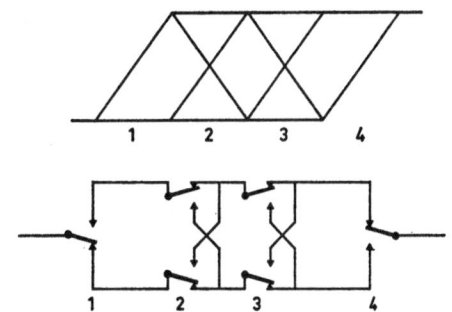

CHAPTER 8

1. 1 0 1
2. 1 1 0 1
3. *a)*

```
            0 0 1 1    0 0 1 0    0 1 1 1
            0 1 0 0    1 0 0 1    1 0 0 0
           ─────────  ─────────  ─────────
            0 1 1 1    1 0 1 1    1 1 1 1
correction             0 1 1 0    0 1 1 0
carry            1          1
           ─────────  ─────────  ─────────
            1 0 0 0    0 0 1 0    0 1 0 1
```

b)

```
   0 1 0 1    1 0 0 0    0 1 1 1
   0 0 1 1    0 0 0 1    0 0 1 0
  ─────────  ─────────  ─────────
   1 0 0 0    1 0 0 1    1 0 0 1
```

c)

```
            0 1 1 0    0 0 1 0    0 1 1 1
            0 0 1 0    1 0 0 1    0 0 1 1
           ─────────  ─────────  ─────────
            1 0 0 0    1 0 1 1    1 0 1 0
correction             0 1 1 0    0 1 1 0
carry            1          1
           ─────────  ─────────  ─────────
            1 0 0 1    0 0 1 0    0 0 0 0
```

4. *a)* minuend   1 0 0 0   0 0 1 1   0 1 1 1
   lend   1 1 1 1
                ───────── ───────── ─────────
                 0 1 1 1   0 0 1 1   0 1 1 1
   borrow   1 0 1 0
                ───────── ───────── ─────────
                 0 1 1 1   1 1 0 1   0 1 1 1
   subtract (compl.) 1 1 0 0   1 0 1 0   1 0 1 1
                ───────── ───────── ─────────
                 0 0 1 1   0 1 1 1   0 0 1 0
   around-end carry 0 0 0 1   0 0 0 1   0 0 0 1
                ───────── ───────── ─────────
   result   0 1 0 0   1 0 0 0   0 0 1 1

b) minuend   1 0 0 1   0 0 1 0   0 1 1 0

     lend   1 1 1 1   1 1 1 1

                  1 0 0 0   0 0 0 1   0 1 1 0

     borrow   1 0 1 0   1 0 1 0

                  1 0 0 0   1 0 1 1   0 0 0 0

     subtract (compl.)   1 1 0 1   0 1 1 1   1 0 0 0

                  0 1 0 1   0 0 1 0   1 0 0 0

     around-end carry   0 0 0 1   0 0 0 1   0 0 0 1

     result   0 1 1 0   0 0 1 1   1 0 0 1

c) minuend   0 1 0 0   0 0 1 1   0 1 0 1

     subtract (compl.)   1 1 0 0   1 1 1 0   1 1 0 1

                  0 0 0 0   0 0 0 1   0 0 1 0

     around-end carry   0 0 0 1   0 0 0 1   0 0 0 1

     result   0 0 0 1   0 0 1 0   0 0 1 1

CHAPTER 9

1. *a)* Gate-lock-out circuit.

  b) The circuit is as on page 307.

2.

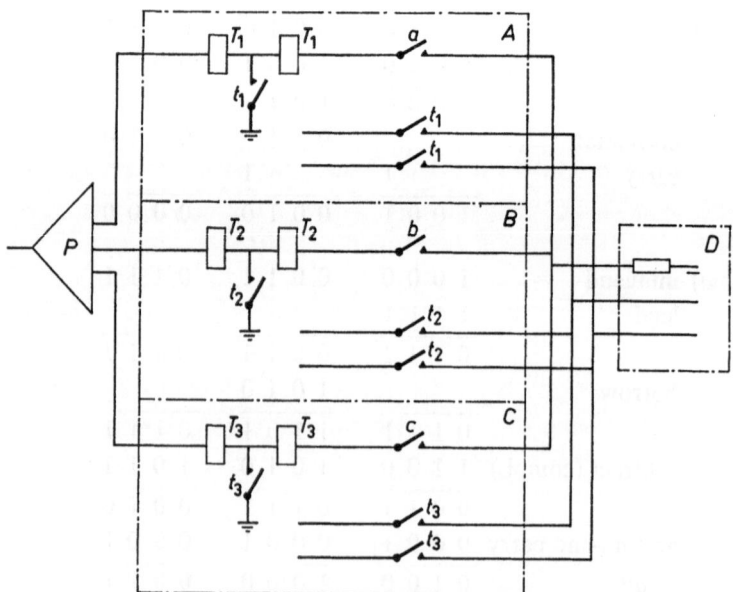

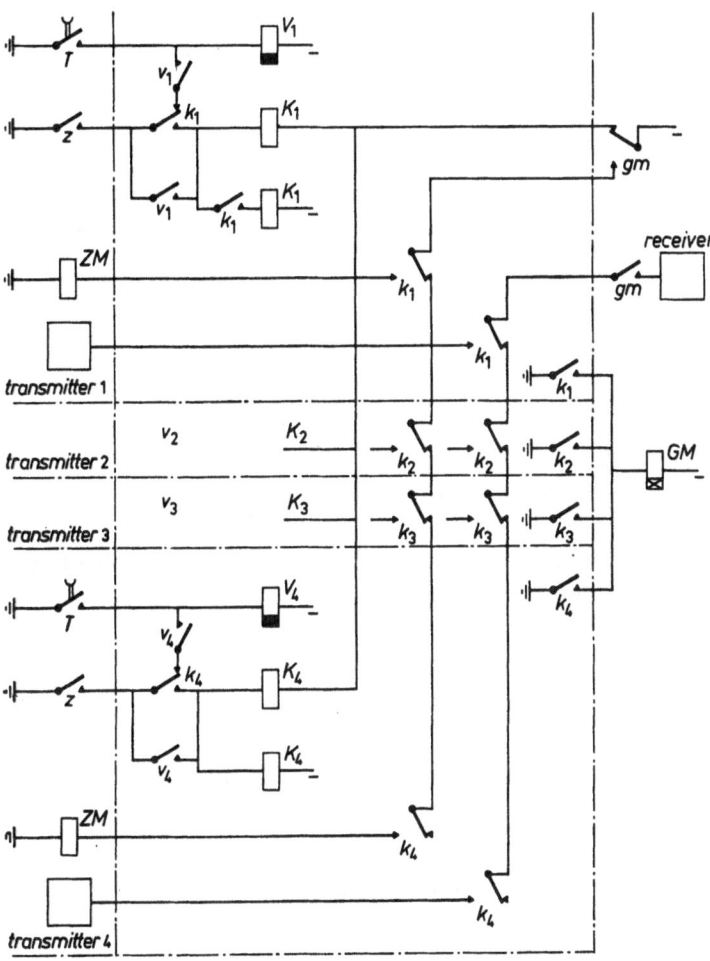

3. *a* and *b*) Circuit as given in the answer to question 1.
   c) The circuit of Fig. 156, Chapter 9.

4. *a*) Relays *A, B, ..., N* must be faster than relay *SL*, because break contact
      units of *SL* are included in the switching-on circuits of *A, B, ..., N*.
   b) Relay *SL* is common to the whole circuit. It is thus important that it
      should be reliable.
   c) By adding a differential relay in series with the windings of *SL*.

5. In order to bridge small time gaps during the making or breaking of
   contact units *a–n*.

6. Because once the relays A–N are energized, they will be connected to minus after the opening of contact unit s, via another contact unit s, and may thus remain energized.

7. a) $R_1 = 800\,\Omega$ $R_2 = 750\,\Omega$
   b) 3 V
   c) $9.6 - 3 = 6.6$ V
   d) No current flows. The diode must therefore be cut off.

CHAPTER 10

1. a) $(m+n)v = 51$
   b) $m \times n = 72$
   c) See theory of Section 10.1.

2. 3 connecting lines give $(m+n)v = 39$ points of intersection and thus still the possibility of internal blocking.
   4 connecting lines give $m \times n = 36$ points of intersection and no internal blocking.
   The version with 4 connecting lines is thus preferable.

3. a) 70, 210, 280, 560, 840, 1000, 1000
   b) $(m+n)v$ in the cases 1, 3, 4, 8 and 12.
   $(m \times n)$ in cases 15 and 20.

4. a) $m \times n$ because this product $= 50$, while $(m+n)v = 60$.
   b) $(m+n)v = 45$, but $m \times n = 50$
   c) no internal blocking.

CHAPTER 11

1. a) Relays.
   b) Selectors are not suitable for the parallel reception of information. Capacitors do not store the information for long enough.
   c) The circuit can be realized with 14 relays. Relay X switches over to the second digit. Relay Y blocks the input after the second digit, and makes "reading" (closing of contact unit k) possible. Relay L remains energized after reading until all register relays, and X and Y, have released. Fresh information can then be admitted.

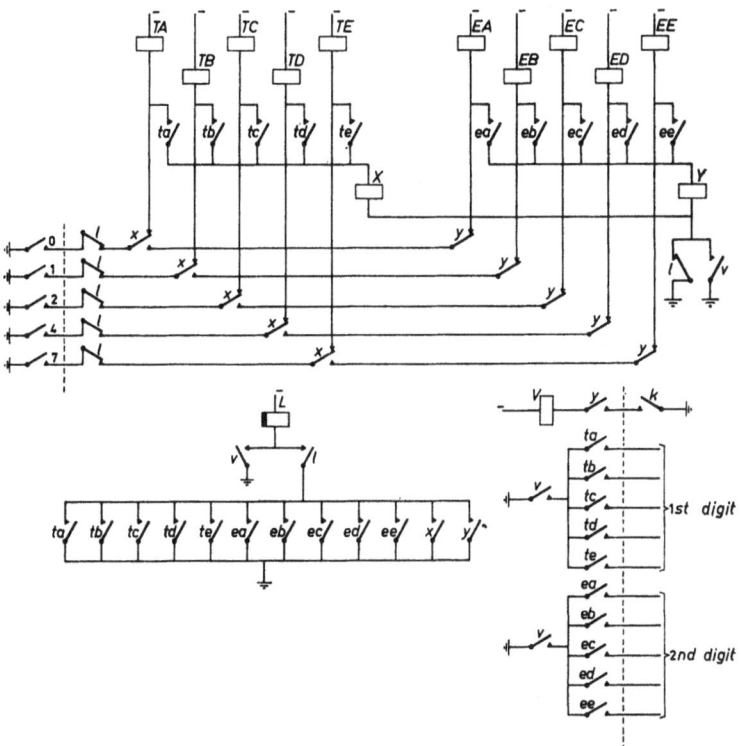

2. *a*) Registering the 1st, 2nd and 3rd digit respectively.

   b) Switching over from the 1st to the 2nd and from the 2nd to the 3rd group of register relays respectively.

   c) Establishes that the last digit has been received, blocks the input and makes it possible for information to be read out (by energizing relay *V* via contact unit *k*).

   d) Relay *V* initiates the reading out, digit by digit. Relay *L* switches on relays *HW*, *TW* and *G* in that order each time relay *V* releases after the reading of one digit.

   Relays *HW* and *TW* connect the output wires successively with the information contact units for the 1st, 2nd and 3rd digit.

   Relay *G* indicates that the last digit has been passed on. Relays *HW* and *TW* are then switched off again.

   The release of relay *L* is retarded, because relays *HW*, *TW* and *G* must be able to operate in the period during which *V* has already released, but *L* not yet. *L* must therefore release slowly, because it is switched off by a contact unit *v*.

3. *a*) Yes.
   b) Because the digit 0 does not switch on the register relays, so that some other means must be sought of indicating that a digit has been presented.
   c) No, because with this code at least one register relay is switched on during the reception of each digit.

4. Relays *X*, *Y* and *Z* form the circuit for counting the three digits to be received. The relays *A*, *B* and *C* act as auxiliary relays in this connection. By means of change-over contact units of *X* and *Y*, the right capacitors for each digit are connected to the input wires. Relay *Z* indicates that all digits have been received; this indication is stored by the switching on of relays *D* and *F*. Contact units of relay *F* block the input and prepare for the output of information (in the circuit for relay *V* and *LA–LD*). A contact unit *d* switches off the 1st and 2nd counter elements (*A*, *X* and *B*, *Y*), as a result of which the 3rd counter element is also switched off. Relay *E* then operates. The function of relays *D* and *E* is to switch off the counter and then to prepare for the output of the three digits. Relay *V* initiates the reading out of each digit.
   Relays *LA–LD* receive the information concerning each digit from the capacitors and passes it on to the wires *a*, *b*, *c* and *d*. These relays remain energized as long as contact unit *k* is closed, i.e. as long as relay *V* is switched on.
   The counter indicates that the last digit has been passed on (relays *C* and *Z* energized), after which relay *D* releases, while relay *E* is still held. This has the result of switching off the 1st and 2nd counter elements, so that relays *C* and *Z* release too. When relay *E* has then released, relay *F* is switched off and the register circuit is again able to accept fresh information.

CHAPTER 12

1. Since two of the six relays are always energized, we have:
   $$f(1) = cf + ef + ce + bd + bc + ac + ae + bf$$
   In $cf + ce + bc + ac$ we have all combinations of one letter with c, apart from cd, so that:
   $$cf + ce + bc + ac = cd'$$
   $$f(1) = cd' + ef + bd + ae + bf$$
   $ef + ae = eb'c'd'$, because e occurs with a and f but not with b, c

and d, so that:

$f(1) = cd' + eb'c'd' + bd + bf = cd' + eb'c'd' + b(d+f)$

This gives the circuit:

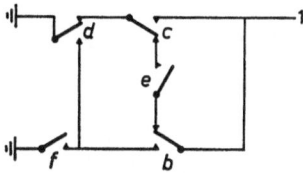

$f(2) = df + ef + de + bd + ab + ac + af + bf$

In $df + ef + af + bf$ we have all combinations of one letter with f, apart from cf, so that:

$df + ef + af + bf = c'f$

$f(2) = c'f + de + bd + ab + ac$

$de + db = da'c'$, because d occurs with b, e and f but not with a and c, so that:

$f(2) = c'f + da'c' + ab + ac = c'f + da'c' + a(b+c)$

The circuit is as follows:

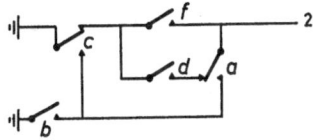

$f(4) = be + ce + de + bd + ad + ae + af + bf$

In $be + ce + de + ae$ we have all combinations of one letter with e, apart from ef, so that:

$be + ce + de + ae = ef'$

$f(4) = ef' + bd + ad + af + bf = ef' + (a+b)(d+f)$

The circuit is as follows:

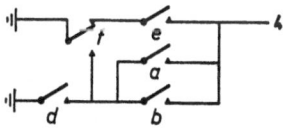

$f(8) = cd + bc + ab + ac + ad + ae + af + bf$

In $ab + ac + ad + ae + af$ we have all combinations of one letter with a, so that:

$ab + ac + ad + ae + af = \underline{a}$

$f(8) = a + cd + bc + bf$

$bc + bf = bd'e'$, because b occurs with a, c and f but not with d and e, so that:

$f(8) = a + bd'e' + cd$

The circuit is as follows.

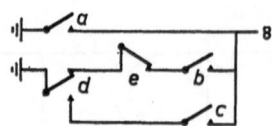

2. $f(A) = 12'4' + 1'24' + 12'4 + 1'24 = 12' + 1'2$

$f(B) = 1'24' + 124' + 1'2'4 + 12'4 = 24' + 2'4$

$f(C) = 1'2'4 + 12'4 + 1'24 + 124 = 4$

The circuit is therefore as follows:

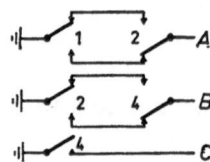

3. *a)*

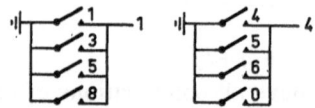

b)

4. *a)*

	A	B	C	D		0	1	2	4	7
1	1	1	0	0		1	1	0	0	0
2	1	0	1	0		1	0	1	0	0
3	0	1	1	0		0	1	1	0	0
4	1	0	0	0		1	0	0	1	0
5	0	1	0	0		0	1	0	1	0
6	0	0	1	0		0	0	1	1	0
7	1	0	0	1		1	0	0	0	1
8	0	1	0	1		0	1	0	0	1
9	0	0	1	1		0	0	1	0	1
0	0	0	0	1		0	0	0	1	1

b)   *c)*

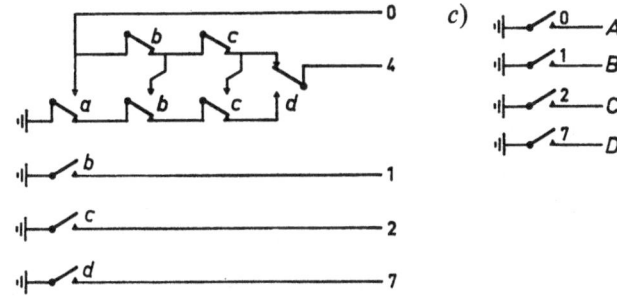

5.

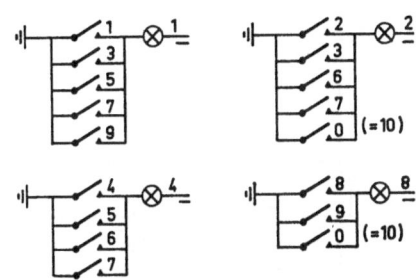

6. $f(A) = 15 + 12 + 26 + 13 + 16 + 56 + 46 + 14 = 1 + 26 + 56 + 46 =$

$\overline{1 + 63'}$ (all digits occur together with 1; 6 does not occur

together with 3)

$f(B) = 12 + 25 + 23 + 26 + 56 + 36 + 24 + 46 = 2 + 56 + 36 + 46 =$

$\overline{2 + 61'}$

$f(C) = \overline{23 + 26 + 13 + 35 + 34 + 16 + 56 + 36} = 3 + 26 + 16 + 56 =$

$\overline{3 + 64'}$

$f(D) = \overline{34 + 16 + 56 + 36 + 24 + 46 + 14 + 45} = 4 + 16 + 56 + 36 =$

$\overline{4 + 62'}$

The circuit is as follows:

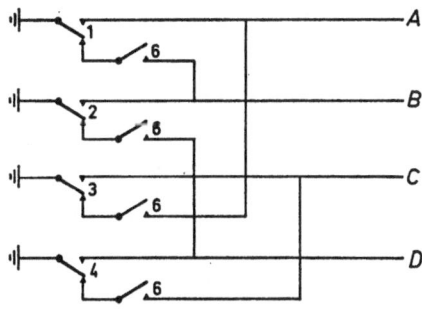

7. $f(1) = ac + ad + ab + ae + be = a + be$

    All combinations with $\overline{a}$ occur, also be.

$f(2) = bc + ab + bd + be = b$

    All combinations with $\overline{b}$ occur.

$f(4) = ac + cd + ce = cb'$

    All combinations with c occur, apart from bc.

$f(5) = de + ae + bd + be + ce = e + bd$

    All combinations with $\overline{e}$ occur, also bd.

    The circuit is as follows:

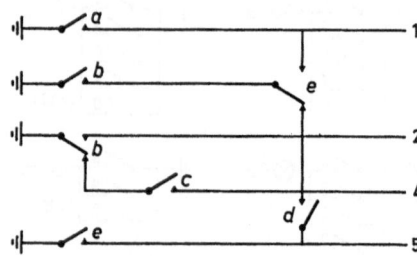

8. Combinations 1, 3, 5 and 7 give: 1 1 0 – –

   Combinations 2, 4, 6 and 8 give: – – 1 0 1

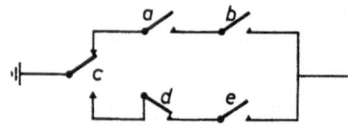

CHAPTER 13

1.

                           TENS   *UNITS*

2.

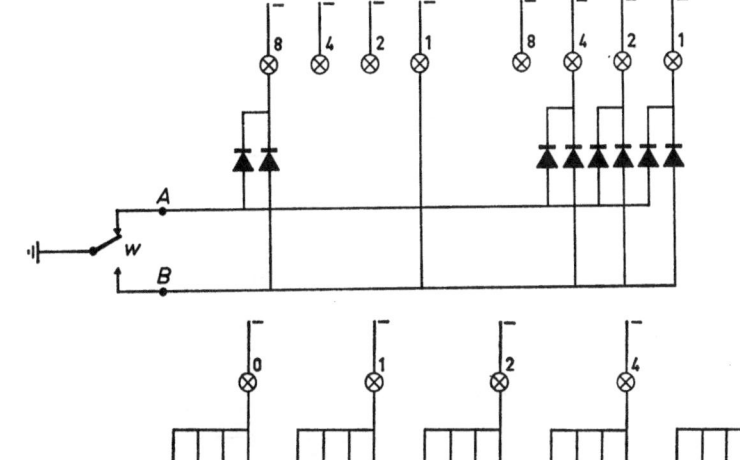

3.

4.

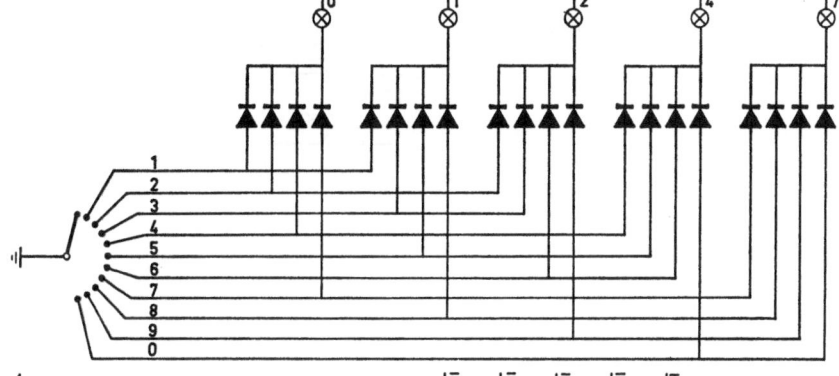

5.

6.

INDEX